June 1997

Science, Vine, and Wine in Modern France examines the role of science in the civilization of wine in modern France. Viticulture, the science of the vine itself, and oenology, the science of winemaking, are its subjects. Together, they can boast of at least two major triumphs: the creation of the post-phylloxera vines that repopulated late-nineteenth-century vineyards devastated by the disease; and an understanding of the complex structure of wine that eventually resulted in the development of the widespread wine models of Bordeaux, Burgundy, and Champagne.

The first part of the book deals with the grapevine and the relations of varieties of vine to the quality of wine. Harry Paul describes the battle to save the French vinifera vine, ravaged by phylloxera in the 1860s, and in so doing demonstrates that the vine itself is central to any history of wine. His discussion includes a portrayal of Montpellier as the center of viticulture.

Bordeaux was the heart of the oenological empire, whose development is the next episide in the story of the modern wine industry. Paul provides an extended discussion of the importance of Louis Pasteur and Jean-Antoine Chaptal to the development of oenology, detailing the role of research in the production of wine in the Champagne, Burgundy, the Languedoc, and Bordeaux. Along the way, he questions the popular idea that the more complex the oenology, the duller the wine. Quite the opposite, he suggests: research has put the science of wine on a solid foundation and made it possible for people to enjoy a greater variety of better wines.

Science, Vine, and Wine in Modern France

Science, Vine, and Wine in Modern France

HARRY W. PAUL
University of Florida

CAMBRIDGE
UNIVERSITY PRESS

Published by the Press Syndicate of the University of Cambridge
The Pitt Building, Trumpington Street, Cambridge CB2 IRP
40 West 20th Street, New York, NY 10011-4211, USA
10 Stamford Road, Oakleigh, Melbourne 3166, Australia

First published 1996

Printed in the United States of America

Library of Congress Cataloging-in-Publication Data
Paul, Harry W.
Science, vine, and wine in modern France / Harry W. Paul.
p. cm.
Includes bibliographical references and index.
ISBN 0-521-49745-0
1. Wine and wine making – France – History. 2. Viticulture – France – History.
I. Title.
TP553.P38   1996
663'.2'0944 – dc20   96-1596
CIP

A catalog record for this book is available from the British Library.

ISBN 0-521-49745-0 Hardback

Contents

Tables and Figures

Acknowledgments

I wish to thank the National Endowment for the Humanities (Grant RH-20971) and the National Science Foundation (Grant SBE-9120629) for supporting my work. A large number of people helped me to finish this project. Louis Latour, Jean-Paul Médard (Moët et Chandon), Betty Smocovitis, Russ Watkins, and Li-Hui Tsai were especially vigilant in showing me how I had strayed from the path of virtue. I am especially grateful to Pascal Ribéreau-Gayon for help and encouragement. I thank Renee Akins, Claire Orologas, Helen Wheeler, Barbara Folsom, and Paige Porter-Brown for "processing the final product."

The following presses and organizations granted me permission to use materials from copyrighted works:

Académie d'agriculture de France: Pascal Ribéreau-Gayon, note in the *Procès-verbal de la séance du 19 janvier 1977.*

Cambridge University Press: Michael Mullins et al., *Biology of the Grapevine,* 1992.

Oxford University Press: *The Oxford Companion to Wine,* edited by Jancis Robinson, 1994.

The Regents of the University of California: A. J. Winkler et al., *General Viticulture,* University of California Press, 1974.

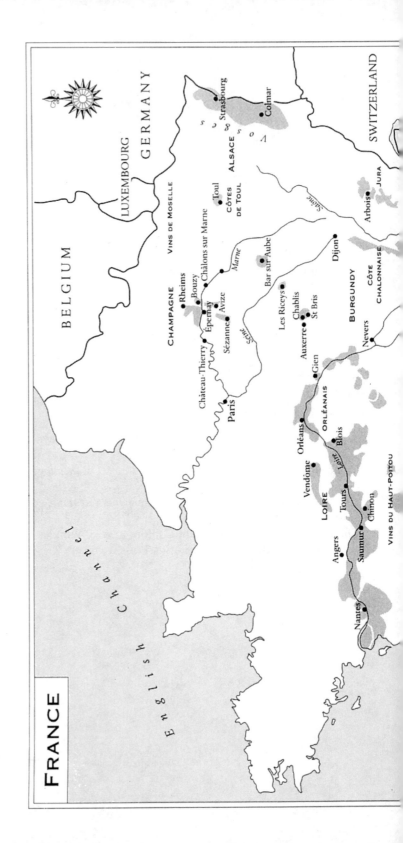

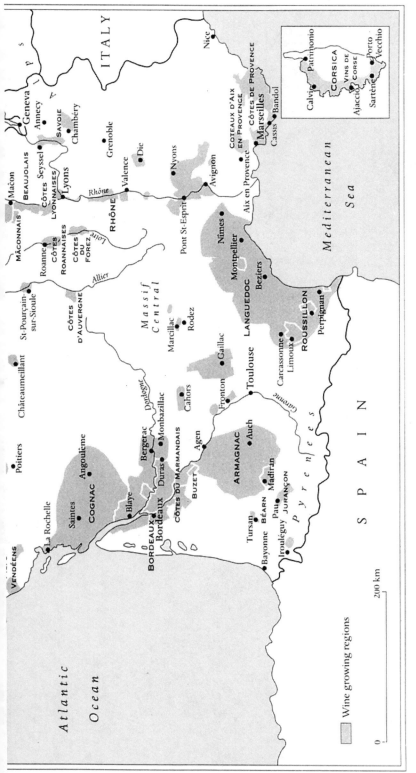

©Oxford University Press. Reprinted with permission of Oxford University Press from *The Oxford Companion to Wine*, ed. by Jancis Robinson (Oxford and New York, 1994).

Introduction

This book deals with the role of science in the French wine world since the Enlightenment. France, historic center of the civilization and science of wine, is the central concern of the book. The Champagne, Burgundy, and Bordeaux (historically, synecdoche for the southwest) have given us great wines, along with an abundant, sometimes self-congratulatory, literature, much of it scientific. This literature is vital for arriving at an understanding of the historical and contemporary dominance of these wine models as well as for an analysis of the role that science has played in their evolution. The main argument of the book may be baldly stated: the modern or post-phylloxeric vine and its wine are the fruit of the sciences of viticulture and oenology, especially institutional science in Montpellier and Bordeaux. This argument is clearest and strongest in the cases of the reinvention of the vine in the late nineteenth century and the rise and long influence of the school of oenology in Bordeaux. Oenologists and viticultural scientists are being bashed these days in some popular wine guides. In a sense, this book is a *historical* counterthesis to the argument that the stronger the oenology, the more uninteresting and duller the wine – even if technically perfect. (But even Robert Parker finds good things to say about oenologists outside California.)

Part I of the book deals with the grape vine and the relation of varieties to quality of wine. Viticultural discourse today is about cultivars (*culti*vated *var*ieties) and clones, the creation of a viticultural science that began its ascendancy over the plant world in the nineteenth century. Salvation of the vines essential for the production of quality wines came from viticultural science in Montpellier and Bordeaux. The slow death of the French vinifera vine began with the arrival of phylloxera in France in the 1860s. Part I sketches the fight to save the old vine along with futile attempts to find substitute hybrid vines. Both efforts finally faded in light of the triumph of the vinifera vine grafted onto resistant rootstocks of American vines.

It is difficult for writers and readers to get passionate about the vine (or about botany and botanists, who have now changed their name to plant biologists and plant scientists in the hope of becoming as notorious as physicists and biologists). The vine has suffered from periodic neglect in both production and scientific literature. But since the mid-nineteenth century the vine has forced us to pay attention to it by catching various frightening diseases and then dying dramatically. The vine has also come to benefit from general recognition of a basic axiom that underlies the production of good wine: quality begins in the vineyard with the plant and its culture. A discussion of the vine, which has its own science of viticulture, must be an essential part of any serious history of oenological research and of wine.

Part II shifts from vine to wine, to the laying of the foundations of oenology by the chemists Jean-Antoine Chaptal and Louis Pasteur, two scientists of great importance in the history of agriculture. Many natural philosophers (later called scientists and more specialized names) have always been interested in plants and agricultural production. As a chemist, property owner, and powerful politician in the Napoleonic empire, Chaptal devoted much of his energy to agriculture, especially the production of wine. Perhaps he deserves his place in the history of oenology more because of his influence on winemaking practices than his research; but he did produce a classic in scientific literature. Pasteur recognized the importance of Chaptal as a worthy predecessor in oenology.

A modern American classic on wine presents Pasteur as a sort of patron saint of oenology, the man who applied the scientific revolution to the wine industry.[1] Pasteur spent much of his career dealing with agricultural problems, especially disease. The prevention and curing of wine diseases came into fashion at the same time (1860s) as the curing of many human diseases and for the same reason: the discovery of the role of specific bacteria in causing animal and human diseases. Pasteur's influence on the study and treatment of wine diseases was perhaps greater than in any other area of microbiology. The idea of curing wine diseases by heating, not original to him, was only part of Pasteur's *Etudes sur le vin,* which appeared in the 1860s. In his microscopic analysis of the composition and characteristics of wine, Pasteur began a whole new approach to the subject. His systematic, experimental team study of wine diseases and of killing bacteria by heating the wine led to a basic change in wine production that united science and technology in a new way. In his work on wine Pasteur worked closely with producers and professional wine tasters. He recognized that science, taste, and production cannot be separated. (It may be partly due to his influence that the Bordeaux school of oenology has usually kept this idea prominent in its teachings.) Pasteur's studies on

---

[1] Maynard A. Amerine and Vernon L. Singleton, *Wine: An Introduction,* 2d ed. (Berkeley, 1976), pp. 22–3.

wine united for the first time the science, technology, and aesthetics of wine in the interests of commerce.

Part III of the book examines the role of research in the production of wine in the Champagne, Burgundy, and the Languedoc. Champagne is one of the greatest commercial successes in the history of wine. Like fine wine, port, scotch, coffee, and chocolate, champagne became popular with well-heeled consumers in the eighteenth century.[2] The Champagne provides some good examples of the collaboration of capitalism with viticultural and oenological research. The special problems associated with the production of champagne, its bottling and storage, as well as the guarantee of its quality, seemed insoluble without the help of scientific research.

The second chapter of Part III struggles with the elusive history of oenology in Burgundy, where the connections between science and winemaking are less evident than in the Champagne and the Bordelais. After the late eighteenth century, the idea of a decline in the reputation of the *grand vin* of Burgundy was not uncommon; at the same time, the reputation of bordeaux improved. In the nineteenth century, red burgundy had the reputation of being a less reliable commercial commodity than red bordeaux, unless burgundy was stabilized through heating, which was Pasteur's panacea for Burgundy's problems. In recent years, better control of production and more attention to sanitation, combined with shipping in appropriate refrigeration, has made it possible to drink good burgundy nearly anywhere. It is tempting if somewhat simplistic to see the explanation for the better reputation of bordeaux, at least in foreign markets, in the rise of oenological science at the University of Bordeaux, whereas the faculty of sciences in Dijon remained surprisingly peripheral to either old or new oenology, at least until after the Second World War.

The last chapter of Part III deals with the Narbonnais school of oenology, little known outside specialized texts, though everyone seems to know about its most notorious subject of research: carbonic maceration or vinification without crushing the grapes – except those that are crushed by the weight of the other grapes! It seems difficult to become well known for oenological research without plugging into high-level mainstream science as Montpellier and Bordeaux did, and it helped even more, as in the case of Bordeaux, to have a reputation for the production of quality wine.

In Part IV the narrative moves to Bordeaux, where we find the greatest success story in the history of oenology. The first chapter treats the development of Pasteurian oenology in Bordeaux. In the late nineteenth century, Pasteur's student, disciple, and collaborator, the chemist Ulysse Gayon, took a position in the faculty of sciences at Bordeaux and became

---

[2] Wolfgang Schivelbusch, *Tastes of Paradise: A Social History of Spices, Stimulants, and Intoxicants,* translated from German by David Jacobson (New York, 1993).

the area's leading agricultural scientist. With his botanist colleague Alexis Millardet, he perfected the famous Bordeaux mixture for the treatment of downy mildew, a fungus disease that threatened to ruin the wine industry in the 1880s. Thus the young oenology could claim some credit in saving a great industry from disaster. The popularity of Gayon's course on grape growing and winemaking helped to establish a formal program designed to cater to agricultural interests. More important, Gayon, chiefly through his work on fermentation, went on to create the oenology of fine wines, which was continued and developed by his students. They extended the Pasteurian paradigm to the limits of its usefulness.

The second chapter of Part IV reveals the development of a new oenology in Bordeaux during the 1920s and 1930s with introduction of the theory of ionization, or electrolytic dissociation, into research on the composition of wine. The theory also entered oenology in Montpellier, where it was the creature of the faculty of pharmacy rather than the school of agriculture. The key figure in Bordeaux was the biochemist Louis Genevois, a prominent if unloved figure in the history of oenology. Jean Ribéreau-Gayon and Emile Peynaud were the most prominent of his students to apply the physical chemistry of solutions to develop a new model of wine in order to understand exactly what goes on in the making and subsequent aging of wine.

Although the evolution of different models of wine was evident in Roman antiquity, it seems to have accelerated in the eighteenth century. Montesquieu noted the existence of many models of Bordelais wines catering to a wide range of consumer tastes. Scientists entered into the business of creating new models of wine for commerce after the Chaptalian revolution. The most widely imitated wine model has been the one created by the Bordelais school of oenology, conveniently symbolized for the media by the peripatetic oenologist Emile Peynaud, who has done as much to promote the fortune of Bordeaux as it has done to promote his. The new model wine, less tannic and less acidic, and therefore drinkable young, may not be to everyone's taste, but it is the wine that has been universally adopted from Piedmont (Italy) to the Hunter valley (Australia) as the best model of marketing success. The second chapter of Part IV reveals the scientific foundations of the new model wine in the work done by Ribéreau-Gayon and Peynaud about a half-century ago.

The third chapter of Part IV deals with the integration of oenology into higher education with the eventual birth of an oenological institute in Bordeaux. Ribéreau-Gayon, assisted by Peynaud, worked in a laboratory of the famous old wine firm of Calvet in Bordeaux. A connection, however tenuous, existed between science and commerce, and it was later extended to education through the faculty of sciences. In moving to the University of Bordeaux in 1949, Ribéreau-Gayon and Peynaud were able to achieve a gradual transformation of oenological practices by training dozens of oenologists, who went into the wine business in one capacity or another. Oenology drew on a wide range of sciences from analytical chemistry to plant physiology. Combined, after the 1950s, with a real technical revolution at the level of production, the use of this new scientific knowledge transformed the vinification of fine wines, and eventually the *petits crus.* The gospel of Bordeaux emphasized that the vinification process must be carried out within the context of general scientific principles, technical know-how, and the artistic sense of the winemaker – wine must be a *Gesamtkunstwerk.*

PART I

Reinventing the Vine for Quality Wine Production

1

Death and Resurrection in the Phylloxeric Vineyard

It is difficult to overestimate the economic and social importance of the vine in French history. In 1865 winemaking was one of France's flourishing industries, producing seventy million hl (hectoliters) of wine worth one and a half billion francs. Between 1882 and 1892 the annual value of all plant production fell from 9.149 to 7.865 billion francs. A large part of this drop in production was due to the general fall in prices during "the agricultural crisis," part of the Great Depression of 1873–96. A wine industry weakened by vine diseases must also be taken into account.[1]

Figures for Bordeaux indicate the damage done to wine production by several devastating diseases. Between 1840 and 1849 the average annual production of the *vins du Bordelais* was 1,646,000 hl; in the 1850s it dropped to 1,183,000 hl. By the early 1870s (1869–75), vats were overflowing, with an annual average production of between 2,500,000 and 3,000,000 hl. In 1875 over 5,000,000 hl were produced; this figure was not surpassed until 1900, when the production of nearly 5,750,000 hl of bordeaux established a record unbroken until 1922, a year producing over 7,000,000 hl. But between 1876 and 1885, phylloxera reduced production to an average annual figure of 1,774,000 hl. It was not until 1891 that production climbed back to nearly 2,500,000 hl, reaching nearly 5,000,000 hl in 1893 (for national figures see Figures 1a and 1b).[2]

Winemaking supported one and a half million vigneron families and as many people in dependent industries and retailing. The department of the Hérault (Languedoc) alone cultivated 180,000 ha (hectares) with revenues of 180 million francs. The vine supplied the state with about one-sixth of its revenues. Wine was the second French export after textiles. The inva-

---

[1] Clive Trebilcock, *The Industrialization of the Continental Powers 1780–1914* (London and New York, 1981), pp. 156–7, puts the final bill for phylloxera at over four hundred million pounds, "approximately 37 per cent of the average annual GDP for 1885–94." He includes phylloxera along with Prussians and *pébrine* (silkworm disease) as the three plagues that hit France in the late-nineteenth century.

[2] Ch. Cocks et Ed. Féret, *Bordeaux et ses vins,* 10th ed. (Bordeaux, 1929), pp. 59–60.

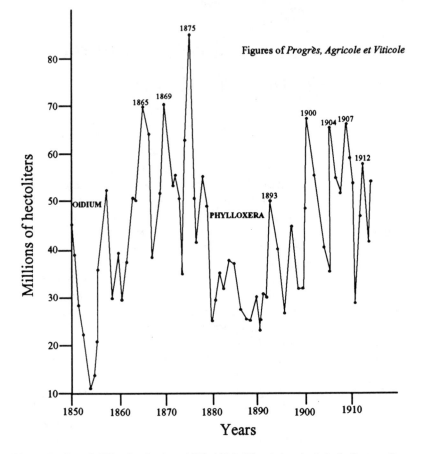

Figure 1a. French Wine Production, 1850–1914 (Algeria not included). *Source:* Jean-Paul Legros and Jean Argeles, *La Gaillarde à Montpellier* (Association des Anciens Elèves de l'ENSAM, 1986), pp. 189, 237.

sion of the vineyards by the phylloxera insect threatened this vast industry, along with its way of life, rural social structure, traditional political arrangements, and urban drinking habits. (Wine and wheat, two key items in rural thought, were of great importance in politics.)

How to save it all became a national obsession, one of the top priorities of politicians and scientists. The road to salvation was littered with good intentions and useless remedies, but in the end the industry survived and in good economic times returned to prosperity. Saving the wine industry was a complex affair lasting nearly a generation. Although it would be an exaggeration to say that salvation came completely from science, it is accurate to say that science played a large role in the reconstitution of the

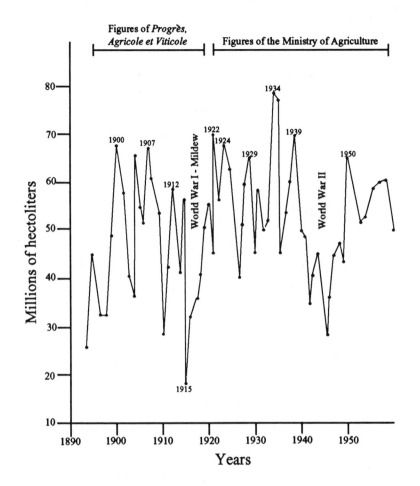

Figure 1b. French Wine Production, 1895–1956 (Algeria not included). *Source:* Jean-Paul Legros and Jean Argeles, *La Gaillarde à Montpellier* (Association des Anciens Elèves de l'ENSAM, 1986), pp. 189, 237.

devastated vineyards, and especially in the preservation of quality wines from *vinifera* vines.[3]

In his superb history of *Vins, vignes et vignerons,* Marcel Lachiver sums up the period 1850–1900 in Straussian terms as "death and transfigura-

---

[3] Mouillefert estimated that 1,950,000 winegrowers supported about seven million people, about one-third of the total population. Pierre Mouillefert, *Les vignobles et les vins de France et de l'étranger* (1891), pp. 20–1, and Pierre Barral, "Un secteur dominé: la terre," in Jean Bouvier et al., *Histoire économique et sociale de la France,* vol. 4: *L'ère industrielle et la société aujourd'hui (siècle 1880–1980),* pp. 378–84.

tion." The most important event in this period was the conquest of phylloxera through grafting. Before the coming of phylloxera, France had about 2,500,000 ha of vineyard. By 1900, out of the approximately 1,730,000 ha in production, about 1,200,000 had been planted with replacement vines (chiefly French scions on American or Franco-American rootstocks) for those dead or dying from phylloxera. One major difference between the pre- and post-phylloxera situations was that 200,000 ha of the reconstituted vineyards included land not used for wine production before: irrigable valleys, maritime sands, and even marshland. Perhaps the most serious challenge was the replacement of 500,000 ha of vines in the calcareous soils of the Charentes, the Champagne, and Burgundy.[4]

Vine Diseases: Mildews, Rots, Phylloxera

The nineteenth century was one of striking increases in agricultural production and also of devastating plant diseases, which science codes in Graeco-Latin nomenclature. The potato and the vine were the most prominent victims of *Mycophyta* – fungi, a ubiquitous group of good and evil eukaryotic organisms that includes mushrooms, molds, rusts, smuts, and yeasts – the agents of disease, decay, brewing, fermentation, baking, some food proteins and vitamins, and antibiotics. Life is hard to define, but it includes a lot of fungi. Among them are the *Peronosporales,* very destructive parasites that include *Plasmopara viticola,* the cause of downy mildew of the vine, and *Phytophthora infestans,* the cause of late potato blight, notorious in the 1840s for provoking starvation, revolution, and emigration. Killer fungi seemed to rule the agricultural world with a reign of terror in the vineyard that was aggravated by the arrival of the plant louse (*Phylloxera vastastrix*) from North America (see Table 1).

Oidium (powdery mildew) showed up in 1846 and, after several years of insidious activity, triumphed in the general disaster of 1854. Oidium does not require much humidity if the weather is hot but not too hot, thirty but not thirty-five degrees centigrade. Mildew can be fought only by preventing it from invading the plant, for it penetrates the tissues. But the fungus that causes oidium can be stopped by treatment of the disease itself, for it affects the herbaceous organs only superficially.[5] Phylloxera (1864) was ably supported by invasions of downy mildew (1878) and black rot (1885). It does not seem that there were any devastating diseases before

---

[4] L. de Malafosse, "La mode en viticulture," *Revue des hybrides franco-américains porte-greffes et producteurs directs* (*RHFA*), no. 22 (1899), pp. 232–6.

[5] The pathogen causing oidium is the fungus *necator* Burr of the genus *Uncinula* Lév.; its class (*Ascomycetes*) includes morels and truffles. For a brief description of the four classes of the taxonomic division fungi (mycota or mycophyta), see M. Abercrombie et al., *The Penguin Dictionary of Biology,* 7th ed. (Harmondsworth, 1980), pp. 195–6.

Table 1. *Common Diseases of* Vitis vinifera

Type	Common Name	Causal Organism
Fungal diseases of leaves and fruit	Powdery mildew	*Uncinula necator*
	Downy mildew	*Plasmopara viticola*
	Botrytis bunch rot and blight	*Botrytis cinerea*
	Black rot	*Guignardia bidwellii*
	Phomopsis cane and leaf spot	*Phomopsis viticola*
	Anthracnose	*Elsinoë ampelina*
Fungal diseases of vascular system and roots	*Eutypa* dieback	*Eutypa lata*
	Esca	Unknown
	Armillaria root rot	*Armillaria mellea*
	Verticillium wilt	*Verticillium dahliae*
Bacterial diseases	Crown gall	*Agrobacterium tumefaciens*
	Pierce's disease	*Xylella fastidiosa*
	Bacterial blight	*Xanthomonas ampelina*
Diseases caused by viruses and virus-like agents	Fanleaf degeneration	Grapevine fanleaf virus (GFLV)
	Leafroll	Unknown, possibly closterovirus
	Corky bark	Unknown
	Rupestris stem pitting	Unknown
Miscellaneous	*Flavescence dorée*	Unknown, possibly a mycoplasma-like organism

Source: Reproduced with permission from Michael G. Mullins et al., *Biology of the Grapevine* (Cambridge, 1992), p. 188.

the invasion of oidium at mid-century. (There were plagues of insects, of course, of which the most serious was the devastating 100 years' war waged by insects in Burgundy after the mid-fifteenth century; Camille Rodier suggests that only phylloxera could have destroyed the vineyards of Beaune.)

Downy mildew ruined grape crops in France (and Europe) from 1878 to 1885. Black rot, caused by the fungus *Guignardia bidwellii,* was the most serious and the most extensive vine disease in the United States, where both wild and cultivated vines east of the Rockies were infected. This was one disease that bypassed California. Black rot first attacks the leaves and then moves on to the fruit. In the United States the rot showed more preference for the leaves than it did in France. A first copper treatment has to be carried out by May 15 if the leaves are to be saved from the attacks that show up around the second fortnight in the month. There is no authentic evidence of the disease in France before 1848.[6] Invasions being usually local, no national disaster resulted. But in 1887 a violent invasion of black rot in the Midi and the southwest was so frightening that vignerons turned from vines grafted on *Vitis vinifera* to direct-producing hybrid vines, which scientists had singled out because of their resistance to diseases.

Grape-growing conditions changed greatly after the post-phylloxera reconstitution of vineyards in the late nineteenth century. Antidisease treatment of the vine, rather rare before the mid-nineteenth century, became a normal part of grape growing. Scarcity of supplies and labor could lead to a catastrophic drop in production, as in the war years 1915–17, when wine production fell to an annual average of 31,500,000 hl, with a low of 20,400,000 in 1915, less than half of the prewar production. Before the 1850s, French vine diseases, generally local in nature, were easily controlled by simple procedures, frequently by removal of the diseased vines. This operated over time as a selection process, producing a nearly disease-free *vignoble*, subject only to disaster from plagues of insects from time to time. A workable alliance between living things existed before the disasters came from the New World.

The grafted vine, taller and more vigorous, resulted in increased production. Immediately before the plague of phylloxera, from 1863 to 1875, average wine production in France was 56,900,000 hl from a cultivated surface of 2,200,000 ha: 25 hl per ha. In the period after the reconstitution of vineyards, from 1899 to 1909, annual average production was 55,500,000 hl from the reduced area of 1,500,000 hectares: 32 hl per hectare. In the decade from 1922 to 1931, the annual average was 56,567,000 hl from 1,440,000 ha: 39 hl per ha. The increase was not due solely to grafting. Changes in cultivation also counted, but there is no doubt that diseases of the vine completely changed growing practices. Before the mid-nineteenth century vines could do without treatment. The new vine could not live without its chemical lifeline. Worse, in order to survive in the new disease-plagued environment, the old vine in particular needed massive doses of chemicals to keep it from succumbing to phylloxera. But the success of the program, chiefly grafting, was so great that in 1929 wine production rose to 81,799,000 hl. High productivity, combined with American prohibition and an economic downturn, led to a disastrous fall in prices.[7]

The drop in production for the period 1908–21 was not entirely due to

---

[6] Pierre Viala, "Le Black Rot en Amérique," *Annales de l'Ecole nationale d'Agriculture de Montpellier* 4 (1888–9): 308–43.

[7] L. Moreau et E. Vinet, *La défense du vignoble* (Paris, 1938), pp. 13–25. In the series *La Terre, Encyclopédie paysanne*, ed. J. Le Roy Ladurie, Secrétaire général de l'Union nationale des syndicats agricoles. Léon Moreau and Emile Vinet were director and director-adjunct of the Station oenologique régionale d'Angers and professors at the Ecole supérieure d'agriculture d'Angers.

the First World War. French vines treated with anti-phylloxera insecticides required relatively large doses of fertilizer to prevent them from being dangerously weakened by the poisons. A new system of "intensive fertilization" with chemical fertilizer developed in the 1880s. By 1884–5 about one-ninth to one-tenth (instead of the traditional one-twentieth) of the Latour estate was fertilized yearly, which required 160 tons of fertilizer, a fivefold increase. The paradox was that keeping the vine alive and productive threatened quality and profits through a high yield. The situation was made doubly paradoxical by the intrusion of the *négociants* (merchants). They wanted French vines with low yields, high-quality wines, and solid profits, but they pushed a policy that gave them lower quality and lower profits. The grafted vine provided a way out for both owner and merchant. Pijassou believes that grafting was a sort of large-scale experiment in which reasonable analytical guides were supplied by the Station agronomique de Bordeaux, headed by the oenologist Ulysse Gayon. Production, sales, and science became necessary if not happy bedfellows.[8]

In spite of some limited local successes in the use of insecticides, phylloxera was generally conquered by grafting the fruit-bearing part of French vines onto the bug-resistant roots of American vines. Other diseases soon attacked the new vine. Intensive cultivation and heavy fertilization of the new vine produced a very vigorous plant, but one more easily invaded by fungus diseases than a weaker vine. Abundant leaves provide an ideal shelter for insects, especially those that attack the grape itself. At the same time, the leaves make it difficult to defend the fruit crop. The larger crops of the grafted vine resulted in bigger and tighter bunches of grapes, a preferred place for the development of insects. The bugs often produced lesions that promoted the onset of *Botrytis* or rot. Humidity encourages mildew; its most dangerous form is gray rot, chiefly the sinister work of *Botrytis cinerea,* a form of which is also the agent of *pourriture noble.* Vital for Sauternes, this rot is disastrous for the production of other wine. Brown rot, though insidious, is more leisurely in its progress. A successive invasion of gray and brown rot could kill a crop easily; even skillful and heavy use of poisons would make it difficult to save the grapes. Nonappearance of mildew could be as dangerous as its appearance if the winegrower was lured into the trap of not treating the vines. Mildew did not appear in 1918 and 1919, but it showed up suddenly in 1920, when it

---

[8] René Pijassou, in Charles Higounet, ed., *La seigneurie et le vignoble de Château Latour. Histoire d'un grand cru du Médoc (XIVe–XVe siècle),* 2 vols. (Bordeaux, 1974), 2:450, 488–9, 496–501. Daniel Jouet, a graduate of the Institut national agronomique and a veteran of the state anti-phylloxera brigade, was appointed manager of Château Latour in 1883, when the estate decided to intensify its attack on the insect. State subsidies were restricted to small property owners (5 hectares or less) who formed anti-phylloxera associations. In 1884–5 Château Latour spent about 22,000 francs, 20% of its administrative budget, to pay for the technical team and the insecticide it injected into the soil.

reduced the crop by one-third to one-half in Anjou vineyards that had been insufficiently treated. It was necessary for the viticultural scientist to keep reminding the winegrower of the need for eternal vigilance in the fight against fungus and insect pests that could infect plants or destroy crops.

Science and Agriculture

Science's understanding of the random behavior of disease was a better guide to action than experience alone. It was not easy to convince all winegrowers of this wisdom, but a disaster or two usually made even stubborn growers more willing to listen to scientific advice than to gamble with nature. The scientist's intimacy with disease made available to the grower a wealth of useful information easily transformed and channeled into better harvests. Specialized knowledge about devastating diseases was a potent weapon in the hands of the grower, especially when coupled with the yield-increasing products that science provided to industry.

The marriage of science and agriculture was not without its stormy quarrels, particularly over the replacement of the European vineyards destroyed by the American plant louse *Phylloxera vastatrix,* a formidable sap-sucking aphid. First, the cost. Between 1868 and 1900 in France alone, 2.5 million ha of vines were uprooted at a cost of about 15 billion francs. The cost of chemical treatments, imports of vines, replanting, and grafting added about another 20 billion francs to the bill.[9] The cost of reconstituting individual vineyards was high, far beyond the resources of many small wine producers. Scientists at the University of Montpellier's school of agriculture advocated replacing dead vines with French plants or scions (the stems) grafted on aphid-resistant American stocks (the rooted portion of the plant), while wine producers in many southern departments insisted that direct production (nongrafted) hybrids were superior in resistance to diseases.

A second major issue in the quarrel was wine quality: hybrid vines produced mediocre wines but were resistant to disease, whereas grafted vines produced good wines but were prone to disease and lived for only about 25 years or less. The producers, armed with the latest science, carried on a long debate that evolved with the history of plant genetics from the botany of Darwin to that of Hugo de Vries and the resurrected Gregor Mendel. One journal of the producers, the *Revue des hybrides franco-américaines,* was well informed on the latest developments in plant genetics and possible implications for vine breeding. The issue was international

[9] Pierre Galet, *Cépages et vignobles de France,* vol. 1: *Les vignes américaines,* 2d ed. (Montpellier, 1988). Galet converts 20 billion francs of the period into 500 billion in 1988.

in scope, with German, Spanish, and especially Italian producers and scientists interacting with the French, who were at the center of the quarrel.

The scientific issue in the debate over direct-production hybrids was inseparable from social structure, for the peasantry and poorer producers were pro-hybrid and the producers of fine wines were pro-grafted. Nor can we separate the issue from its cultural context. Choice of vine depended on the consumer's acceptance or rejection of the taste of wine from two types of vine: the peasant palate could tolerate hybrid wine, while the bourgeois palate could accept only *vins de cru*. A clear case of distinction, Bourdieu-style.

Echoes of the quarrel over the grafted vine can be found today in the discussion over the planting of vine clones, which are produced free of diseases and can be selected to give a low yield of high-quality fruit but may be genetically susceptible to some diseases and may also produce a less complex wine than grafted vines. This is another typical paradox of scientific progress, which in this case is clearly preferable to the alternative of disease, and even death, for the vine.

Perhaps for the first time in its history, science found itself in a quarrel where being right or wrong could have serious economic consequences. The intellectual challenge of the problems to be solved and the new economic importance of science gave a great boost to agricultural science and the funding of its institutions by governments at national and regional levels. The sciences relevant to agriculture had themselves changed considerably during the nineteenth century. It is useful to look briefly at what scientific resources were available for the battle against plant diseases, especially phylloxera.

The Plant Sciences in the Nineteenth Century

In the second half of the nineteenth century, the sciences dealing with plants moved from observation-description into a more experimental phase.[10] Linnaeus (1707–78) would have been at home with most of the research and publications; yet this remark is misleading. Botany, which often had impressive adjectives added to it to attest to its scientificity, had become a highly developed, well-established, flourishing discipline, one of the sciences in which France was a leading research and teaching power. Botany was a key science in the Muséum national d'histoire naturelle, the faculties of science, and institutions like the Jardin botanique de Nancy. Its

---

[10] Increasingly complex changes in the nature of research on plants has produced an evolution of labels for packaging the old botany as transformed by new sciences, i.e., botanical sciences, plant biology, and plant sciences. See V. B. Smocovitis, "Disciplining botany: A taxonomic problem," *Taxon* 41 (August 1992): 459–70.

fortunes were promoted by a network of local and national organizations, including numerous Linnean societies – that of Bordeaux was well known. The most famous society was the Société botanique de France, publisher of a distinguished bulletin.

Botanical research, much addicted to the microscope, laid bare the structure and function of plant parts, detailed plant anatomy and tissues, outlined plant physiology, including nutrition, and began to explain plant reproduction. Experimentation in the laboratory, the greenhouse, and the field became part of the botanical empire. Many of the big names in nineteenth-century botany were French: Boussingault, Brongniart, Chatin, Duchartre, Dutrochet, Millardet, Planchon, Naudin. Even though specific scientific knowledge of the vine was poor when the phylloxera crisis came, the advanced state of botanical science was a big advantage in the battle to save the vine. Some of the best-known researchers on phylloxera came out of nineteenth-century classical botany. Their knowledge and skills was to be rudely tested by the complexity and unpredictability of the most difficult of plants.[11]

Mendelism, however important it would become in agriculture, was unknown in schools of agronomy. The cultivation, grafting, and hybridization of the vine were highly developed in both practice and theory. Until at least the last quarter of the nineteenth century, specific scientific knowledge of the vine remained at a low level. Phylloxera changed all that, although scientists had become slightly more interested in vines as a result of the other "American diseases" – the great blight of oidium in the 1850s, for example. Still, treatment of most plant diseases or of vines attacked by insects was a simple practical matter not requiring much technology and less science. Phylloxera was the first disease that required a mobilization of part of the scientific community to "conquer" it.

On September 25, 1871, the Académie des sciences pondered the most important problem it had ever faced, a national agricultural disaster, the unstoppable destruction of French vineyards by the insect *Phylloxera vastastrix,* an event pregnant with incalculable economic, social, and political consequences. The significance of a scientific problem may be judged by the prominence of the people who try to solve it or, in this case, are chosen to solve it. After hearing reports on the destructive advance of the insect in the Midi, the Academy asked Jean-Baptiste Dumas to chair a commission that would do research on means of fighting this plague. By 1875 the Commission supérieure du phylloxera also included Milne-Edwards,

---

[11] Pierre Duchartre, *Rapport sur les progrès de la botanique physiologique* (1868), pp. 400–1, concludes that eighteenth-century French botany was superior to that in other countries and that the highest level of research was maintained in the nineteenth century – "l'un des brillants fleurons de notre couronne scientifique" – but by mid-century botany was as *international* as it was national.

Duchartre, Blanchard, Pasteur, Thenard, Boulez, and later, Cornu, Risler, and Balbiani. The faculties of science, the Collège de France, and schools of agriculture – all representatives of high-level science – were mobilized for battle. The Academy also chose scientists as its own delegates to the commission: Balbiani, Duclaux, Cornu, Boutin, Maurice Girard, Millardet, Mouillefert, and Rommier; initially, three scientists in zoology, botany, and chemistry were charged with working in infected areas. The ministry of agriculture funded the research of the delegates and other fighters. The government expected results from the Academy, and it got them, but not a final solution to the problem. In its battle against phylloxera, the Academy sometimes coordinated its efforts with those of provincial scientists, particularly in Montpellier. The scientific campaign, like the disease, was national.

Montpellier Takes On Phylloxera and Other Enemies

Jules-Emile Planchon, botanist and doctor of medicine, professor in the faculty of sciences and the faculty of pharmacy in Montpellier, claimed to have discovered the American aphid sucking the roots of vines at Château de Lagoy, near Saint-Rémy, when he, Félix Sahut, and Gaston Bazille, backed by the Société centrale d'agriculture, were on an investigative expedition in the heavily infected Rhône valley in 1868. Sahut was a property owner, scientist, professor, and author of viticultural works. Bazille was a big property owner in the Hérault, a horticulturist, a senator after 1879, and a member of the Commission supérieure du phylloxera (a group of politicians and scientists). Sahut claimed that he found the insects on the vine roots and pointed them out to Planchon, who was later forced to admit Sahut's priority. Sahut, who was clever enough to pull up the roots of the living plant, was the first to see the moving insects – the Christopher Columbus of phylloxera, Vigier called him. Planchon carried out the identification and confirmation process through the machinery of university science, proceeding with the unseemly haste of the scientist who knows he is on to something big.

So the answer to the question who discovered phylloxera all depends on how one defines discovery.[12] The discovery followed a publicity route characteristic of modern science: an article in the paper *Messager du Midi* to inform the public of the importance of the discovery and of the discoverers, then a note in the *Comptes rendus de l'Académie des sciences,* and

---

[12] Félix Sahut, *Un épisode retrospectif à propos de la découverte du phylloxéra* (Montpellier, 1899). Planchon had died over ten years before; see Sahut, *Discours prononcé aux obsèques de . . . Planchon* (Montpellier, 1889), an exercise in politeness that says nothing about phylloxera. For years after this revelation Foëx kept attributing the discovery to Planchon alone: educational institutions share glory reluctantly.

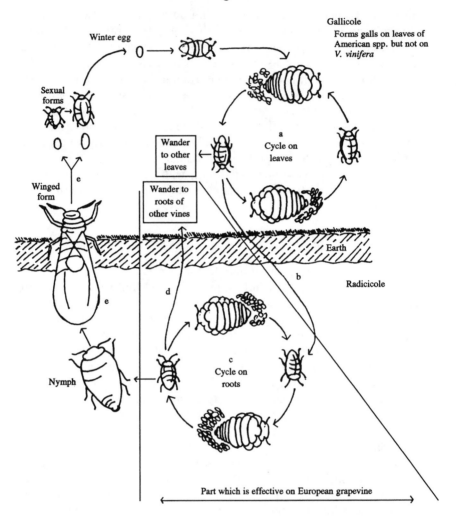

Figure 2. Diagram of the Life Cycle of Phylloxera (*Daktulosphaira vitifoliae*). Based on the diagram in A. J. Winkler et al., *General Viticulture* (1974), p. 540 (Copyright 1974 The Regents of the University of California).

finally entry into the world of literature. Baptized as *Rhizaphis* (root aphid) for its presentation to the Academy of sciences, this root louse soon became known as *Phylloxera vastatrix,* although some scientific purists insisted on using *P. vitifolii* (see Figure 2). A literary text endowed the insect with reality in 1874, when Planchon, a vain man with words, announced its existence in the *Revue des deux mondes;* he forgot to mention the other two members of the team. Sahut was not amused to find that Planchon gave him and Bazille no credit for the discovery of the insect.[13]

Succumbing to the temptations of minor bureaucratic power, Planchon became director of the school of pharmacy and also of the Jardin des plantes; he was perhaps too versatile to make a really basic contribution to the scientific study of phylloxera. He studied the biology of the insect but did not write any book comparable to the *Etudes sur le phylloxera vastatrix* (1878) by Maxime Cornu of the Muséum. Nor did he make any biological discovery about the insect comparable to that made by E. G. Balbiani, whose study on *Le phylloxera du chêne et le phylloxera de la vigne* (1884) laid bare at least part of the bizarre sexual life of *P. vastatrix*, including parthenogenetic reproduction as well as production from eggs laid by sexuals.[14] Planchon had spent the years 1844–9 at London's Kew Gardens as Sir William Hooker's assistant. His greatest expertise was probably in botany. After spending a few months in 1873 roaming through American grape-growing regions, meeting American entomologists, and visiting nurseries and vineyards, he returned home loaded with vines and wines. His report on American vines, the "first synthetic study" of the subject, established three categories of these American vines according to their resistance to the insect. Planchon became a staunch advocate of the use of resistant American roots: the cure came from the disease, the solution from the problem.[15] A solid basis was established for a great scientific debate.

From 1877 until well into the twentieth century, the most powerful

---

[13] The term *phylloxera,* from the Greek for dried leaf, was suggested to Planchon by a Parisian entomologist (Signoret) because the insect's mechanism of destruction resembled that of the oak tree phylloxera, whose pricks in the leaves dry them up. The vine phylloxera (insect family *Aphididae* of the order *Homoptera*) sticks its rostrum in the root tissue and sucks its cells dry. Other learned (Latin) and erudite (Greek) name-calling went on, but more suitable systematic names fortunately never caught on, or we might have to stutter out *Dactylosphaera* or *Daktulosphaira vitifolii* (syn. *Viteus vitifolii*). Botanists still indulge in this residual classicism, a mystery for the masses that scientists believe essential for scientific clarity.

[14] In 1874 the Academy of Sciences delegated Balbiani, who was professor of comparative embryogeny at the Collège de France, and Cornu to do a study of phylloxera. Some promising experiments were carried out near Montpellier to test Balbiani's recommendations of destroying the insect's winter eggs, but in an age of primitive insecticides the labor-intensive process never caught on. See the article on Balbiani in the *Dictionary of Scientific Biography* (*DSB*), Charles Coulston Gillespie, editor-in-chief, 18 vols. (New York, 1970–90).

[15] J.-E. Planchon, *Les vignes américaines: Leur culture, leur résistance au Phylloxera et leur avenir en Europe* (Montpellier, 1875); see the excellent history of phylloxera by Roger Pouget, *Histoire de la lutte contre le Phylloxéra de la vigne en France* (Paris: INRA, 1990), pp. 53–5. The best general book on phylloxera is Gilbert Garrier's *Le Phylloxéra: Une guerre de trente ans 1870–1900* (Paris, 1989). In English, the fine book by George Ordish, *The Great Wine Blight* (London, 1972), is still indispensable.

instrument of propaganda in favor of the Americans was *La vigne américaine, sa culture, son avenir en Europe,* a journal published by J. E. Robin and V. Pulliat and edited by Planchon. This powerhouse had the collaboration of leading members of the viticultural establishment, presidents of agricultural societies, departmental professors of agriculture, and politician–vineyard owners. The big names included Alexis Millardet, G. Foëx, C. Saintpierre, head of the Ecole d'agriculture in Montpellier, Oberlin, mayor of Beblenheim (Alsace), and Louis Vialla, president of the agricultural society of the Hérault. By 1880 the group had expanded to include Gaston Bazille and a gaggle of foreigners – four Italians, three Germans, one Austrian, one Spaniard, one Portuguese, and one American – among whom were some top people in viticulture, including A. Blankenhorn; Dr. Baron Dael von Koeth, president of the International Commission of Ampelography;[16] Rudolphe Goethe, editor of the *Rheinische Blaetter für Wein, Obst und Gartenbau;* Cerletti, head of the school of viticulture in Conegliano and editor of the Italian review of viticulture and oenology; Munoz J. del Castillo, professor at the institute of Logrono; the count of Villamaior, head of the University of Coimbra; and G. E. Meissner, partner of Isidor Bush of St. Louis, whose nursery was one of the Missouri suppliers of rootstocks to France.

The wine industry, especially the small segment of it obsessed with quality, has always been closely connected to the networks of social, political, and scientific-educational power. Most winemakers in the Hérault were small-property owners, uninterested in the science and politics of the introduction of the American vine, an affair of the big estates and the railway companies. The local politicians, in cooperation with the Ecole d'agriculture de Montpellier, undertook a campaign using the printed word and free courses in grafting to spread the American gospel.[17]

In his introduction to the first number of *La vigne américaine,* Planchon noted that the study of American varieties of vine in the south and west of France had entered into an experimental phase. This had a prosaic meaning for Planchon: facts had more value than theoretical discussion. Of course, the facts had to be carefully observed, prudently interpreted, compared, and judged in context. The indispensable source of this viticultural knowledge would be trials in the vineyard, large-scale field experimentation. The role of science would be to supply method, order, precision. Agronomists often conceive a passion for Isis because she seems to be the ideal housekeeper. *La vigne américaine* would offer serious experimenters an open forum for courteous discussions; it would not be an arena

---

[16] The International Ampelography Society is located in Minneapolis.
[17] W. Ian Stevenson, "La vigne américaine, son rayonnement et importance dans la viticulture héraultaise au XIXe siècle," in *Economie et société en Languedoc-Roussillon de 1789 à nos jours* (Montpellier, 1978), pp. 78–84.

for irritating polemics. Two assumptions made by Robin and Pulliat underlay this publishing ideology. First, the impotence of all systems of defense of the vine that were based on the strategy of absolute destruction of *Phylloxera vastatrix,* and, second, the imminent ruin of all the great French vineyards. Planchon did not believe that the true rootstock value of an American variety could be established without a large-scale experiment. A friendly relationship between scientists and politician-proprietors facilitated this type of experimentation: Montpellier's school of agriculture propagated American plants for the departmental councils of the Midi to distribute to growers.[18]

Viticulture and oenology in Montpellier was divided into two houses: the university, especially the faculty of pharmacy but not excluding the faculty of sciences, and the Ecole nationale d'agriculture, La Gaillarde. (This name refers to the estate of 22.5 ha – Domaine de la Gaillarde – acquired in 1872 by the department of Hérault and the city of Montpellier; it was the original property from which the school started its astounding territorial expansion, especially in the twentieth century.)[19] The Ecole nationale d'agriculture (ENA), transferred from La Saulsaie (near Lyon) in 1872, was the first scientific school of agriculture in the Midi. Its destiny was inevitably linked to oenology, and especially to viticulture. The first five years were difficult, but under the leadership of Camille Saintpierre (1876–81), the school became an important center of scientific agriculture, supported by local money and political power.

Saintpierre, owner of the fine domain of Rochet, was one of the class of professor–property owners. Professor at the medical school, professor of technology at La Gaillarde, and secretary of the Société centrale d'agriculture de l'Hérault, Saintpierre conceived of the school as a center of research: teaching the rudiments of scientific agriculture to locals, however necessary and desirable, would not be of any importance outside the Hérault; and Saintpierre was an ambitious man. But a regional research project could be of national, even international, significance, and this was certainly true in the case of fighting phylloxera by grafting French plants onto American roots. Saintpierre became a convert to the American gospel.

Yvette Maurin has pointed out the religious dynamic that inspired the

---

[18] Once the vineyards were reestablished, *La Vigne américaine* evolved into just another solid, dull viticultural journal dealing with routine matters like the weather, insects, pesticides, fertilizers, and fraud, as well as hybrids. In 1908 its editor was G. Battanchon, an inspector in the ministry of agriculture and honorary president of the Lyonnais Society of Viticulture. J. Roy-Chevrier, producer, author, viticulturalist, was also on the administrative and editorial committee.

[19] *La Gaillarde* has metamorphosized into ENSAM, Ecole nationale supérieure agronomique, which corresponds better to its leading role in making Montpellier "un pôle d'attraction en matière d'agronomie méditerranéenne et tropicale" (Jacques Chirac, minister of agriculture, 1973).

school to solve social problems by the application of science, source of social and political progress. Most of the professors, including Planchon, were protestant. La Gaillarde became the "foyer américaniste," acquiring a collection of American varieties of vines that soon became the biggest in France – 400 varieties by 1889 and up to 3500 later. "L'Université du Phylloxera," some called the school. Collaboration with the Société centrale d'agriculture, where science and property met, was ensured through interlocking personnel monopolizing the presidency and vice presidency: Saintpierre, Gaston Bazille, and Louis Vialla. The school taught the new grafting techniques necessary for viticultural workers and property owners engaged in the great renewal. La Gaillarde was the world center of the new viticulture.[20]

Saintpierre died in harness at the age of forty-seven in November 1881, on his return from an oenological congress in Conegliano, Italy. Gustave Foëx, thirty-seven years old, became the new director. A graduate of the agricultural school of Grignon, Foëx established a laboratory of viticulture and introduced a course in the subject, both firsts in France. He kept comparative viticulture for himself, but after 1886 he let his collaborator, Pierre Viala, teach general viticulture. In 1890 Viala escaped to a chair at the Institut national agronomique in Paris. La Gaillarde was simply too small for two prima donnas of viticulture. In 1886 Foëx's *Cours complet de viticulture* began its long career as a classic European text of scientific viticulture. Foëx and Viala published numerous works on ampelography, including American vines. All of this was vital professorial activity not without interest to some enlightened men of property, some of whom were also well-known scribblers on viticulture.

American vines were introduced for the production of rootstocks for grafting without paying much attention to the relations between soil, climate, and plant. Soon the nonsuccess of the American plants became as important an issue as phylloxera. Hybridizers and scientists soon concentrated on this complex problem, and no one more successfully than Alexis Millardet of the University of Bordeaux. Of course Montpellier's bigger viticultural operation produced a greater quantity of high-level science than Millardet. Foëx studied the chlorosis suffered by the vines planted in calcareous soil, and Viala was to move the problem far along the road to solution after his visit in 1887 to the United States, where he looked for vines in soils similar to the killer soils in Cognac and Champagne country. Foëx and Viala also studied fungus diseases, old and new. To show that it believed in its own science, La Gaillarde produced and gave away more than two million American plants. This was one of the benefits of close

---

[20] Yvette Maurin, "Société et Ecole d'agriculture de Montpellier devant le phylloxera," in *Economie et société en Languedoc-Roussillon,* pp. 57–9.

relations between the school and producers; some of the school's leading scientists were themselves owners of big wine-producing properties and enjoyed close relations with regional *notables*. They were all part of the great rootstock experiment.

Montpellier: A Center of Official Science

By the end of the nineteenth century there had developed an entity that its enemies scorned as "official science," the research and its applications done chiefly by professors in institutions of higher learning and pushed by bureaucrats in the ministry of agriculture. After the failure of chemical warfare to eliminate phylloxera and the demonstration of impressive success in the program of grafting French vines onto American rootstocks, the powers of science allied with the government to encourage winegrowers to replace their dead vines with grafted ones. Grafting had the approval of the school of agriculture at Montpellier as well as the ministry of agriculture. The government's well-baited financial incentive helped.

Who were the leaders of this official science? What were its aims and accomplishments? Montpellier's *Revue de viticulture*, which proclaimed itself to be an "organ of the agriculture of viticultural regions," was often cast in the sinister role one of the most powerful instruments of "state viticulture," a publication supposedly capable of exercising ruthless power in the scientific arena. Its praise or condemnation of works dealing with wine and viticultural and oenological subjects probably had a major impact on the fate of viticultural publications. Members of the editorial board of the *Revue de viticulture* were viewed as men of great power, "sommités officielles." In 1898 these men of scientific power could be easily identified: Pierre Viala, an inspector general of viticulture and professor of viticulture at the Institut national agronomique de Paris-Grignon (INA); F. Couvert, also professor at the INA; Ravaz, professor at the Ecole de Montpellier; and B. Chauzit, departmental professor of agriculture in the Gard and director of the department's agricultural laboratory in Nîmes. Viala and Chauzit were also identified on the cover of the *Revue de viticulture* as members of the viticultural elect ("propriétaire-viticulteur"). By 1913, fourteen of the sixteen big guns of oenology and viticulture comprising the editorial board belonged to the class of "propriétaire-viticulteur."

The two most powerful scientists in French viticulture were Viala and Ravaz. Who were they and what were the sources of their power?

Pierre Viala, classified by friends as a cold and reserved man of the Midi with a sensitive nature, finished his studies at the Ecole de Montpellier in 1881. One of his *maîtres* was Gustave Foëx, director of the school. It was the time when the fight against phylloxera had clearly failed. That failure was a clear challenge to an ambitious young man like Viala, who saw that ignorance was the chief difficulty: little was known about the physiology of the vine and the nature of its resistance to its enemies. Research on these matters required a strong scientific background, especially in methods of study and controlled techniques. So Viala went back to school for a doctorate in natural science.[21] After doing several years of work with the botanist Charles Flahaut, head of the Institut de botanique at Montpellier and one of the founders of botanical cartography, Viala finally earned his degree at the Sorbonne in 1891. In a biological study of host–parasite relations, he developed an original method of sowing seeds to prepare pure cultures and invented a gadget for observing the biological evolution of the parasites. His work in phytopathology was extended from the laboratory to the vineyard with the help of fellow winegrowers. He won a prize from the Académie d'agriculture for his book *Maladies de la vigne* in 1885.

Then came an official mission to the United States. In 1886 the committee on phylloxera of the Charente-Inférieure, in a moment of desperate inspiration, turned to geology to look for similar soils in the United States where there might be wild vines immune to both phylloxera and chlorosis. The Société centrale de l'Hérault organized a scheme for French viticultural societies to fund the project. The clear popularity of this move convinced a reluctant ministry that democratic science should be funded. In 1885 the head bureaucrat (Directeur de l'agriculture), Eugène Tisserand, chose Viala to search for areas in the United States with calcareous soil for a resistant vine suitable for the Charentes. Léon Degrully and his journal *Progrès agricole* provided financial support for Viala's trip. His discoveries were *Vitis berlandieri, Vitis cinerea,* and *Vitis cordifolia.* Viala found, in Belton, Texas, the object of Charentais desire, *Vitis berlandieri,* flourishing in a soil where other vines succumbed to chlorosis. After many disappointing experimental marriages with both American and French plants, a successful French partner was found for this vine. Their generally happy union led to the successful repopulation of many of the Charentais vineyards.[22] *V. berlandieri* (also known as *V. aestivalis* and *V. monticola* Buckley) had first been revealed to the scientific world in 1834 by the Belgian-Swiss botanist J.-L. Berlandieri, who found it in Texas. Elevating this vine to the status of a new species in 1880, Planchon dedicated it to

[21] Montpellier, the capital of botany, has a distinguished scientific tradition in botanical research. The Jardin des plantes, founded in 1593 in the reign of Henri IV, is the oldest botanical garden in France. Auguste Pyrame de Candolle (1778–1841) spent some time there as a professor of botany in the early nineteenth century. A new Institut de botanique was founded in 1959 and a Centre d'études phytosociologiques (CEPE, funded by the Centre national de la recherche scientifique) was established in 1961. Jean-Paul Legros and Jean Argeles, *La Gaillarde à Montpellier* (Montpellier, 1986), pp. 252–8.

[22] A. Verneuil, "La reconstitution en terrains crayeux," *Revue de viticulture (RV)* 39, no. 25 (1913): 211.

Berlandieri. The vine was introduced into the Var by Dr. Davin, who drew attention to its ability to survive in calcareous soil. Viala studied the species with botanical rigor. No startling novelty here, but often what counts in science is not what one finds but what one makes out of it. Viala made a great deal out of *V. berlandieri.*[23]

On his return to Montpellier, where he became Professor of Viticulture in 1886, Viala mounted a scientific assault on phylloxera. Planchon had pointed out that American vines had brought the disaster and could possibly save French viticulture as well. But the vines used were often diseased and hard to acclimate. Worst of all, they often produced grapes and wine that tasted foxy. Viala used his own two vineyards of Cournonterral and Lavérune, which he had inherited from his parents, to carry out a series of experiments on resistance to major diseases in 400 varieties of vines grafted on different rootstocks. Ten years of laboratory and field research provided the basis of certainty on which Viala proceded to rebuild the vineyards of France.

For many years Viala was also the technical director of the Forceries de la Seine, with its four hectares of vines and fruit trees in 104 hothouses. It was an ideal place for experiments on artificial fertilization and the influence of various soils on plants. Viala never ignored agronomic issues in his botanical research, but his superiority over many others working on vine diseases was that he actually did solid, original scientific work in plant pathology. This was clear in the debate over the possible modification of vines by grafting with a consequent influence on the quality of wine. Viala showed that grafting caused neither morphological nor physiological modification of the characters of the vine's leaves, flowers, fruit, stalks, and roots to a higher degree than modifications produced accidentally in nature by variation of bud or shoot. The bottom line for the grower was that the grafted vine was as good a plant as the ungrafted vine. Nor was the wine it produced inferior to that of the ungrafted vine. Not everyone agreed with Viala.

In 1890 Viala became professor at the Institut national agronomique. He was elected to the rural economy section of the Académie des sciences in 1919. His election to the Chamber of Deputies, which probably increased his influence in the viticultural world, gave a great boost to the idea that it was the duty of the state to fund scientific research. Science and patriotism were important elements of viticulture. In 1890, on his visit to the vineyards of Anjou, Viala warned against planting hybrid direct producers to replace the vines of French blood that produced famous wines: it would be antipatriotic to sacrifice the legitimate fame of a high-

---

[23] Pouget, *Histoire de la lutte,* pp. 76–82, emphasizes the nonimportance of "La mission de Viala en Amérique," pointing out that *V. berlandieri* had been long introduced and used in French hybrids. See also Galet, *Cépages et vignobles de France,* 1:94.

quality French product. The title of a talk he gave to the 36th division of the French army in 1916 is a masterly summary of a certain French ideology based on wine – *bifteck* and *frites* used to be less important in this system of beliefs. "Le vin, première richesse du sol de France, sa défense contre les Boches, son avenir après la guerre, son rôle contre l'alcoolisme."[24]

In the academic genealogy of La Gaillarde, Foëx was the *maître* of Viala, who was the *maître* of Ravaz. After receiving his diploma from Montpellier's Ecole nationale d'agriculture (ENA) in 1880, Ravaz went to work in the laboratory of viticulture with Foëx and Viala. A scientific coup followed. In 1885 he and Viala discovered black rot, a new American horror. Ravaz became a great expert on the biology of the vine, attaining a profound knowledge of its diseases and an equally profound knowledge of practical viticulture. After two years of scientific servitude in Montpellier, he was rewarded in 1888 by being chosen by the Comité de viticulture de Cognac to study the reconstitution of the vineyards of the Charentes. Reconstitution in the Charentes was based on the model of Montpellier: experimental fields for demonstration, distribution of American cuttings, teaching people to graft, lectures, and publications. To fight against the threat of chlorosis, he recommended the *V. berlandieri* that Viala had brought back from the United States and the 41 B (the hybrid *V. chasselas* × *V. berlandieri*). In 1897 he became professor of viticulture at La Gaillarde and then director of the ENA in 1919.

A famous teacher, Ravaz inspired thirty-four graduating classes of the school. One of his most famous students was Lucien Sémichon, who became an oenological power in Narbonne. His work in ampelography made him famous enough to attract even Germans – and he kept up with research in Germany. His knowledge of the resistance of vines in different soils was unsurpassed. An authority on hybrids, American vines, diseases – especially mildew and its evolutionary cycle – he made classic contributions to phytopathology and phytopharmacy. A cofounder of the *Revue de viticulture,* Ravaz left it in 1912 to become codirector with Léon Degrully of the *Progrès agricole et viticole,* with which he had been collaborating more and more since he and Viala had politely but professionally squabbled in 1901 over an obscure vine disease.[25] The international reputation of La Gaillarde was in large part the work of Ravaz.[26]

[24] See articles by Emmanuel Leclainche, Henri Hitier, Mario Roustan, E. Schribaux, Jean Perrin, and Paul Marsais in *RV* 86 and 88 (1937, 1938).
[25] Relations between Viala and Ravaz cooled after 1901, when they disagreed over the cause of cracking in vine (*la gélivure*). Both were wrong on this physiological phenomenon resulting from the exhaustion of an overly productive plant. Legros et Argeles, *La Gaillarde,* pp. 178–9.
[26] Robert Dunges, editor of *Der Deutsche Weinbau,* a student of Ravaz's, said farewell with a very friendly obituary. See the article on Louis Ravaz by D. Vidal, Jean Branas, and Sémichon in *RV* 86 (1937).

Agricultural science was developing into a laboratory science without losing its essential connection with field reality, the practice of science *en grande culture*. The fight against phylloxera shows the close and necessary interaction between laboratory pot science (experimenting on plants in pots) and large-scale field experiments. Montpellier was in the forefront of this powerful laboratory development in agricultural science. The usefulness of the laboratory was not immediately evident to growers, and even a scientist or two. It took a long, hard fight to prove the success of the new approach. What was the pot science war all about?

Method and Reality in the Laboratory and the Vineyard

Viticultural laboratory research of the pot-science variety became the subject of considerable discussion in 1897–8. At stake for the agricultural scientist was the heart of his research world, the value of laboratory science. A common criticism of laboratory research made by the practical man in the vineyard was that the small-scale research done on plants in the laboratory had no practical value because of the totally different situation of the same plants existing in a natural state, though this was seen by some as an artificial cultured state. A classic example of this quarrel of the pot men with the vineyard men was the disagreement between Ravaz and Millardet over the putative lack of resistance to phylloxera of vine 33A2, which Millardet had bred and touted as highly resistant.

Ravaz reported having found "numerous voluminous nodes" on the roots of vine 33A2 grown in the school's nurseries. Millardet could hardly deny his fellow scientist's observational powers, but he did refuse to admit that the experiment showed that his plants could not resist the phylloxera louse. Some observers thought that the property used for experiments in Montpellier was highly unsatisfactory for growing vines. Even before phylloxera, French vines generally did poorly there because of the very chalky soil, which was also excessively humid as a result of poor drainage. But the Cornucopia vine, suited to the soil there, did well, even though its resistance to phylloxera was low. The problem was the poor adaptation of Millardet's vine to that soil. Some observers found it difficult to understand why the school carried out its experiments in such abnormal conditions. Millardet seemed justified in refusing to accept the experiment, while harsher critics of Foëx claimed that it had not been conducted in a very scientific fashion.

Was the reputation of Bordeaux threatened by Montpellier? This is one layer of interpretation we can peel off the exchange. "Exaggerated Claims for Resistant Vines Made by One of Bordeaux's Scientists, Claims Leading Scientist of Montpellier" – or so one might have read in some version of the *New York Times* science section had it existed. Millardet carefully avoided a direct challenge to Ravaz's research results by accepting them as valid only for the bad conditions in which the vines grew in Ravaz's pots, but he also saved his own scientific work by rejecting Ravaz's research as inapplicable to normal vine culture.

Paul Gouy, in his *Revue des hybrides franco-américains, porte-greffes et producteurs directs* (*RHFA*), supported Millardet and attacked Foëx. In 1898, Gouy, "viticulteur à Vals près Aubenas" in the Ardèche, founded this journal, with the help of "nombreuses Notabilités viticoles," to deal exclusively with the theoria and praxis of viticulture. Pierre Galet points out that the journal played "a big role" during the next half-century. Because disease, especially phylloxera, had wiped out the viticulture based on the old French vines, Gouy announced that his journal would be the organ of the new viticulture. In its period of imperialistic neutrality between 1907 and 1913, the journal appeared as *La revue du vignoble,* but it returned to the original honest title in 1913.[27] André Perbos, a viticulturist and hybridizer at Villeneuve-sur-Lot, became editor in 1910. In 1937, during a period of hybrid expansion, the journal metamorphosized into *La vigne moderne.* (It died in 1953; the hybrid empire would fade away a generation later.)[28]

Gouy went beyond Millardet's critique to reject Ravaz's thesis on the transmission of the characters of parent plants to hybrids. Ravaz argued that the result of crossing any vine with *V. vinifera* would be a significant drop in the hybrid's resistance to phylloxera. Not necessarily so, replied Gouy, using the curious example of successful interbreeding between different ethnic groups as an analogy against the theory of weakening of resistance to phylloxera in vines. Producers believed in facts as much as professors did, perhaps more. So it is not surprising that a satisfied and cocksure Gouy would note the *fact* that all sorts of combinations resulted from hybridization, which left the choice of reproductive strategy to the producer. The selection of the most resistant hybrids could be made by the vine breeder.

Rather than accept what they castigated as the theoretical, fact-ignoring discourses of the professors, Gouy and the men of practical science would follow the laws of botanical science and established agricultural practice. Gouy and the *RHFA* were interested in making a general argument about what kind of experimental science was of importance to agricultural producers. Nearly ten years after a vast European experiment had started "en grande culture," where millions of grafted hybrids were under observation,

[27] The *Revue du Vignoble* advertised itself as the "Organe de la viticulture nouvelle" dealing with "Culture, hybridation, oenologie, commerce extérieur des vins et eaux-de-vie."

[28] French hybrids are are still around and are called French hybrids; some nominalists in the United States now want the name changed to reflect the latest genetic breakthroughs in "hybridizing *vinifera* with with the Scuppernong-type rotundifolia of our southern states." Philip Wagner, "Wagner's Question: Better Name for Hybrids?" *Wines and Vines* 59 (1978): 54.

no general serious weakening of vines had been detected. The growers were also satisfied with the vines. This was hailed as a double demonstration against science based on "la culture en pots"; so much the worse for science if its results did not agree with the real world of commercial production.[29]

As news of these Olympian quarrels drifted down, there was a certain confusion among winegrowers about what to do. To follow a wrong or even doctrinally defective school could be a ruinous decision. Some people thought that the confusion created by science could only be cleared up by more science. To remedy the confusion over the real value of the Franco-American direct-production hybrids most in vogue, the Société centrale d'agriculture de la Haute-Garonne called on the Conseil géneral of the department to obtain help from the state (Paris) in establishing an experimental vineyard. The society thought that experiments – presumably carried out "en grande culture," not in pots – would answer questions about resistance to phylloxera, bad weather, and fungus diseases, as well as questions about the fertility of the vines and the quality of their wines. Certainly, experimentation could and would answer questions, but the society was unable to recognize that the existing confusion arose from the conflicting evidence provided by experimentation and its interpretation.[30]

It seems that no viticulturalist had heard of Pierre Duhem's demolition of the Baconian idea of an *experimentum crucis,* but that was for physics, and the practitioners of the life sciences have always believed that God will give them a degree of certainty denied to physicists. No one has yet written a book on how the laws of biology lie.[31]

There was little sign of doubt in Ravaz's magisterial reply to Millardet. Nor did Ravaz mention the gloating and complaining of Gouy and his cronies, although their criticisms received incidental refutation in the reply.[32] Ravaz's conciliatory approach to Millardet recognized the classic character of Millardet's first research on the resistance of vines to phylloxera. He even praised the competence and good faith of the creators of such interesting hybrids, whose resistance to phylloxera, he hastened to add, had not yet received any solid proof. This was why Ravaz placed himself in the camp of hopeful skeptics who did not condemn the Franco-Americans on a priori grounds, for it was possible that some of these vines that were very resistant to phylloxera existed. Ravaz linked the value of the reasons for the choice of hybrid vines to the value of the methods of

[29] "La méthode de M. le professeur Ravaz pour la détermination des résistances phylloxér-iques," *RHFA,* no. 8 (1897).

[30] Ibid., p. 169.

[31] Cf. Nancy Cartwright, *How the Laws of Physics Lie* (Cambridge, 1983).

[32] Ravaz, "Sur la résistance phylloxérique. Réponse à M. Millardet," *RV,* no. 8 (1897): 688–94.

selection that were used. Methods have to be judged according to the results produced, including, in this case, some authentic bad vines. Different methods produced different realities. Ravaz was less than impressed by the results obtained by the hybridizers.

Ravaz's urgent concern in 1897 was to find the right vine for the right soil, and that meant finding a good method of selection. In his study, Millardet had covered the same territory; Ravaz even followed the order used by Millardet. But a patina of politeness covered serious differences of opinion. First, Ravaz criticized the ostensibly fertile method of studying new varieties in the vineyard, large-scale experimentation – *en plein champ, en grande culture.* The weak point of this method was simply that it put the cart before the horse in assuming that there already existed numerous plantations of the varieties in question; in a way it assumed what had to be proved. It is not surprising that it rendered contradictory results. An egregious example: a few years earlier the Société des agriculteurs de France carried out an inquiry on the principal American varieties cultivated as rootstocks and direct producers. The responses were favorable to the Othello, while it was recognized that the Jacquez was very valuable as a rootstock. A year after this extensive inquiry there was general agreement in the vine world on striking both the Jacquez and the Othello from the list of resistant varieties.

Now it was not easy to tell if a vine could resist phylloxera, although it was easy to tell if it did not when it died; the problem was what to conclude when it was slightly affected by the insect. One could be certain about the resistance of a vine only when one knew that it did not have any resistance to phylloxera. So Ravaz concluded that the contradictory results given by the large-scale method of experimentation made it useless as a method of selection of resistant vines. Worse, the frequent progressive deterioration of vines from phylloxera over a period of several years made it difficult to come to any conclusion about long-term results. The interpretation of observations made in open fields was uncertain at best.

On the basis of Gouy's exegesis of Ravaz's work, one might conclude that Ravaz was a hard-line defender of pot science. But a careful reading of Ravaz soon establishes that this is the exaggerated accusation of a determined opponent, not the sober estimation of an objective reviewer. Not all pot experiments are equal. Ravaz clearly pointed out that isolated pot cultivation of each new variety in the presence of phylloxera did not give good results. The insect was fickle, avoiding some pots, frolicking in others, avoiding some nonresistant plants while damaging the roots of plants known to be resistant. The puzzled scientist could only look for another method of research.

What method? The comparative method seemed initially to promise better results because the variety of vine to be tested for resistance to phylloxera was placed next to a control plant whose resistance was known. One could always compare the plants in the pot where the infestation succeeded, without worrying about the results being falsified by noninfestation in certain pots that should have been infested according to the rules of the experiment but perversely ignored by the insect. Millardet had used this method; Ravaz had derived results from it for three years. Like similar methods, it was flawed because of the curious fact that it was a great way of developing nodosities on roots. Worse, the most resistant plants developed the most nodosities. True, this meant that these lesions were not very serious; Victor Gazin had shown that this was true even in vineyards. What the method showed was the phylloxeric receptivity of the radicles of the plants, which was interesting but of no use in determining the resistance of a vine. More often than not, the least resistant vines were those whose radicles were least attacked. So the receptivity of the radicle to phylloxera was irrelevant to attempts to answer the question concerning the resistance of the plant to an insect whose complex behavior was far from known.

The ingenious experiments of Millardet at Villat-de-Vic (Montagnac), whatever their results, seem to have been vitiated by his choice of the *riparia* vine as the control plant. In his experiment, Ravaz found that when Millardet's 33A2 vine was grafted onto an infected *riparia gloire de Montpellier,* the phylloxera remained for over three years only on *riparia*'s radicles, which were loaded with nodosities, whereas the radicles and roots of the other vine remained uninfected. One could conclude that the 33A2 was the resistant vine and the *riparia* the nonresistant vine, exactly the reverse of what Millardet had concluded. Ravaz lived in a time when scientists had the luxury of believing in the possibility of objectivity secured through repeatable experiments in which experimenters got exactly the same results. He logically concluded that it was the method that was at fault, not that he needed to reexamine the nature of the scientific enterprise itself.

The unpredictable behavior of *Phylloxera vastatrix* forced Ravaz to rethink his approach to discovering the most resistant vine that would also produce the best wine. He was unwilling to believe that Millardet could at his convenience always direct the insect to perform like a method actor in the right place at the right time. Nor did he accept the optimistic hypothesis that several defective methods would control one another, producing in Keplerian fashion a sum of errors that led to the right answer in the end. Defective methods had been useful in indicating mistakes to be avoided, but the inability to specify the resistance of a vine had made any rigorous selection impossible. Ravaz's fertile mind therefore dreamed up a new method, which produced such alarming results for Millardet's work that Millardet subjected it to a serious critique in the *Revue de viticulture* itself. Of course, Ravaz was unshaken by the fact that Millardet thought that his experiments were flawed.

What was Ravaz's new method that put Millardet's Bordelais science in doubt? First, there was "an indisputable fact," meaning a fact "verified by competent observers" after Ravaz had made his claim: the lesions on radicles are of no importance, or nearly no importance – in science it is crucial to be as precise as possible. Millardet agreed. Second, the receptivity of the radicles to phylloxera had nothing to do with the resistance of the plant to the insect. This may have been a less indisputable fact, but it was well established and made the first fact more important. Third, it was important to know only the receptivity of the roots (the organs on which the insects formed the tuberosities), which according to both Millardet and Ravaz were the only serious site for lesions. To obtain these lesions with certainty, an arrangement was needed to make the insect attack the roots. As all species of vine carried tuberosities, there was no point in just examining those on one plant: the plant with an unknown resistance had to be compared under a microscope with one of known resistance. Thus the more resistant plant would be known.

The triumphant advance of Ravaz's argument seems at this point to need only Q.E.D. for its conclusion, except for the man in the vineyard. This was, after all, pure pot science. What about real vineyards? No problem, argued Ravaz, for the method could also be applied in the field, although there was no obvious reason for doing so. Pot culture furnished all the cases that one would meet in practice; so science done "en grande culture" or in big field experiments would have no advantage over that done in pots. If one is able to control the type of soil used, fertile or poor, one can control the vigor and development of the vine, thus obliterating any possible distinction between a vine grown in a pot and a vine grown in the field, at least from the scientific point of view. But soil was unimportant for Ravaz's experiments unless it was an obstacle to the multiplication of phylloxera. This comparative method could establish whether a given vine was equal, inferior, or superior in resistance to the control vine, whose resistance was known.

Ravaz went on to score a few more points against his critics, including a denial of an important assumption of Millardet that he eliminated the radicles (with their nodosities), thus depriving the plant of its principal protection against the insect. But he ended on a conciliatory note. He was not judging Franco-American hybrids to be unusable just because his experiments showed that they were more susceptible to phylloxera than the Americans were. It was not always necessary to have vines as highly resistant as *cordifolia* × *rupestris* and *riparia,* for in certain soils a lower degree of resistance would be acceptable. What counted for Ravaz was that knowing the precise degree of resistance of the hybrids would allow him to predict the chances of success if vines were planted under certain conditions. And this knowledge was of immense importance in making a decision on the type of vine to plant in a given type of soil in certain situations. Ravaz also pointed out that his work was based on the first research done by Millardet on vine resistance to phylloxera. Ravaz piously observed that being classic did not diminish the value of that research, and by this observation certainly did diminish its contemporary value even as he established its historical significance.

The quarrel over pot science was part of the larger debate over grafting French *vinifera* vines to American rootstocks. As the nineteenth century drew to a close, it became clear that grafting was the best long-term weapon against phylloxera. Arriving at this conclusion was a long, painful process for scientists and growers, especially the producers of quality wines. What were the apparently successful alternatives to grafting that allowed so many growers to delay for so long in adopting grafting as the program of salvation for the vineyards?

Fighting Phylloxera by Flooding Vineyards

After phylloxera hit France it was soon noticed that the vines of the sandy soils of the Mediterranean littoral did not succumb to the insect. While scientists quarreled about the explanation of the inability of the insect to attack the vines in sandy soils – the movement of water in the sand, especially near tidal waters, seems to destroy the larva and the eggs – producers extended their holdings into soils having a clay content of less than 3 percent, the ideal soil having at least 60 percent siliceous rather than calcium sand. Only a few areas outside the Languedoc conformed to this model: the Landes, the Vendée, the Ile de Ré, and the alluvial river sands of the Rhône valley. None were producers of high-quality wine. Fortunately for the Graves and the Médoc, some of their vineyards had a high enough sand content in their soils to slow down the spread of the insect, if not enough to kill it.[33]

Another successful form of protection that became obvious to growers strong in analogical reasoning was drowning the pest in flooded vineyards, a procedure called the Faucon system, named after Louis Faucon, the owner of 21 ha at Gravison (Bouches-du-Rhône). In 1875, after five years of flooding and fertilizing, Faucon produced 2480 hl of wine compared with 925 hl in 1867, a pre-phylloxera year when the vines were not fertilized, and 45 hl in 1868, the first year of invasion. The duration of flooding the vineyard with 20 to 25 cm of water varied from 35 to 40 days in the fall to 45 to 50 days in the winter. Faucon obtained permission to take

---

[33] See Pouget, *Histoire de la lutte,* pp. 42–4, and Ordish, *The Great Wine Blight,* pp. 94–6.

water from the canal des Alpines. Vines recovered after four years of flooding with heavy fertilization and produced high yields up to 100 hl per ha and more.

Flooding became an important factor in wine production. By 1883 an estimated 20,000 ha of submerged land was producing between one and a half to two million hl of wine, worth more than 45 million francs. By 1896 nearly 40,000 ha were being flooded in the Bouches-du-Rhône, Gard, Hérault, Aude, and Gironde (only the *palus* or riverbanks), all departments where water could be pumped into the vineyards for about fifty days. Although the Faucon system was less expensive than chemical pest control, it was still limited in use because of the amount and cost of the water required. Production costs were also increased by a greater susceptibility of immersed vines to fungus diseases. Canals, including the canal du Midi for the Hérault and Aude, were the chief source of water. State authorization and financial help, amounting to nearly two and one-half million francs in 1880, were necessary for the diversion of water from the canals. If grafting had not become the major weapon in the fight against phylloxera, up to 100,000 ha could have been flooded to produce eight to nine million hl of wine worth 250 to 300 million francs, including 20 to 25 million in taxes and an equal amount for transportation. The Faucon system is an important part of the history of the national effort to stop phylloxera.[34] But it has not captured the historical imagination like the use of insecticides, a program that had the advantages of enjoying major scientific support and being the preferred treatment in the best vineyards.

Chemical Warfare against Phylloxera

It seems reasonable to classify the use of carbon disulfide (CS_2) to kill the phylloxera insect as the first large-scale (and often excessive, not to mention overconfident) chemical assault on agricultural pests. CS_2, in use as a powerful industrial solvent since its discovery in 1796 by the German chemist Lampadius, is highly toxic, volatile, and flammable. After the 1850s, CS_2 was used as an insecticide against weevils in grain silos. Paul Thenard was perhaps the first to use it against the great vine pest. He was the chemist son of a much more famous chemist and owner of the Domaine Thenard in Givry. In an experiment on two estates in the Gironde, Thenard had no trouble killing the insects, but the death of the few vines

---

[34] Pouget, pp. 44–6; Ordish, pp. 75–9; and Louis Faucon, *Guérison des vignes phylloxérées* (Montpellier, 1874), and *Mémoire sur la maladie de la vigne et sur son traitement par le procédé de la submersion,* published in the Mémoires présentés par divers savants à l'Académie des sciences, 1874. Faucon spread the wet word to winegrowers through the *Messager agricole du Midi.*

treated was an unintended consequence. The vines rose again the next August, too late to prevent the spread of the bad news that killed the experiment.

The chemical treatment required 150 to 200 kg of carbon disulfide per hectare to be injected into the soil around the vine.[35] Hundreds of thousands of hectares of land were injected with CS_2, some areas for decades; yet the viticultural literature is silent on what we would call environmental impact. One is tempted to explain this silence by fitting it into the notorious lack of concern over the use of poisons that was characteristic of scientific agriculture. The spraying of copper sulfate on vine leaves to kill fungus diseases did stimulate a debate, however; so it may be that the consciousness of the danger of poisoning the soil just took a longer time to develop because of the obvious difference between burying a poison in the soil and putting it on the visible fruit-producing plant. Not to mention that some of the most politically influential people in France depended on the poison for profits and production.

Railroad companies, which profited from the transport of wine and insecticides, soon concluded that it was in their interest to form an alliance with scientists against phylloxera. A.-F. Marion, a natural scientist of the faculty of sciences in Marseille, was among the first to carry out successful large-scale experiments on Provençal vineyards in the 1870s. His research was supported by the Compagnie des chemins de fer de Paris (PLM), one of the big profit makers in transporting products for the viticultural industry, especially carrying *gros rouge* from the Midi to the North. Marion had the right to use a special letterhead of the PLM company: "Service spécial pour combattre le phylloxera."

The railway companies were generous with subsidies and free travel to scientists who worked on insecticides. But when Valéry Mayet, a professor at the school of agriculture in Montpellier, was included in a list of free travel requests for people doing work for the phylloxera commission, the PLM company declined to issue him a card because Mayet was pro-Americanist. Company policy was to help only those researchers opposed to the introduction of American vines and therefore necessarily supporters of the use of insecticides. The commission granted subsidies to Mayet anyway. As in the case of Foëx, this backing of Mayet indicated that the commission supported all scientifically legitimate research on phylloxera, whether or not it was profitable to railway companies.

Marion had no illusions about unifying differing scientific interpretations of the biological phenomenon; the fact was that the data furnished guidelines for agricultural practice. This curiosity of science, which ex-

---

[35] Marcel Lachiver, *Vins, vignes et vignerons: Histoire du vignoble français* (1988), pp. 423–34.

plains part of its astounding success, is often overlooked by historians concentrating on the froth of scientific squabbles.[36] Careful studies of effective doses of the poison, of soils, humidity, and temperatures, led scientists to a sufficiently precise knowledge of the action of the insectide to keep in production many thousands of hectares that would have quickly succumbed to phylloxera.

After 1875 the Commission supérieure du phylloxera, presided over by J.-B. Dumas, chemist and man of influence in governmental circles, recommended the use of CS_2 over all other procedures. The PLM supported application of the chemical and research on its use. An injector technology was developed. It is difficult to say if the winegrowers profited as much as the manufacturers of the chemicals, who also gained further profits from selling extra fertilizer needed by vines bombarded by a toxic substance. In the late 1880s about 68,000 ha were being sterilized with thousands of tons of CS_2, at an annual cost of about 300 francs (450 the first year) per ha. After 1880 American varieties with insufficient resistance to phylloxera were also treated. Some scientists were happy with another scientific triumph for mankind; a few modest ones thought that a scientific problem had been solved and that the details of application could be left to industry and bureaucracy.

Dumas solved a major problem in the application of the insecticide when he came up with the idea of using sulfocarbonates (alkaline monosulfides combined with carbon disulfide) in order to overcome the disadvantage of the evanescent nature of carbon disulfide. Potassium sulfocarbonate, for example, would release poisons slowly and supply needed potash to the plant at the same time. This was a sort of synergism of "weed and feed." Dumas pointed out that these sulfocarbonates could only hold back the invasion: there was no universal cure for phylloxera. And the price of application was high because of the need for heavy equipment and lots of water. It was definitely the rich man's insecticide. But then so was CS_2. Poor winegrowers, and some not so poor, ruined by the latest science or by fraudulent products, could only turn to a hybrid vine immune to phylloxera but producing a drinkable liquid.[37]

The threat of phylloxera to local and regional economies led to the organization of departmental commissions of experts and growers. These organizations specialized in studying and combating the insect in a way that was beyond the capacity of more general organizations such as the Société centrale d'agriculture de l'Hérault.

[36] René Pijassou, "Les temps difficiles, 1880–1920," in Higounet, ed., La seigneurie et le vignoble de Château Latour, 2:455; A.-F. Marion, Traitement des vignes phylloxérées par le sulfure de carbone. Rapport sur les expériences et sur les applications en grande culture effectuées en 1877 (1877).
[37] Pouget, Histoire de la lutte, pp. 33–42, and Ordish, The Great Wine Blight, pp. 80–102.

Henri Marès, a prolific writer on the vine, its diseases and treatments, presided over the anti-phylloxera departmental commission of the Hérault. Departmental field experiments with alkaline sulfocarbonates began in 1872 after laboratory trials had been completed. One hundred and forty trials had been completed by 1873, but Marès was skeptical about the efficacy of insecticides, perhaps because he then believed that the disease was less serious than generally accepted. By 1876, when phylloxera had made impressive progress, Marès, recognizing his mistake, praised the use of sulfocarbonates. In the experiments, which included control vines, many substances were tested, including cow urine (rated as no. 3 in effectiveness), and key variables (age and variety of vine, soil) were noted. In the end it was clear that in the Hérault only one or two insecticides could seriously fend off the insect, generally for a limited time. Other departments soon imitated the pioneering Hérault.

The rapid spread of phylloxera, inspiring fear of a threat to the national economy, led the minister of agriculture to create the national Commission supérieure du Phylloxera in 1871. As the government hoped that the disease could be stopped by science, Dumas was made president of a mixed group of scientists, bureaucrats, and politicians, many of whom were landowners. Dumas, permanent secretary of the Academy of Sciences, was perhaps the most powerful scientist in France. It was also logical to appoint a chemist as head of a group that was looking for an extermination scheme. The commission did come up with a killer technology, but the poison did not stop the insect. Although Planchon was a member of the commission, no one seems to have thought much about the study of the plant itself and the degree of its resistance to the insect. Again, this was not unreasonable, for government and industry had not been wrong in hoping for scientific solutions to other economic problems when it had asked Pasteur to look for cures for wine and silkworm diseases.

Fully occupied with seeking cures for other diseases, Pasteur did not come up with any good practical idea for fighting phylloxera. After the Academy of sciences learned that the insect could not exist on roots covered with mycelia, Pasteur observed that it might be possible to use fungi as a biological weapon against phylloxera. It turned out that the roots, killed by the mycelia, were of no interest to the insects, who were looking for food rather than avoiding an enemy. Pasteur also came up with the idea of infecting the insects with the corpuscles of sick silkworms, but his plan of biological warfare was never experimentally tested. Presenting to the Academy a note by Maxime Cornu and Charles Brongniart detailing their observations of small flies killed by the parasite fungus *Entomophthora,* Dumas noted that similar observations had been made in other countries, though no one had taken up Pasteur's challenge to find a fungus to infect the phylloxera aphid. Pasteur became president of the commission in 1885. But no one ever got close to controlling the phylloxera aphid by infecting it with a deadly spore.[38]

Dumas talked enthusiastically about the potential of research on fungi because he saw it as one of the few possible sources of a cure for phylloxera that would win the enormous governmental prize of 300,000 francs. Some people wanted to abolish the prize, which seemed to have been useless as a stimulus to discovery. No one ever won the prize. Many people thought that it should have gone to the Bordeaux grower L. Laliman, but this enthusiastic importer of American vines was also included among those blamed for introducing the pest to France. He seems to have been the first to come up with the idea of saving the vines through grafting. Laliman made many pathetic attempts to get the award. Scientists could not overcome their belief that someone outside the scientific fraternity was incapable of solving a major scientific problem.

Dumas's commission received many quack remedies for phylloxera. Urine was a common element in several of them, perhaps because it fertilized the plant into a brief final fling of growth. A piss prize should have gone to a concoction of horse-urine syrup. E. Maumené advocated planting thyme, asparagus, or beans near the vines, to drive out the insects, though he could personally vouch for only the thyme's success in doing so. The miracle remedy was a glass of water from the fountain in Lourdes.

In 1878–9 the commission was transformed from a research organization into a consultative and administrative body to guide the agricultural bureaucracy in its fight against phylloxera. Dumas and his scientific cronies believed that the scientific problem had been solved: carbon disulfide and sulfocarbonates would kill off *Phylloxera vastastrix* – not the last chemical illusion in agriculture.

And not for the last time would botany have the smug satisfaction of proving chemistry wrong, as it became increasingly clear after 1880 that the chemical fix could at best be only a temporary expedient. Yet for years an unfortunate exercise of scientific and politico-administrative power discouraged grafting on American rootstocks. Legal blocks were set up to the importation and introduction of American vines in specified zones classified according to the level of infection.[39] This is a case where one of our clichés of historical explanation, "Paris versus the provinces" (read the Hérault), has some obvious value. At the time, the position of Paris did not look totally unreasonable to a chemist or a desperate grower or politician. And, after all, the century had seen some striking chemical successes, even in agriculture, and they were by no means at an end. In

---

[38] Maxime Cornu et Charles Brongniart, *Sur une épidémie des insectes diptères causée par un champignon . . .* (Extrait du *Bulletin hebdomadaire de l'Association scientifique,* 2d ser., no. 3, n.d.; but the second series, 1–14, is dated 1880–7).

[39] Pouget, *Historie de la lutte,* pp. 33–9, 48, 59–62.

this particular disaster, the traditional approach did not work. Whigs will be glad to learn that scientists were finally not only able to change their minds but also to take the lead in finding a workable permanent solution to the problem as well as a useful temporary expedient.

One of the areas in which phylloxera was most difficult to conquer by planting grafted vines provided one of the best successes for Dumas's insecticide, or at least for the scientific experiment testing it. The phylloxera committee of the Cognac, presided over by Edouard Martell, famous cognac producer and member of the Chamber of Deputies, collected more than the 8000 francs needed annually to support a researcher chosen by the Academy. Montpellier's school of agriculture, which provided a researcher to do experiments for the departmental commission of the Hérault, was the model to be imitated. For Cognac the Academy chose Pierre Mouillefert of the school of agriculture of Grignon; Maxime Cornu of the Muséum d'histoire naturelle was in charge. Recognizing that the oily skin of the aphid protected it against liquids, they looked for the most effective gas-producing insecticide. Mouillefert tested seven categories of potential insecticide: fertilizing, neutral, and salt substances, essences and plant products, empyreumatic products (coaltar oil, petroleum), and sulfur products (carbon disulfide, potassium sulfide, and so on).

Experimenting first on infected roots in tubes and flasks, as recommended by Dumas, Mouillefert reduced to fifteen the number of products that would kill the insect without killing the vine. Then experiments in pots and field, especially in the vineyard, left four or five products giving general satisfaction. The process of elimination produced increasingly restrictive statements until, in the final section of the research report, published by the Academy, Mouillefert reached a conclusion that was sobering if satisfying (particularly to Dumas). Only one category of substance produced significant results against phylloxera: differently based sulfocarbonates, especially of potassium and sodium. The problem with carbon disulfide was that the liquid evaporated too quickly, before the gas had time to go through the soil to kill all the insects, and repeated applications killed or damaged the vine. The ideal would have been carbon disulfide in combination with another substance to prevent quick evaporation.

Dumas's great scientific contribution to the fight against phylloxera had been to find this combination in sulfocarbonates. And scientists engaged by Dumas, who was chosen by the government and the Academy to direct the battle, found that his potassium sulfocarbonate was the most successful insecticide in laboratory, pot, and field trials. The satisfaction of the pro-insecticide men was summed up by Cornu and Mouillefert: "Science has fulfilled its mission, and it's now up to agriculture to accomplish its mission," by perfecting the use of potassium sulfocarbonate *en grande culture.*

Mouillefert and the engineer Félix Hembert helped winegrowers by developing a total system (including a steam engine) for applying this insecticide, all for 6000 francs. Labor costs added another 310.5 francs per ha. Big property owners could afford the system, and small ones could too if they formed cooperative groups. Mouillefert was also among those eager to take science to the people with the organization of a national popular society for the treatment and reconstitution of vineyards infected by phylloxera. In this ideal world, science, business, and small producers lived in mutually beneficial harmony.[40]

A certain skepticism existed in the wine business concerning recommendations by the Académie des sciences. One of the leading vine and wine publications noted that "M. Dumas has read many things to the Academy without slowing up the advance of the terrible louse at all."[41] B. Cauvy, a professor at the school of pharmacy in Montpellier, unjustly accused Dumas of promoting Pasteurian ideas that did not work and, more serious from Cauvy's viewpoint, of excluding from the list of state-recommended insecticides all sulfocarbonates except those that were potassium-based. One of the country's enterprising professors, Cauvy manufactured his own calcium-based sulfocarbonate. In 1875 a circular of the ministry of agriculture recommended the use of alkaline sulfocarbonates against phylloxera.

Many of the attacks on Dumas were launched in 1874, early in the campaign against phylloxera, when ignorance and tension were at a high level, but as late as October 1878, Maumené wrote a letter to Dumas stating that potassium sulfocarbonate had totally failed. Maumené was also a critic of Pasteur's theory of fermentation. He did not have a distinguished university career. It would be unfair to say that it was as distinguished as his science, but like many of the critics of the scientific establishment, he failed to recognize that great scientists are sometimes right.

The only effective large-scale chemical treatment of vines over long periods was carried out on estates producing commercially recognized high-quality wines. In the Médoc some *premiers crus classés* were treated until 1914 and not replanted on American roots until after the war. In Burgundy the old vines of certain famous vineyards were kept productive even later: Romanée-Conti did not produce a grape crop on a reconstructed vineyard until 1952. Grafted vines, introduced very slowly for the French *grands crus,* replaced the vines that died or else were planted in small plots; the pace picked up after 1900, essentially finishing in the 1920s. This procedure had the great advantage of permitting comparative testing of the

---

[40] Maxime Cornu et Pierre Mouillefert, *Expériences faites à la station viticole de Cognac dans le but de trouver un procédé efficace pour combattre le phylloxera* (1876). "Extrait des Mémoires préséntés par divers savants à l'Académie des sciences," 2d ser.

[41] *Journal viticole. Revue de la vigne, des vins et des spiritueux,* Sept. 5, 1874.

wine from grafted vines with that from nongrafted vines (*les plants francs de pied*). The quality of the wines did not vary with the vines, or if it did, no one could tell.

In the Médoc, generous use of carbon disulfide kept the plant louse at bay in certain terrains. Bordeaux's chamber of commerce estimated that in 1881 one-third of the Bordelais vineyards was destroyed and one-third was under attack. Whatever the accuracy of this guess-estimate, the chamber could hardly have erred in predicting an economic disaster for the region, except for producers who could afford to use insecticides. As late as 1904 the Médoc still had 66,000 hectares of ungrafted vines; it already had 70,000 of grafted vines. Grafting in Bordeaux was delayed by the opposition of some *négociants* to American roots. Motivated by caution and ignorance, they feared a possible drop in wine quality and therefore in profits. Up to about 1914 French *vinifera* vines outnumbered grafted ones at Château Latour. The First World War, with its shortages of chemicals and labor, encouraged grafting, which led to the dropping of the clause in the wine-supply contract that prohibited grafting. By 1920–1 the entire crop was being produced on grafted vines.[42] This experience was typical of big estates in the Gironde.

The use of insecticides proved an effective weapon until grafting could demonstrate that the wine of the grafted wine was no different in quality from that of the old ungrafted *vinifera* vine. That effort required a considerable scientific campaign led by Montpellier and aided by an agricultural bureaucracy, which was itself slowly converted to the grafted gospel.

[42] Pijassou, "Les temps difficiles, 1880–1920," in Higounet, 2:471, and Archives de l'Académie des sciences, no. 1635.

2

Scientific Programs for the Spread of the Grafted Vine

Because the phylloxera aphid destroyed the root system of the ungrafted *vinifera* vine, it was necessary to graft scions or the fruit-producing part of European vines onto pure American or hybrid Franco-American resistant rootstocks. At first it seemed that the replanting of a vineyard was a simple matter depending on time, labor, and capital. In reality, the reconstitution of the vineyards proved difficult because the rootstocks had to be able to survive in different soils, to be compatible with the varieties of *vinifera* vines to which they were grafted, and to be capable of giving the desired yield. So vines had to be crossed in various combinations. The crossings were the subject of a great experiment, or rather many experiments, big and small. In view of the time and expense required to complete these risky trials, the use of ungrafted, disease-resistant American vines or direct-production hybrids seemed an easy way out. But the wine from these vines was vastly inferior to that from straight *viniferas* and French vines grafted on American and hybrid rootstocks.

The grape vine has engaged in an orgy of multiplication of species since it came into existence many millions of years ago. There are now about 5000 varieties of vines covering nine and one-half million hectares in the world: hybrids cover about 700,000 ha and the rest are European vines from *Vitis vinifera*.[1] The leading varieties of vines used for wine production include both the black and white grenache, the carignan, the black merlot, the cabernet sauvignon, the semillon, the chenin, the riesling, the aramon, the Müller-Thurgau, the colombard, the black gamay, the black pinot, the cabernet franc, and the syrah. Some varieties are more notorious than others.

---

[1] Ordinary language fails to convey the complex reality of the plant world. "From the technical point of view, the *cépage* cannot be considered as a *variété,* for it does not reproduce identically by seed," but only by various vegetative means. The botanical term *cultivar* is also a little different and "corresponds to a clone coming from a seed, multiplied then by vegetative means so that all the descendants are identical." This means that some *cépages,* like the Alicante Bouschet, are true cultivars if they come from artificial crossing. Pierre Galet, *Cépages et vignobles de France,* vol. 2: *L'ampélographie française* (Montpellier, 1988), pp. 2–3.

The genus *Vitis* is one of 18 genera in the family of vines *Vitaceae,* which includes over 1100 species and represents more than 3000 taxa. (A taxon is the name of a taxonomic group in a formal system of nomenclature.) J.-E. Planchon, one of Montpellier's great experts on the vine, divided the genus into two sections: *Euvitis* (called *Vitis* in the international code of botanical nomenclature of 1961) and *Muscadinia.* More important for plant breeding is the chromosomic barrier that makes it difficult to cross-fertilize vines between the two sections: *Vitis* has a chromosome number ($2n = 38/n = 19$ pairs) different from that of *Muscadinia* ($2n = 40/n = 20$ pairs). In sexual reproduction the two sets of chromosomes unite but remain constant for each species, whether plant or animal.[2] Even botanists have become excited by the vine (see Figure 3).

Botanists have been squabbling over *Vitis* for a century and a half. An English school created confusion by collapsing the genera into one: *Vitis.* The French school pursued a Cartesian policy of understanding by division: Planchon constructed six genera out of the *Vitis* that Tournefort had defined at the end of the seventeenth century. Recently *Vitis* had 108 species – some fossils and some doubtful – but 60 species were well identified. A basic point is the identification of 33 American and 27 Asian species. The fourth edition of Gustave Foëx's great text in 1895 listed nine species in the section *Vitis;* today Pierre Galet's treatise lists 11, among which are *Labruscae, Cinerae* (includes *V. berlandieri*), *Cordifoliae* (includes *V. monticola*), *Ripariae* (*V. riparia* and *V. rupestris*), and *Viniferae* (*V. vinifera* and *V. silvestris*).[3] All species of the section *Vitis* can be crossed to produce direct-production hybrids (nongrafted plants) or rootstocks to be used in grafting (see Tables 2 and 3).

In the 1870s and the 1880s scientists tried to cross the two sections because of the high resistance of *Muscadina* (*V. rotundifolia*) to parasites. They got no useful results. The more different parent plants are genetically, the higher the probability that the offspring will be sterile – mule plants, one might say. Nowadays it is known that the specific genetic and

---

[2] Variations depend on random alignments, during meiosis, of chromosomes that are divided into short lengths or units of the material of inheritance (DNA), genes. Walther H. Muller, *Botany: A Functional Approach* (New York, 1963), p. 192.

[3] I hate to spoil this botanical clarity by giving popular and often identical names for different species, but here are a few examples that show how really well known these vines are – by different names:

V. labrusca: fox grape, swamp grape, skunk grape, etc.;
V. berlandieri: mountain grape, sugar grape, Spanish grape, etc.;
V. monticola: sweet mountain grape, sugar grape, etc.;
V. riparia: frost grape, river grape, June grape, August grape, etc.;
V. rupestris: July grape, sand grape, sugar grape, bush grape, etc.;
V. rotundifolia: muscadine grape, fox grape, Bull grape, Scuppernong, etc.

See Galet, *Cépages et vignobles de France,* vol. 1: *Les vignes américaines,* 2d ed. (Montpellier, 1988).

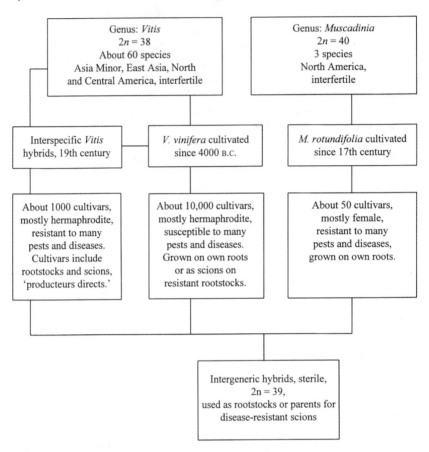

Figure 3. Summary: History and Present Status of Grapevine Breeding (1988). Modified from Alleweldt and Possingham (1988); reproduced with permission from Mullins et al., *Biology of the Grapevine* (Cambridge, 1992), p. 212.

cytoplasmic barriers are weaker in proportion to distance from the first generation. (F_1 is the first filial generation, and $F_1 \times F_2 = F_3$, etc.) Because the barriers are not so great as used to be believed, exogamous marriage promises to open up a new way for the improvement of wine through genetic manipulation. Grafting *Vitis* on *Muscadiniae* (or vice versa) is still a viticultural challenge. Fortunately all the species of *Vitis* can be grafted onto one another. It is not easy to see how the vine would have survived without this sexual compatibility.

For over a hundred years France was "one huge rootstock trial." Many breeders of the thousands of varieties of vines believed in the possibility

Table 2. *Classification of* Vitis *Species (Partial Listing)*

Series	Species and Synonym	Origin
I. Candicansae	*V. candicans*	North America (East)
	V. champinii	North America (East)
	V. doaniana	North America (East)
	V. simpsonii = *V. smalliana*	North America (East)
	V. coriacea = *V. shuttleworthii*	North America (East)
II. Labruscae	*V. labrusca*	North America (East)
	V. coignetiae	Asia
III. Caribaeae	*V. caribaea* = *V. tiliaefolia*	North America (South)
	V. blancoii	North America (East)
	V. lanata	Asia
IV. Arizonae	*V. arizonica*	North America (West)
	V. californica	North America (West)
	V. girdiana	North America (West)
	V. treleasei	North America (West)
V. Cinereae	*V. cinerea*	North America (East)
	V. berlandieri	North America (East)
	V. baileyana	North America (East)
	V. bourgeana	North America (South)
VI. Aestivalae	*V. aestivalis*	North America (East)
	V. linecumii	North America (East)
	V. bicolor = *V. argentifolia*	North America (East)
	V. gigas	North America (East)
	V. rufotomentosa	North America (East)
	V. bourquina	North America (East)
VII. Cordifoliae	*V. cordifolia*	North America (East)
	V. rubra = *V. palmata*	North America (East)
	V. monticola	North America (East)
	V. illex	North America (East)
	V. helleri	North America (East)
VIII. Flexuosae	*V. flexuosa*	Asia
	V. thunbergii	Asia
	V. betulifolia	Asia
	V. reticulata	Asia
	V. amurensis	Asia
	V. piasekii	Asia
	V. embergeri	Asia
	V. pentagona	Asia
IX. Spinosae	*V. armata*	Asia
	V. davidii	Asia
	V. romaneti	Asia

Table 2. *(cont.)*

Series	Species and Synonym	Origin
X. Ripariae	*V. riparia = V. vulpina*	North America (East)
	V. rupestris	North America (East)
XI. Viniferae	*V. vinifera*	West Asia and Middle East

Source: Reproduced with permission from Mullins et al., *Biology of the Grapevine* (Cambridge, 1992), p. 21, based on Galet (1967).

Table 3. *Some Well-Known Cultivars of* Vinifera *Grapes*

Cultivars for Red Wines		Cultivars for White Wines	
Name	Origin	Name	Origin
Aramon	Southern France	Chardonnay	Burgundy,
Cabernet franc	Bordeaux, France		Champagne, France
Cabernet sauvignon	Bordeaux, France	Chenin blanc	Loire Valley, France
Carignane	Southern France	Colombard	Charente, France
Gamay	Beaujolais, France	Traminer	Germany
Grenache	Rhône Valley, France	Riesling	Germany
Merlot	Bordeaux, France	Sauvignon blanc	Loire Valley, France
Mataro	Spain,	Semillon	Bordeaux, France
	Southern France	Sylvaner	Germany, Austria
Nebbiolo	Northern Italy	Müller-Thurgau	Germany
Pinot noir	Burgundy,	Emerald Riesling	Davis, California
	Champagne, France	Ugni-Blanc	Italy
Syrah (syn. Shiraz)	Rhône Valley, France		
Ruby Cabernet	Davis, California		
Carmine	Davis, California		

Source: Reproduced with permission from Mullins et al., *Biology of the Grapevine* (Cambridge, 1992), p. 28.

of producing a rootstock for general use, an idea that Lucie Morton called "the myth of the universal rootstock." This viticultural miracle has never occurred, and growers have had to select from a few dozen rootstocks, depending on soil, climate, and the type of vine they want to graft. Three American species "provide a good base for understanding most commercial rootstocks: *Vitis berlandieri, V. riparia,* and *V. rupestris. V. berlandieri* × *V. riparia* provides over 20% of the rootstocks in French vineyards. It

Table 4. *Grapevine Rootstocks: Some Long-Established and Newly Bred Cultivars*

Rupestris St. George (du Lot)	*V. rupestris*
Riparia Gloire de Montpellier	*V. riparia*
3309 Courderc	*V. riparia* × *V. rupestris*
101-14 Millardet et de Grasset	*V. riparia* × *V. rupestris*
Schwarzmann	*V. riparia* × *V. rupestris*
99 Richter/110 Richter	*V. berlandieri* × *V. rupestris*
140 Ruggeri	*V. berlandieri* × *V. rupestris*
1103 Paulsen	*V. berlandieri* × *V. rupestris*
5BB Teleki, selection Kober	*V. berlandieri* × *V. riparia*
SO$_4$	*V. berlandieri* × *V. riparia*
420A Millardet et de Grasset	*V. berlandieri* × *V. riparia*
Ramsey	*V. candicans* × *V. rupestris*
Börner	*V. cinerea* × *V. riparia*
Vialla	*V. labrusca* × *V. riparia*
Ganzin 1 (AxR#1)	*V. vinifera* × *V. rupestris*
41B Millardet et de Grasset	*V. vinifera* × *V. berlandieri*
Fercal	*V. vinifera* × *V. berlandieri*
039-16	*V. vinifera* × *Muscadinia rotundifolia*
043-43	*V. vinifera* × *Muscadinia rotundifolia*

Source: Reproduced with permission from Mullins et al., *Biology of the Grapevine* (Cambridge, 1992), p. 30.

is also widespread in Italy and practically the only hybrid rootstock in Germany" (see Tables 4 and 5).[4]

There were five categories of direct-production hybrids gotten by crossing different species of vines.[5] Most of these hybrids were a cross between two American vines or American and French vines. These binary hybrids were genetically simple, but complex hybrids were soon produced, including as many as six species. These complex creatures could include American, French, Asian, and Russian vines, as well as their combinations. The first hybrids were from seeds; so the father was unknown. Artificial crossings soon ended the mystery: the pollen of the variety chosen as father was

[4] Lucie T. Morton, "The Myth of the Universal Rootstock," *Wines and Vines* 60 (1979): 24–6. Galet, *Cépages et vignobles de France*, 1:233, 305, summarizes the virtues and vices of the standard group of artificially obtained hybrids used for rootstocks: *berlandieri-riparia, berlandieri-rupestris, vinifera-berlandieri,* and *vinifera-rupestris.* The *V. berlandieri,* usually used as the mother in spite of its very small flowers, is a hermaphrodite difficult to castrate; *V. riparia* is the male whose pollen is used to impregnate the *V. berlandieri.* André Créspy, *Viticulture d'aujourd'hui* (Paris, 1987), p. 19, has also created a convenient table showing the principal qualities of the breeder vines used to obtain rootstock.

[5] *Hybrides producteurs-directs, producteurs-directs,* or *plants directs* are the French terms.

Table 5. *Grapevine Breeding: Principal Sources of Resistance*

Character	Species
Abiotic Stress	
Winter hardiness	*V. amurensis, V. riparia*
Lime-chlorosis	*V. vinifera, V. berlandieri*
Salinity	*V. berlandieri*
Fungal Diseases	
Downy mildew	*V. riparia, V. rupestris, V. lincecumii, V. cinerea, V. berlandieri, V. labrusca, M. rotundifolia*
Powdery mildew	*V. aestivalis, V. cinerea, V. berlandieri, V. labrusca, M. rotundifolia*
Botrytis bunch rot	American species
Bacterial Diseases	
Crown gall	*V. amurensis, V. labrusca*
Pierce's disease	*V. caribaea, V. coriacea, V. simpsonii, M. rotundifolia*
Nematodes	
Meloidogyne spp.	*V. champinii, M. rotundifolia, V. longii*
Xiphinema spp.	*V. rufotomentosa, M. rotundifolia*
Insects	
Phylloxera	*V. riparia, V. rupestris, V. berlandieri, V. cinera, V. champinii, M. rotundifolia*

Source: Reproduced with permission from Mullins et al., *Biology of the Grapevine* (Cambridge, 1992), p. 211.

put on the ovary of castrated hermaphroditic flowers of the mother variety or directly on the ovary of female flowers. In the naming of hybrids, convention decrees that mom comes first, followed by the name of the father. With pure rootstocks there are two possibilities: *riparia* × *berlandieri* or vice versa – for example, depending on who gets whom pregnant. Beyond these simple binary hybrids, any degree of bastardy is possible, depending on the mixed origins of the rootstocks crossed: *riparia* × *rupestris* × *vinifera* would provide a tertiary combination. Sexual reproduction or artificial crossings can lead to complexity in the genetic composition of hybrids.[6]

[6] On this topic see the magisterial treatment by Pierre Galet, *Cépages et vignobles de France*, vol. 1: *Les vignes américaines*. I have drawn heavily on his science in my summary of the sex life of hybrids.

The ancient hope was that these hybrids would live and produce directly on their own roots – hence their name – in contrast to the grafted vine, whose rootstock came from one or more American varieties and its scion, or top grape-bearing part, from a variety of *Vitis vinifera* of French blood.

It was only after 1880 that direct-production hybrids became important, with the creations of the famous hybridizers Couderc, Castel, Seibel, and so on. Hybrids were big business, often in places like Denison, Texas, where T. V. Munson created many new varieties using native American grapes. Munson's input into the reconstitution of French vineyards earned him membership in the Legion of Honor in 1888.[7] The aim was to produce an American leafage and root system in a plant that would bear fruit tasting like that of the *vinifera* vine. Good hybrid rootstocks were created even if the direct-production hybrids were too American (foxy) in taste to replace French vinifera vines.[8] Complex hybridizations, or crossings between hybrids, produced varieties of considerable resistance to diseases that also bore satisfactory fruit – the wine was another matter.

Official science, led by Montpellier, conducted a relentless campaign to promote grafted vines and certain rootstocks – some discovered, some invented. Final victory by Montpellier against its opponents came after many battles over both the right rootstock for specific soils in regions producing quality wine and over the quality of wine from direct-production hybrids. It is to these clashes that we now turn.

Cognac: The Berlandieri War in the Charentes

In the seventeenth century the English and Dutch had become enamored of French *eaux-de-vie,* thus provoking a change in the viticulture of the area from the cultivation of grapes that gave good white wine to grapes that gave poor white wine good for distillation. The *crus* of the Grande Champagne and the Petite Champagne became part of the gastronomic aesthetic and the taste of luxury.

Reconstituting vineyards in the Champagne, Burgundy, and especially the devastated Cognac-producing region of the Charentes, was a great accomplishment of official science. Phylloxera appeared in the Charentais vineyards in 1876, but its progress was slowed by the use of insecticides. From the beginning, the replanting of vineyards with American rootstocks was a striking failure in the Charentes. A key test for the new viticulture was the *berlandieri* war in the Charentes, whose calcareous and very friable

---

[7] Thomas Pinney, *A History of Wine in America from the Beginnings to Prohibition* (Berkeley and Los Angeles, 1989), pp. 409–10. Munson left his own literary monument to his work: *Foundations of American Grape Culture* (Denison, Tex. 1909).

[8] On "Fox Grapes and Foxiness," see Pinney, *History of Wine,* appendix 1, pp. 443–7, the best analysis of this ambiguous term, which presumably refers to the taste of labrusca grapes, some varieties of which contain the ester methyl anthranilate, often blamed for the *goût foxé.*

soils were murderously chlorotic for rootstocks. Although it experienced some initial defeats, pro-grafting official science went on to victory or, if it could not win – the opinion of its opponents – it redefined the nature and limits of the victory over phylloxera that was possible in the production of fine wine.

The great hybridizer Couderc had created a successful line of hybrids, but he did not succeed any more than Millardet in creating a hybrid rootstock that could survive both phylloxera and chlorosis, the killer twins of all vine replacements in the Charentes up to that time. Comparable in effect to the young women's anemic disease of the same name, chlorosis weakens its plant host. The prevention of the formation of the green pigment chlorophyll, giving the plant a pale-yellow coloration, is the result of the inability of the roots to absorb enough iron from the soil to satisfy the large demands of the leaves of the grafted plant.[9] The reconstruction of the *vignoble Charentais* was the greatest challenge to viticultural science.

It was too much of a challenge for Couderc to resist, especially after his hybrids had been declared useless for the Charentais by the Station viticole, representing official science. Gouy presented his fellow Ardèchois as "a Benedictine looking for the characters that had disappeared from a palimpsest," interested only in unraveling the enigmas of nature, snatching off masks, unveiling secrets, searching for truth. To test his hybrids in an experimental Charentais vineyard, Couderc bought the estate Tout-Blanc, twelve hectares in Marville. After expanding the estate, he tested 22,000 different hybrids, which grew with a success that astounded all who saw this splatch of green in an arid white countryside. No chlorosis then. Was official science wrong?

It seems that this is more a case of Couderc's being right. Couderc brought seeds from Aubenas to plant at Tout-Blanc so that the roots would, he hoped, acquire resistance or adapt physiologically to phylloxera in the killing Charentais soil. The plants that grew could then be used as roots for grafting on the desired varieties for the region. The key concerns in breeding such plants were affinity, adaptation, fertility, and fertilizer requirement. Couderc had had his failures and mediocre successes in breeding vines, but he had ruthlessly eliminated those vines in trying to create the best possible collection – we might say genetic pool – for vine replacement. Gouy thought that the direct-production vines were most promising and attractive. By comparison, the *rupestris du Lot* and the hybrids of the Ecole de Montpellier, except for 333EM (cabernet sauvignon impregnated by the pollen of the *berlandieri*), were mediocre. Even this 333EM, although 20 percent more productive, was not preferred to the

---

[9] Pouget, *Histoire de la lutte,* p. 72. The culprit: "les sols calcaires riches en ions bicarboniques."

41B because growers were suspicious of the "prominent tuberosities" on its roots.[10] Official science lost again.

Once Couderc had shown what could be done, he became uninterested in pursuing business in the Charentais. Mildew, phylloxera, and chlorosis were conquered, at least in theory. The challenge of black rot remained. Couderc established the Champ d'expériences de Cadoret to try to defeat the new enemy. Meanwhile Ravaz and his station went on with their research operation, backed by big cognac producers and the resources of official science, until the wine and cognac industries could be reestablished. *Sitzfleisch* is as important as talent in solving many scientific problems.[11] Unlike Ravaz, to whom the government and the cognac establishment looked for a definitive defeat of phylloxera, Couderc could claim victory with a half-solution.

Alexis Millardet also took up the challenge of the Charentais soil. No effort shows more clearly than Millardet's the significance of botanical science in the reconstitution of the vineyards. Between 1862 and 1866 he had studied with famous German botanists (Hofmeister, Bary, and Sachs) in Heidelberg and Freiburg-im-Brisgau and received his doctorate in natural science at Strasbourg in 1869.[12] Millardet's doctoral thesis in botany, on the male prothallium in vascular cryptograms, was distinguished by an obsessive emphasis on heredity and an acceptance of the role of variations in producing new forms. In his faculty courses he praised the fertility of the "scientific method" in Darwin's works.

Like Pasteur and Darwin, his research heroes, Millardet kept in close touch with science applied to production. He had an extraordinary knowledge of research on an international scale, including a special contact with American botanists and nurserymen. The *Journal d'agriculture pratique* carried many of his articles. Millardet believed that the resistance of American vines to phylloxera was a strictly hereditary property, dependent on whether or not they descended directly from resistant species. Resistance was not a Newtonian absolute but a relative property dependent on a variety of environmental factors; hence the importance of establishing a scale of resistance for different vines, a guide for the perplexed winegrower.[13]

---

[10] Galet, *Cépages et vignobles de France,* 1:202–3.

[11] P. Gouy, "Les nouveaux porte-greffes en sols crayeux. Le champ d'expériences de M. Couderc à Tout-Blanc," *RHFA,* no. 1 (1898), pp. 20–2. Ravaz returned to the ENA in Montpellier in 1897. His successor, J.-M. Guillon, followed bravely in the research steps of the master.

[12] W. Hofmeister was famous for his research "on the life cycle and reproduction of the cryptograms and the homologies of their reproductive structures." His *Comparative Studies of the Cryptograms* appeared in 1851. Anton de Bary also spent his life following fungi, analyzing their life cycle. Ernst Mayr, *The Growth of Biological Thought* (Cambridge, Mass., 1982), pp. 216–17.

[13] A. Millardet, *Le prothallium mâle des cryptogames vasculaires* (thèse de botanique, Doctorat ès sciences naturelles, Strasbourg, 1969); *La botanique, son objet, son importance*

Millardet became a professor of botany at Bordeaux in 1876. He did a great deal of work in pure botany but is best known for his work on the use of hybrids and grafted vines in the conquest of phylloxera and the fungus diseases. In his experimental vineyard of Marville, near Cognac, he grafted vines of the southwest onto his sturdy hybrid no. 41B, the result of years of crossing *V. chasselas* with *V. berlandieri*. Millardet was part of official science, but that of Bordeaux, which seemed far enough away from the Rhône and the school in Montpellier for Gouy to look upon it benignly. Millardet was not "an armchair viticulturist."

The need to move with dispatch against the American louse was appreciated by commerce. The Station d'études viticoles, created at Cognac, was headed by Cornu and Mouillefort, authors of classic studies on the insect's behavior, the damage it caused, and treating sick vines with potassium sulfocarbonate. Some vineyards were able to continue producing cognac and to ensure export for a while. Because production required the existence of large stocks for aging and blending, the cognac business had a great commercial advantage over wine areas, where the product was sold on a yearly basis. But a solution had to be found quickly. The direct-production hybrid was no good for producing high-quality cognac. Martell distilled 150 casks from direct producers in the Grande Champagne. The liquid was not without commercial virtue, though clearly inferior even to the *eaux-de-vie à terroir* of the Charentes. Everyone nervously awaited the results from the two experimental stations at Crouin and Marville.[14]

Yet the grape production of the hybrid vines guaranteed the stability of the harvest. The money saved by skipping chemical treatments that would have been necessary for the *vinifera* vine ensured that hybrids flourished in the Charentes. In 1920, J. de Fayard, head of the Station viticole de Cognac, warned winegrowers of the urgent need to get rid of the Noah in the Charentes because the brandy it produced had nothing in common with even the poorest *eau-de-vie* of the region.[15]

Several of France's leading viticultural scientists took up the challenge of finding vines that would grow in the Charentes and produce quality cognac, and with encouraging results. Millardet's venture was more modest than that supervised by Ravaz, but it was still a large-scale, successful experiment. In 1894 Millardet acquired Le Parveau, a property of 47 hectares 2 km from Cognac, in the heart of Grande Champagne, where the level of calcium carbonate in the soil varied from 40 to 66 percent. It was a stiff challenge. Millardet used his *franco-berlandieri* 41B as a rootstock

---

(Montpellier, 1872); *La question des vignes américaines au point de vue théorique et pratique* (Bordeaux, 1877); *Histoire des principales variétés et espéces de vignes d'origine américaine qui résistent au phylloxera* (1885).

[14] *RHFA,* no. 100 (1908), pp. 664–7.
[15] *RV,* no. 53 (1920), pp. 402–4.

for grafting *folle blanche, colombard, blanc-ramé,* and *saint-émilion,* the traditional vine varieties for *eaux-de-vie* of high quality. In 1899 he installed a distillery at Le Parveau. After Millardet died (November 1902) a son carried on, even adding another three hectares of *saint-émilion* grafted on rootstock 420. His production of cognac was 2200 hl in 1905, earning praise for its "exceptional quality." Although Millardet was a professor in Bordeaux, his gamble in Cognac was private. So it was not a triumph for official viticulture in the sense that the term was used by its enemies. Montpellier's Ravaz would have to be as successful as Millardet to avoid the ironic gibes of Gouy and the pro-hybrid camp.

A serious private interest in viticultural research existed in the Cognac area before the professors landed in Marville. Wine merchants and the big producers of *eau-de-vie* formed a well-funded committee (Comité de viticulture de Cognac) to begin research on the value, qualities, defects and culture of grafted vines. Ravaz, the head of the committee, believed that the challenge of the Charentes could be met by transferring the techniques that had worked so well in Montpellier. In 1893 the committee welcomed the state as its collaborator. This marriage transformed the venture into a viticultural research establishment baptized as the Station viticole de Cognac. Ravaz, to whom fell the job of organizing and directing the station, believed that it was unique in France. Eventually the station, supported by the Hennessy firm, was given the use of Le Parveau as its experimental vineyard. It was the Comité de viticulture de Cognac, presided over by James Hennessy, also a senator, which took the decision in the late 1920s to replace the Marville vineyard, notorious for vine 41B.

Ravaz was not long in showing the power of science. Growers had made few attempts to reconstitute dead vineyards because they thought that American rootstocks could not survive in the highly chalky subsoils of the Charentes. Ravaz boasted that he was the first to demonstrate that this opinion was wrong: only the nature of the topsoil was important. Diffusion of this information led to the spread of grafted vines in areas considered impossible before Ravaz had spread the good word. True, Planchon, Millardet, Davin, and above all Viala, Ravaz's colleague and sometime collaborator, had shown the ability of the *berlandieri* to grow in the Charentes – and optimists thought that the problem was solved – but the *berlandieri* could not easily be propagated by cuttings. So it was necessary to create hybrid rootstocks like the *berlandieri* × *riparia* and *berlandieri* × *rupestris.* The experiments done by Ravaz at the Marville station formed the basis of his recommendations. Because of the effective propaganda carried out by the station and its allies, Ravaz's work had a far greater influence than Millardet's. This was not entirely a fortunate thing, according to the enemies of the Ecole de Montpellier.

The introduction of *V. berlandieri* into France stirred up a nasty controversy. Bad results in areas like the Auvergne and the Yonne earned the vine low marks in an enquiry conducted by the association of Agriculteurs de France, which found the vine interesting only as a basis for hybridization. Supported by the agricultural establishment, the vine's success was really assured by the absence of serious competitors. The American vine withstood phylloxera, chlorosis, and denunciation by the hybrid camp. Gouy found the *berlandieri* more pleasing to the administration than to the viticultural public. But the *berlandieri,* the *riparia,* and the *rupestris* made up the new trinity proposed for winegrowers by the pontiffs of the higher levels of state viticulture, who believed the *berlandieri* to be the only resource available for calcareous soil. Once Viala had found a viable vine for the Charentes, bureaucrats and politicians quickly started to take advantage by associating themselves with success and taking credit where none was due.

The *Revue des hydrides franco-américains* (*RHFA*), relentless in its opposition to the *berlandieri,* argued that *vignerons* did not see in this vine the rootstock qualities accorded it by the professors. Its fatal faults were that it was very difficult to select from the numerous varieties of *berlandieri,* some of which were poor; that the vine was hard to reproduce from cuttings; and that it was not a plant that easily accepted grafts. There was even a rumor that problems in bearing fruit made it of dubious economic value. In 1898 Gouy was able to cite official science, speaking through the voice of Guillon, then director of the Station viticole of Cognac, who did not think that it was then practical to encourage the spread of the *berlandieri.* Guillon's study of the crosses between that vine and both *V. riparia* and *V. rupestris* did not justify the enthusiasm with which they were being promoted. The *berlandieri* could serve only through hybridization: the 420 *Millardet* (*berlandieri* × *riparia*) and other hybrids did well. For Gouy the *berlandieri* could have only a small role in *la grande culture.* The hybrids of *berlandieri* would participate with the hybrids of *rupestris* and others in the reconstitution of the calcareous and chalky soils of areas like the Charentes, the Saumurois, the Yonne, and the Marne.[16]

In a spoof, "Le baptême du *berlandieri,*" published in November 1898, the *RHFA* admitted the necessity of the *berlandieri* only in ultracalcareous and very chalky soils – a suitable *vinifera* × *berlandieri,* like the 41B of *Millardet* (*chasselas* × *berlandieri*), would be required. In fact, the *RHFA* reminded the potentates of official science (Foëx and Ravaz) that the 41B *Millardet* was widespread in the Charentes because of its resistance to phylloxera and chlorosis. In less alkaline soils, the "américo-américains à demi-sang *berlandieri*" were inferior to certain Franco-American combinations without the *berlandieri* element: the 1202 *Couderc* and the *aramon*

---

[16] "Les *berlandieri* et leurs hybrides," *RHFA,* no. 7 (1898), pp. 156–9.

× *rupestris de ganzin* no. 1 (now being attacked in California by phylloxera), which were then touted as better than the *American × berlandieri* even in very calcareous soils. This analysis was remarkably well informed, even if a little harsh on the professors.[17]

Over a decade later, in 1913, the *Revue de viticulture* published an exhaustive analysis of the reconstitution of vineyards in chalky soils. The article was doubly significant. First, the author, A. Verneuil, president of the Fédération des viticulteurs charentais, could not be suspected of ideological sympathy for the *berlandieri*. Second, the article appeared in the *Revue de viticulture,* formerly the organ of propaganda for the professors and the viticultural establishment, at least according to Gouy and his supporters. It was clear that opposing sides in the *berlandieri* war had by this time arrived at some sort of truce, although the wider conflict over hybrids had not ended.

Verneuil limited his remarks to the Charentes. As the Charentes contained the most chlorotic soils, Verneuil could argue a fortiori that any vine able to survive in the Charentes could survive anywhere, even with very high calcareous levels. In the most chalky soils of the Charentes there were several plants that could be used for production: the pure *berlandieri,* the *rupestris du Lot,* and Franco-American hybrids, which survived better than the Americo-Americans, capable of surviving in only mildly calcareous soils. Numerous experiments had eliminated plant after plant until a limited number of *franco × rupestris* (*aramon × rupestris de Ganzin* and the 1202 *mourvèdre × rupestris de Couderc*) and *franco × berlandieri* (the 333 *cabernet × berlandieri de l'Ecole de Montpellier* – 333EM – and the 41B *chasselas × berlandieri de Millardet et de Grasset*) showed their capability of being grown commercially – *en grande culture.*

In 1913 the preferred rootstock was the famous 41B of Millardet and de Grasset, which had replaced the 1202 as the favorite. The advantages of the 41B were its resistance to phylloxera, its adaptation to calcareous soil, its regularity of production, and the ease with which it could be reproduced by cuttings and also grafted. It did require well-prepared soil and lots of fertilizer in order to avoid the lack of vigor it had first shown in the Charentes; properly pampered, it would produce a full crop in its fourth year. Of course the 41B was not the ideal plant, for in the rainy years 1910 and 1911, bad years for chlorosis, even grafts that had previously flourished on the 41B were seriously afflicted in soils where the calcareous level

---

[17] "Le baptême du *Berlandieri,*" *RHFA,* no. 11 (1898), pp. 287–8, which drew on a rhapsody in the *Indépendant des Pyrénées-Orientales* on a baptism that took place at Alénya près Elne, with Foëx and Viala present. A different view is put by George Barnwell, although he gives no sources, in his article "A New Rootstock Is in the Pipeline," *Wines and Vines* 66 (1985). "In Europe A × R never became popular. There it was suspect on the grounds that it was not sufficiently immune to all the existing strains of phylloxera and it did not thrive on the higher-limed soils of Europe."

exceeded 40 percent. Fortunately, such high levels were rare in the Charentes. Other plants could be more widely used, for example the 333EM,
which had been dropped by the nurseries because of the many tuberosities
that developed on its roots, but it showed itself to be resistant anyway to
phylloxera and was just as good as the 41B in calcareous soils. It also
grafted well. Much of the Charentes had changed to other types of farming, but cognac makers also had a future.

And what of the much debated pure *berlandieri?* Although in the awful
year 1911 Verneuil had seen it suffer from chlorosis in very calcareous
soils, he still rated it superior to the 41B. Its serious drawback was that it
was difficult to reproduce. This difficulty led to a switch to *franco-berlandieri* hybrids like the 41B. There was also the natural hybrid 19–62 (*berlandieri* × *inconnu*) of the Millardet and de Grasset collection, just as good
in the Charentes but difficult to reproduce and therefore little used, although it was more vigorous than the 41B. All factors taken into account,
the 41B was the winner. Verneuil was diplomatic enough not to say it, but
of course it was what Gouy and the *RHFA* had long been arguing. On the
other side, the *Revue de viticulture* and the Ecole de Montpellier had come
to see the virtues, at least provisionally, of the use of hybrids in some situations.

In 1935 Viala asked J.-D. Vidal, director of the Institut de recherches
viticoles Fougerat (Bois-Charente) and a collaborator of the *Revue de viticulture,* to write an article on the results of fifteen years of research on the
survival of vines in the area. Vidal himself owned property with a 20-year
collection of hybrids. Verneuil's observations on the 41B were confirmed:
in a succession of cold and rainy years it became sensitive to chlorosis.
Depending on circumstances, the *berlandieri* × *colombard,* the *cabernet* ×
berlandieri (333 EM), and the 19–62 (*berlandieri* × *inconnu* of Millardet)
were better for the Cognac area. Between 1906 and 1910, and again from
1925 to 1933, a large number of growers noticed that the 41B lacked vigor,
dropped in resistance to chlorosis, and bore less fruit. The Société de viticulture de Cognac decided that new combinations better than the 41B had
to be developed. Hence the decision in 1928 to create the experimental
vineyard at Le Parveau.[18]

When the study by Verneuil is compared with another piece on phylloxera that had been written a generation before, the most striking difference
may be the degree to which the precision and confidence of science had
conquered viticulture. A vague discourse indicating only that a soil is calcareous has been replaced by statements specifiying the percentages of
limestone in topsoils and subsoils. Variations in percentages are noted for

---

[18] J.-L. Vidal, "Orientation nouvelle des porte-greffes pour terrains calcaires et crayeux," *RV*
83 (1935): 117–24. Fougerat was a pharmacist who left "une magnifique propriété" to
Angoulême that ended up as the Institut de recherches viticoles Fougerat.

areas separated by only a few meters, a piece of information vital in explaining successes and failures that would otherwise be inexplicable and give rise to much fruitless argument. Rainy springs in the Charentes made the 40 percent limestone soil of the Charentes more chlorotic than soils with up to 90 percent in Belton, Texas. The percentage of specified plants surviving in soils of different percentages of alkalinity could be given with reasonable assurance.

The best rootstocks and the best plants for different soils could be identified with certainty. The *rupestris* and the 1202 were found to be poor stocks for grafting the *folle blanche* and the *colombard,* although the *rupestris* welcomed being grafted to the scion of the *saint-émilion,* more commonly called by a name showing its Italian origins, the *ugni blanc,* which spread from the Midi to the west of France and became the key grape for the making of Cognac and Armagnac. Grafts made on the 1202 stock were also more susceptible to *coulure* (dropping of flowers or small berries after flowering), which made it necessary to graft varieties resistant to that disease. It was possible to give a ranking of rootstocks for different soils, in absolute terms – a stock for all soils and seasons – for those uninterested in the details that viticulturalists loved to inflict upon the producer. The bottom line in this case was the advice to use vine 41B.

The use of hybrid rootstocks was finally accepted even by the professors who longed for pure varieties. Each year brought new hybrids, some of which might turn out better than the 41B. So the producer, especially after being disappointed in several solutions to problems, refused to become dogmatically attached to one variety, except as the best solution for the moment. Verneuil declared that research remained as open to finding new hybrids with rootstocks whose success was certain in soils of 50–60 percent alkalinity as it was open to finding a good practical method to assure the reproduction of the pure *berlandieri* by cuttings. It is just as well to remind ourselves, however, that cognac production was saved by science – that is, experimental viticulture in cooperation with commerce and government.

Viala's Program to Save the Viticultural Patrimony of Burgundy

Life would be possible without cognac and even champagne, both recent inventions in the long history of the consumption of alcohol. But it is difficult to imagine how dull life would be without bordeaux and burgundy. One of the great grafting triumphs of the Americanists was the reconstitution of the vineyards of Burgundy. No question of planting direct-production hybrids there. At a meeting of winemakers in Beaune in 1891, Couderc emphasized that his talk about hybrids was only for people *not* from the Côte-d'Or. Even to talk about direct producers for *la Côte* would be blasphemy. But the Burgundians still had to be convinced that grafting would both save the pinot noir and maintain wine quality. Meanwhile winegrowers put their hope in saving the vine by using insecticides. The Bordelais model of salvation provided some grounds for hoping that this strategy would succeed.

It was widely accepted that insecticides initially gave good results nearly everywhere in Burgundy. Of course, the bitter medicine of reconstruction of the vineyards had eventually to be swallowed. The prescription was delivered with professorial authority by Viala in September of 1891 at Les. conférences viticoles de Beaune, organized by the Société vigneronne de Beaune, an influential organization with over a thousand members. On the platform for the occasion were the chosen ones among winegrowers and professors, including Lyoën, president of the society, Bouchard, Latour, Margottet, Peneveyre, Gayon, Ravaz, and Roy-Chevrier. Viala announced that, instead of giving a magisterial lecture, he would just talk as winegrower to winegrower, ignoring the scientific side of things. (Viala's father was a winegrower.) He then gave a good general scientific lecture, declaring that science renders us great services – science the servant of mankind – although pure scientific speculation must be ignored in the domain of the application of facts. This was a refreshing change from the usual professorial line that application flows from theory. Viala advertised the hard empirical line in a world suspicious of science: only certain, acquired facts would lead him to the conclusions that must be drawn from what he had seen.

Viala claimed that preconceived ideas, deliberately used as a rallying flag, had too often hindered the reconstruction of vineyards. Scientific objectivity decreed that personal predilection for American vines or for the use of insecticides should have no role in determining which method to follow in saving the vine. The idea of two camps (one *sulfuristes,* the other *américanistes*) was absurd, for the only reality was a situation in which insecticides would work or in which only grafting would work. Viala agreed that insecticides would be effective in all Burgundian terrains if treatment were started at the beginning of the invasion or while the vine was still vigorous. The vine could live on "indefinitely" with its regular dose of poison. So the old vines that gave Burgundy its quality wines and represented a centuries-old capital, "les vignobles des grands crus et des grands ordinaires," could be conserved. Only near the hilltops, areas producing *vins ordinaires,* would both the use of insecticides and reconstruction be difficult.

Viala's 1891 speech set forth a complete if general program for handling the spread of phylloxera in Burgundy. The big question was what to do when the vines were eventually killed by the disease. It was even an issue in newspapers with a certain influence in the politico-scientific world. Viala's irony was perhaps appreciated by the sardonic Burgundians. Leaving for later the controversial issue of the cultivation of the grafted vine, Viala dealt first with a more important question: would the grafted vines produce wines with the primordial qualities of *bouquet, finesse,* and *arôme* that distinguished Romanée, Richebourg, Corton, Beaune, Chambertin, and Montrachet?

The reply was an emphatic yes from the professor-vigneron. Proof? First, a theoretical comparison on essentially a priori grounds: the wines produced from grafted vines are of better quality than those from non-grafted vines. A general rule in fruit production is that grafting hastens maturity and improves quality. Grafted trees produce tastier and sweeter peaches and pears than non-grafted trees. From this general fact, one could make an a fortiori argument, but Viala, sticking grimly to his role of vigneron, did not succumb to the temptation. Some loose tongues had spread the rumor that the sap *(sève)* of the American plants – Viala was not sure what they meant by *sève* – transferred its bad taste and even bad odors to the French scions. Without going into the complex and fascinating system of sap circulation in plants, Viala dismissed this interesting if erroneous idea as plain nonsense.

Accurate viticultural facts gave Viala the basis on which to declare with absolute certainty that the quality of the *grands vins* produced by grafted French vines would be the equal of its classic quality. For those who doubted the scientific dogma proclaimed in the vigneron vernacular, there were witnesses and documented proofs. Many growers from the Beaujolais at the conference could testify that the vines they had grafted eight to twelve years before were giving wines just as good as or better than pre-phylloxera vines of the same age. For those who doubted that an argument concerning the noble, delicate pinot could be based on the performance of the plebian, robust gamay, Viala cited evidence from areas that produced better and certainly more historic wines than the nine *crus* (now ten, with the recent addition of Régnié) of the Beaujolais. Experiments done on wines in the Blayais, on some *grands crus* in the Médoc, and with châteaux in the Bas-Médoc showed that the grafted vine had maintained the quality of wine in the Gironde. And some said that the wine had improved – this conventional comment on grafting at least showed that the pro-grafting Americanists believed in the power of positive thinking, one of the vital elements of wine tasting.

Similar cases could be cited for the Libournais and also the vineyards producing *grands vins* in the Saint-Emilionnais. Closer to home, Burgundians could consider the example of reconstitution in several famous vineyards of the Hermitage, where an insecticide was also used, Côte-Rôtie and Côtes du Rhône, including Châteauneuf-du-Pape. The only vineyard in the Hérault with a claim to distinction was Saint-Georges-d'Orques, whose vines, grafted for 12 to 16 years, still produced red wines justifying its old designation of *cru réputé*. There were no exceptions to the maintenance or improvement of quality. So even the vines of the Clos de Vougeot could be grafted without fear of losing part of the national patrimony.

Viala spent so much time on the issue of quality because of the recent excitement generated in wine-producing areas by the widespread arguments that wine from grafted vines was inferior stuff. The botanist Lucien Daniel was the chief culprit in this affair. Viala did not give Daniel any free publicity by naming him. Convinced that Daniel was wrong, Viala just ignored him.

After this impressive rhetorical rout of the anti-grafting forces on tasting grounds, Viala turned to a more serious problem, that of culture or cultivation of the vine. First was the basic issue of rootstocks, of which there were only a small number to consider: *V. riparia, solonis, jacquez, rupestris,* and *vialla*. The *rupestris* and the *vialla* would play only a very secondary role in the reconstitution of the Côte-d'Or, unlike in the Loire-Inférieure and Maine-et-Loire, where the *rupestris* was the dominant rootstock, and in the mildly phylloxeric granitic and shale soils of the Beaujolais, where the *vialla* was equally important. In the one-tenth of the soil of Burgundy that was classified as calcareous, the *jacquez* and the *solonis* would suffer a chlorotic death. In 1891 Viala was awaiting the final results of Ravaz's experiments in the Charentes, but he and Ravaz were confident that it would be easy to reconstitute at least some of the vineyards in Burgundy's calcareous soils. The rootstocks for replanting with grafted vines were available when needed.[19]

A serious question remained: would the "procedures and systems of cultivation" in Burgundy have to be modified with the use of grafted vines? Would these modifications lower the quality of the wines? Viala answered at length. There were four basic issues in the culture of the vine. First, the depth to which the earth should be plowed or dug: no change was required in procedure, except that the subsoil must not be disturbed if it was calcareous. Second, type and quantity of fertilizer: no change was needed in the relatively copious amounts used in Burgundy. Rumor had it that American vines required more fertilizer than the French and produced more fruit, thus leading to a decline in the quality of the wine. An absolute mistake, said Viala optimistically, also adding that the *rupestris* and the *berlandieri* grew in very poor soils in the American wild. Third, what distance should separate vines? The Société vigneronne de Beaune had done a study concluding that closely planted vines give better wine than vines widely separated. Viala corroborated this observation. It had been assumed that the

---

[19] Viala also preached the gospel of grafting in the Loire. See "La reconstitution des vignobles dans la Loire-Inférieure [Atlantique]," *Progrès agricole et viticole (PAV)* 16 (1891): 290.

Americans would need more space than the French – at the beginning of reconstruction in the Midi, the distances were even exaggerated. But the vines grew just as well in the old system of first plantation, even for the *grands crus:* 15,000 to 18,000 plants per hectare. The basics of cultivation would remain pretty much the same for the grafted vine as it was for the traditional ungrafted *vinifera.* This was a reasonable opinion at the time.

The last issue Viala addressed on vine cultivation was that of the traditional layering (*le provignage* or *le marcottage*) of vines – putting a plant's branch or shoot into the earth to strike root while still attached to the plant. It could be done, as Foëx showed in practicing annual layering on pinots grafted on taylors, but it was not advised. Layering was best for feeble vines, except in the Champagne, where the aim was to get plants close to one another, thus promoting the spread of roots in the less calcareous topsoil. In a northern climate it was also an advantage to grow the grapes near the ground to profit more from the heat reflected from stones. Given the strong vigor of American vines strengthened by fertilizing, layering was superfluous.

Viala concluded that the cultivation of the American rootstock was no different from or more difficult than that of the French vine. Science had banished another popular myth. Burgundy would still produce burgundy, even on grafted vines. After the conquest of Burgundy, science could follow phylloxera farther north to the last viticultural frontier, in the Champagne.

Science, Capitalism, and Politics in Champenois Grafting

In 1890 phylloxera was discovered in the Aisne, on the departmental border of the Marne. Administrative and political powers in the department joined forces with most of the proprietors in the Champagne to defeat the enemy. An Association syndicale promoted the heavy use of carbon disulfide, while the departmental politicians, in a reasonable political act catering to the ignorant, prohibited the entry of American vines into the Champagne. Raoul Chandon, president of the Association antiphylloxérique d'Epernay, pointed out to the prefect of the Marne that the prohibition served no purpose. The American louse had already arrived and, more important, the roots of imported vines could be easily sterilized by immersion in hot water. Logic eventually prevailed. A few years later (1905–6) the Conseil général of the Marne was distributing disinfected plants in forty communes.

Moët et Chandon, led by Raoul Chandon, played an important role in the fight against phylloxera in the Champagne. The company property was kept in good condition by two successive applications to the soil of 300 kg of CS_2 per ha, carried out at 15-day intervals after the harvest. The slow advance in the Champagne gave Raoul Chandon the time to note the influence of CS_2 on vines of different ages and in different soils. The chemical was encouragingly efficacious in light, permeable soils, but it did often damage old vines. Two basic questions had to be answered. Could the traditional growing methods of the Champagne accommodate grafting, and would grafting modify the products of this great industry? Raoul Chandon went to the Midi to study the work of reconstitution carried out under the guidance of the Ecole d'agriculture de Montpellier. Back in Epernay he eventually established more than 100 experimental plots to determine the best way to preserve the classic varieties of vine in the Champagne.

The Montpellier connection was important. Raoul Chandon, who also published on phylloxera, joined the board of editors of the *Revue de viticulture,* and he gave a laboratory job to Manzade, a graduate of the Montpellier's school of agriculture. Manzade's mission was to find the best type of grafting for the Champagne. The Ecole de viticulture of Moët et Chandon, as the establishment came to be called, became a model school that trained an army of grape growers in the art of grafting. The most widely used rootstocks in the Champagne were *berlandieri* hybrids and *riparia* × *rupestris,* although other less desirable varieties (hybrids of *solonis,* of *rupestris,* and of *riparia,* etc.) were used by those who did not have access to the knowledge and money of Moët et Chandon. As late as 1911 fearful but optimistic *négociants* were still advising growers to replant with French vines. Replanting picked up after World War I and went on throughout the 1920s, "the heroic period of Champenois reconstruction," although 1921, 1922, and 1930–40 were years of very poor wine sales. The production of great wine requires long-term, heavy investment in knowledge and resources.[20]

Official science now seems to us to have been immensely successful. Its grafting program seems in retrospect to have been the only possible solution to the problem of how to save French vines and their quality wines. The success of the solution appears to us to have been reasonably well demonstrated. Why was this official viticultural science so controversial, so vigorously contested by large numbers of intelligent viticulturists as well as by large numbers of small wine producers? In particular, why did they insist on the virtues of direct-production hybrids? This nearly hundred-year-old fight over these controversial vines is the subject of the next chapter.

[20] *Bulletin du laboratoire Moët et Chandon* 1 (1898): 5–13, 69, 90, and 3, no. 2 (1909), issue on Comte Raoul Chandon de Briailles; Etienne Henriot, "De Champagne, à Monsieur Raymond Cordier," *RV* 85 (1936): 370–2; and Georges Chappaz, *Le vignoble et le vin de Champagne,* vol. 2 of *Vignobles de vins fins et eaux-de-vie de France* (1951), pp. 97–114.

3

Direct-Production Hybrids: Quality Wines?

The Hybridists Take on the Grafters and Big Science

From the viewpoint of winegrowers, the science of the Montpellier professors was not very useful because it failed to provide vital, precise, detailed information. Because Montpellier's professorial science had committed itself to giving practical advice to growers on the reconstitution of vineyards, the complaint is not so absurd as it might appear to later historians. Growers were eager to obtain reliable information about Franco-American hybrids (both grafted vines and direct producers) and new rootstocks that were superior to pure American roots in difficult chalky soils as well as new direct-production hybrids. These direct producers were becoming more and more popular with small as well as big producers because of their reputation for resistance to the devastating black rot disease that was seriously lowering grape production in the late 1880s. Many producers found that the traditional sources of information (viticultural congresses and societies, journals and newspapers, and even occupants of departmental chairs of agriculture) all failed to provide the complex knowledge required by French viticulture.

In the program he proclaimed in the first issue (1898) of the *RHFA* (*Revue des hybrides franco-américains*), Paul Gouy condemned the spokesmen of official science as poor purveyors of new scientific information. Most of the departmental chairs of agriculture and their institutions, either hostile to or suspicious of the new vines, were not inclined to advise growers to plant them, even on an experimental basis. Nor was the domain of the printed word much better in encouraging the spread of hybrids. Two manuals had just been published on the direct producers, but nothing yet existed on hybrid rootstocks.[1] And hybrids still needed a periodical publication for the quick dissemination of the latest news.

Gouy condemned the strong opposition to hybrids in some of the periodicals controlled by official science, which continued to recommend the

---

[1] E.g., J. de Bouttes, *Etudes de viticulture nouvelle: Les nouvelles hybrides,* 2 vols. (published by the author, 1897–1900).

65

planting of grafted vines. Other publications treated hybrids just inciden-
tally or ignored them altogether. The few favorable journals were inade-
quate because they were local or regional in nature, thus necessarily nar-
row in scope. Even their generality was a problem: avoiding specialization,
they tried to cover the wide front of viticulture, sometimes even all agriculture. So the burning issue of the day, the value of Franco-American hybrids, was not seriously covered by any publication. Gouy concluded that
as a journal did not exist for this vital subject it was necessary to invent
one.

No doubt many were surprised that the good news on hybrids came out
of Vals près Aubenas in the Ardèche. But the political artificiality of the
department enclosed part of one of "les grands vignobles de France," the
Côtes-du-Rhône. Aubenas lay between the northern vineyards (Condrieu,
Château-Grillet, Côte-Rôtie, Saint-Joseph, Hermitage, Crozes-Hermitage,
Saint-Péray) and the southern vineyards (Beaumes-de-Venise, Gigondas,
Châteauneuf-du-Pape) of the Côtes-du-Rhône. Although near the mediocre Côtes-du-Vivarais, Aubenas was ideally located for communication
with powerful growers of the Rhône and even of the Mâconnais, the Beaujolais, and Burgundy. In addition to the support of well-known wine producers, Gouy had the backing of two of the best-known producers of hybrids. The well-advertised presence of both Seibel and Couderc in
Aubenas made that small town "the most active center of hybridization in
France." Aubenas was ideal as the Rome of the good news to be spread to
the neglected small producers by the *RHFA,* because the town had access
to better information and documentation on hybrids than many larger
towns.

Gouy worshiped three hybridizers: Couderc, Castel, and Terras.
Georges Couderc (1850–1928), who voyaged through several institutions
of higher learning, spent two years at the faculty of medicine in Montpellier, where he got to know Planchon.[2] On the death of his father in 1876,
he returned to Aubenas to run the family estate, Champfleury, and after
1881 formed an army of hybrids for his battle against phylloxera. Couderc
was one of the first names that the British government thought of when it
needed information at the turn of the century on planting American vines
in Australia. The engineer-hybridizer, as the Montpellier professors called
him, was popularly flattered as "the pope of the new viticulture." P. Castel,
also famous for his hybrids, bred his vines at Carcassonne in the Aude, a
department important in producing the twentieth-century Languedocian
wine flood. At Pierrefeu in the Var, Marcien Terras, known for his paint-
ings of the region, produced among his many hybrids the Alicante Terras
no. 20 (*Alicante Bouschet* × *rupestris-monticola*), widely used in French

---

[2] Pierre Galet, *Cépages et vignobles de France,* 1:211.

vineyards. Gouy included cuttings from these hybridizers among the items sold from his plantings and collections. The production of hybrid vines had become an important local business activity.

In his journal Gouy was determined to avoid the imperial pedagogical tone of many publications, where the editor played professor, speaking ex cathedra to an audience imagined to be attentive and docile. The journal's function should more resemble that of a salon, a circle, or a club, open to everyone who wanted to discuss the problems of the new viticulture – "a sort of intellectual picnic." Gouy knew the importance of attracting the elite of the viticultural public to the Bacchic spread. Before starting the journal he informed a hundred "notabilités viticoles" of France's wine country about his plan. Happy with the project, nearly all wrote back, with more than fifty promising collaboration either by providing information or writing articles. Several notables accepted the desirable burden of serving on the editorial committee. Gouy summed up with admirable concision the aims destined to seduce a large enough public to justify publication of a very specialized journal. "I hope that this publication will become . . . a tribune for the masters of wine-growing science and practice, a school of mutual teaching, a source of useful information for wine growers, nursery men, agricultural associations, and an auxiliary for the other viticultural journals, with which our special limits prevent us from competing."

Science, practice, school, information, publication – key words of enticement for a reading public eager to take the road of progress to the city of profit. The word *research,* presumably a monopoly of the mandarins of official science, was not used. Not that Gouy was ashamed to advertise the approval of his journal by some well-known mandarins of viticultural science: Millardet of the faculty of sciences in Bordeaux; Franc, departmental professor of agriculture in the Cher; doctors Crolas and Cazeneuve of Lyon; and Paul Gervais – a reasonably strong viticultural phalanx by any criteria. Gouy was proud to announce support from the marquis de Vogüé, a big name in Burgundy, president of the Société des agriculteurs de France. The marquis pontificated that hybridization would be the way to salvation for three-quarters of French vineyards.

Of all the support received by the *RHFA,* including that of individuals and numerous regional organizations – Union de Sud-Est, Union Beaujolaise, Union des syndicats de la Drôme, and so on – it seems that Gouy most deeply appreciated that given by A. Grimaud, editor of the bulletin of the Agriculteurs d'Annonay et du Haut-Vivarais. Grimaud hoped that the *RHFA* would thus forge a connection with the small wine producers, who had become sour and cynical as a result of twenty-five years of disappointment in the remedies offered for reviving their vineyards. Both science and commerce had peddled too much optimism. Peasant confidence had to be based on a sober reality, in this case the possibility of planting vines that would make winemaking economically viable once more in the new world of American diseases, especially black rot. Hybrid wine production also had a social significance. One of the targeted groups of readers of Gouy's journal was the small wine producer dependent on direct-production hybrids. A. Jurie gave an explicit social interpretation of the hybrid gospel in recognizing the direct producer as "a social work aimed at small proprietors, sharecroppers, and farmers of disinherited areas."[3]

Perhaps it was a recognition of the difficulty of continuing with French plants grafted on American rootstocks that led *La vigne américaine,* the chief instrument of propaganda for the grafted vine, to publish an apostatical article by B. Paillet, a member of the Comité d'études et de vigilance du Lot, pointing out that several of Couderc's direct producers resisted fungus diseases very well. These hybrids provided a striking contrast to the grafted French vines, which gave a relatively poor crop because they were so susceptible to vine diseases, especially black rot. Not only was the wine from grafted vines worth less than that from hybrids – a curious, dubious observation – but the grafted vines had to be doused with *saloperies* (chemicals) in order to survive. Gouy generously recognized that the learned editor of *La vigne américaine,* Battanchon, was not completely hostile to new direct producers; still it was only with some sorrow that the publication recognized facts upsetting its old theories. It is hard to tell whether Gouy was more pleased with the new support for hybrids or with the discomfort of a supporter of an opposing theory. In any case, *La vigne américaine* was not an organ of "la science officielle," at least not to the degree of following the rule of putting literature on hybrids on a scientific index.

Gouy's journal was able to give a considerable boost to the spread of hybrids. Take the case of the Couderc 4401, a cross between the *chasselas rose* (*V. vinifera*) and the *rupestris.* The first issue of the *RHFA* strongly recommended this phylloxera-resistant vine, even for the Mediterranean climate, whose dryness could make the famous plant louse a worse threat than it was elsewhere. The vine also resisted mildew, oidium, and, to a considerable degree, even black rot without being sprayed with a copper sulfate solution. So much for the vine; what about the wine? The difficulty in replacing destroyed vineyards was that good vines often produced bad wines and bad vines produced good wines. The wine made from the Couderc 4401 was classified as a beautiful deep red in color, having an honest taste that was in no way foxy, but only of ordinary quality.

The subject of new hybrids, which were being invented in astounding profusion, would have kept Gouy permanently bickering with the viticul-

---

[3] A. Jurie, "Encore le 4401," *RHFA,* no. 22 (1899), pp. 226–9.

tural mandarins if the enemy had deigned to do battle with their self-announced opponents. The reserve, even hostility Gouy alleged, of the powerful *Revue de viticulture* concerning new hybrids worried and irritated believers in their future.[4] Hybridizers had a distorted view of science, though the view was not without a certain insight into the conservative nature of the scientific enterprise. Gouy saw the official scientific suspicion of hybrids as "absolutely logical, conforming to all precedents," but also thought that official science eventually ends up following the public, the "public viticole" in this case. This point of view was certainly excessively flattering to this public, Gouy's readers, but then this is the essence of successful journalism. Practice evolves faster than scientific teaching. Perhaps this is the point that Gouy struggled to make in observing that several holders of the chairs in agriculture were members of the Académie des sciences, implying that it was a mausoleum of scientific ideas.

Gouy noted that official viticulture – he meant Dumas and company – had begun the battle against phylloxera by prescribing the exclusive use of insecticides while condemning the grafting of French plants onto American roots. As pumping the highly toxic carbon disulfide and sulfocarbonates into the ground did not get rid of phylloxera, the establishment rallied to grafting, but the grower had already moved on to using new rootstocks and planting direct producers. Again, Gouy ignored the complexities of the arguments concerning hybrids, especially those over their use in good wine country. There was little evidence that hybrids could produce anything other than a poor *vin ordinaire*. Winegrowers had become used to making progress by going their own way, which, like most roads to progress, was mined with unpleasant surprises.

The Hybridist Demonization of Foëx

Gouy, who had flattered himself into believing that his ability to write well set him apart from the camp of official science, advertised his preference for the effervescent concoctions of writers like J. Roy-Chevrier over the standard prose of academic science. "Science doesn't have to be heavy and pedantic in order to be both profound and practical."[5] Although it was nonsense to accuse everyone in the camp of official science of producing bad prose – some of the great stylists of the age were scientific pundits or

---

[4] In 1898 Gouy defended the "authoritative pamphlet" of Joseph de Bottes (owner of the Château de Flamarens in the Tarn), *Etudes de viticulture nouvelle. Les nouvelles hybrides à production directe résistant au blackrot et aux maladies cryptogamiques,* 2 vols. (Flamarens, par Lavaur, 1897–1900). See Gouy, "Les nouveaux cépages, résistance phylloxérique et adaptation," *Bulletin de la Société des viticulteurs de France* (March 1898).

[5] See, e.g., J. Roy-Chevrier, *Voyage au pays des vignes jaunes et au pays des vignes bleues* (Villefranche, 1897). The yellow vines were the diseased vines of the Charentes; the blue, the copper sulfate–sprayed French varieties of the southwest.

renegades, all employees of the educational establishment – Gouy was not being entirely unfair in lamenting the mediocrity of professorial and bureaucratic prose. Foëx, mandarin of Montpellier, provided a convenient target. In October 1898 Gouy ridiculed Foëx's recent report on Franco-American rootstocks written for the Service du Phylloxera.[6] Foëx's offenses were against the French language: banal thoughts, loose style, incoherence, pathos, and bad grammar. But was Foëx's science perhaps sturdy enough to survive despite its crimes against the language of Buffon?

No, gloated Gouy, for even worse was Foëx's approach to viticulture based on a simplistic theory. Grafting French vines on a few different American rootstocks was the unique means of viticultural salvation preached in the Ecole de Montpellier. Foëx began his hybridizations at the Ecole de Montpellier in 1878 and acquired a certain notoriety for his well-known crossings of *Vitis berlandieri* with *V. riparia* and *V. rupestris.* In Gouy's world of hybrid profusion, this was an administrative conception of the "art viticole": three pure varieties and a half-dozen mixed varieties to serve as rootstocks for French *vinifera* vines. The direct-producer hybrids created by Couderc played little role in Foëx's research program. No matter what their success, the Franco-American hybrids were suspect. They derived their good qualities from *Vitus vinifera,* but crossing the hybrids with the *vinifera* vine caused them to lose the resistance they inherited from the American parent. So these hybrids were inferior in resistance to plants grafted on pure American roots. This was the "thèse classique" of Foëx.

Gouy's counterattack appeared to be more scientific than the scientist's science. Did not Foëx know that "the nature and aptitudes of the radicular system[7] of the wild parent" can be inherited along with other botanical characters, like fertility, from the *vinifera* parent? This happens regularly with some plants when generations are numerous enough. Then heavy irony: "This language, familiar to all scientists and all practitioners, would be unintelligible to the Inspector General of French viticulture and the former director of the School of Montpellier." Gouy's attack provoked some welcome hilarity at the staid Congrès oenologique, which included a less staid *Foire aux vins,* at Toulon. Gouy's modest opinion of his intemperate blast was that it was a simple pinprick that much deflated "the inspectorial wineskin."

Foëx was one of a small army of professors and inspectors grinding out *la littérature grise* of reports as well as articles and books on a schedule

---

[6] In the *Compte rendu des travaux du Service du Phylloxera pour les années 1895–97, RHFA,* no. 10, (1898), pp. 238–40. Foëx was also an Inspecteur général de la viticulture pour les départements de l'Est et du Sud-Est.

[7] The radicle is the part of the plant's embryo that develops into its root system, the root of the embryo of seed plants.

more conducive to bad prose than brilliant science. He had the misfortune to be singled out for a stinging attack on his research, his scientific knowledge, and his literary style. Of course it is true that anyone who takes a position on a controversial scientific subject is bound to be attacked by someone in an opposing camp, whether the subject is phylloxera, natural selection, or the extinction of dinosaurs. Nonliterary factors are not irrelevant to the discussion. Foëx was one of the great men of Montpellier's school of agriculture, La Gaillarde. For Gouy and his friends, the Ecole de Montpellier was the notorious source of the most rigorous criticism of the cultivation of direct-production hybrids. Foëx had been the head of the Ecole: to show that the "dread commander" was really an incompetent was therefore not a bad strategy to follow in attempting to destroy his case against hybrids.

Foëx had used the *Revue de viticulture,* the powerful propaganda weapon of his camp, to cast doubt on the value of existing hybrids. Perhaps his crime was that the hybridists viewed him more as traitor than enemy. He had been among the first to share the hopes and illusions that direct producers had originally inspired. As early as 1875 he had begun crossing European and American vines in order to improve the Americans. Several of his creations were as good as anything on the market in 1898. But far better hybrids were needed. There was little hope that the wines of direct-production hybrids would ever inspire the hyperbole that the language of *dégustation* had developed for describing the gustatory qualities of the wines from indigenous *vinifera* vines. Foëx was not optimistic about the possibility of solving scientific problems quickly, especially in the case of the search for practical and commercial applications.

Foëx took a similar approach to the search that he and his colleagues carried out for rootstocks possessing the double virtue of resistance to phylloxera and good adaptation to chalky soil: much remained to be done. He was right; nor was he merely engaging in the tropistic defense of a research program that had been made redundant by someone else's success. The *RHFA* took pleasure in commenting ironically that this confession by Foëx was an admission that the professor and his collaborators had failed in their work on direct-production hybrids and had achieved only a mediocre success with rootstocks. The relative lack of success of the school in marrying American and *vinifera* vines made it an easy target for Gouy's barbs, especially in the case of the *Tisserand,* named in honor of the director of agriculture. Nature, perversely ignoring sponsorship by a *haut fonctionnaire,* failed to give this creation the necessary fertility and resistance to phylloxera. (The "*rupestris* mission" had been launched in 1893 at the Congrès viticole in Montpellier with Tisserand on the stage.)

A new research strategy adopted by Foëx as the school's director and his collaborators aimed at producing hybrid rootstocks for difficult soils.

The chief success here was no. 34EM (*V. berlandieri* × *V. riparia*), not by any means a failure; but growers preferred the creations of other researchers and breeders, notably Millardet, de Grasset, and Couderc. And shortly after Gouy's attack on Foëx's failures, there appeared in the journal an article on the plantings and the 12,000 vines in the collections of Castel at Paretlongue (Aude). Gouy believed that the vineyards of the Languedoc provided a model with lessons for all of France, including the Ecole de Montpellier.

The school seemed to be following a different strategy, concentrating on laboratory testing of vines in pots and transferring the results to the field for verification. Ravaz's pot science cast considerable doubt on many optimistic opinions held by hybridizers. Serious doubts about basic research strategies could be exposed in discussions and quarrels about the value and epistemological foundations of pot science and *grande culture* science. But more important for the producer was the practical question of what was best to do in the vineyard, for a wrong decision might bring financial disaster. Farmers have a reputation for being suspicious of science even if they love its chemical and biological products. When a professor of the Institut national agronomique noted the resistance of farmers to innovation, the journal *L'Agriculture nouvelle* explained their suspicion of experimental agriculture as a protective device against charlatans, who are as common in the countryside as in the city. The journal justified its title, however, by approving of experimenting in a rational way with the new methods and applying the modern theories on a small scale. That was the road of progress leading to fortune.[8] In spite of the skepticism of the hybridizers, pot science would be an important part of the new agriculture.[9]

Confidence and Doubt in the Hybrid Camp

Meanwhile the hybrid bandwagon rolled on. Scientific doubts about the value of hybrid vines were countered by the enthusiasm and empiricism of breeders and growers, partners in promoting polymorphous promiscuity in the hybrid world. At the Conférences viticoles of 1898, played to an audience of the Société de viticulture du Rhône, the revelations and debates continued. The "distinguished orator" J. Roy-Chevrier gave a "Report on Direct Producers." This skillful speaker and scribbler, opera writer, poet, was Planchon's cousin and "a true disciple of Couderc." A

---

[8] *L'Agriculture nouvelle*, no. 71 (August 1892).
[9] Foëx-bashing remained a popular sport in Gouy's journal for several more years. In 1899, Foëx's treatise *Comment devons-nous reconstituter nos vignobles?* was gleefully trashed as a summary of technical and cultural ideas that had been prevalent twenty years before. *RHFA*, no. 16 (1899), pp. 77–81, and no. 22 (1899), pp. 221–3.

preacher of the hybrid gospel for a over a decade, he promised an imminent success for a direct-production vine that would be as immune from the astringent criticism of the professors as it would be from the American diseases. Even this modest success would be a matter of many local vines, for there was no universal variety for all viticultural regions. So the Conseil général de la Haute-Garonne had made the right decision in allowing direct producers to be tried in numerous experimental fields; the general council of the Haute-Saône showed its closed mind on hybrids by excluding direct producers from its list of experimental vines. Hybrids stirred up controversy.

Discussion of Roy-Chevrier's report had at least one interesting moment when Ernest Leenhardt of Montpellier begged winegrowers not to divide into two hostile camps, a tendency he saw exacerbated by the attacks of Roy-Chevrier on Foëx as one of the leading theoreticians of grafting and therefore one of the leading opponents of the spread of direct producers. Roy-Chevrier replied that he too wanted an agreement between French winegrowers but that the harmony must be based on a judicious eclecticism excluding neither grafting – the basic dogma of the agricultural Ecole de Montpellier – nor hybridization (including direct-production hybrids). Until Foëx and his school recognized the need for a dual program, the two groups would continue to exchange insults disguised as the latest results of experience and experiment.

A. Jurie, a "very distinguished hybridizer from Millery" (Rhône), made clear the reasons for so much confusion and squabbling.[10] In an article bristling with originality, Jurie had not only attacked the sacred vine of the hybrid men but even added insult to injury by arguing, on theoretical grounds, that a good direct producer could not exist because of the difference between *vinifera* and American vines in their wild state.[11] Essentially, the strategy of the *RFHA* in replying to this salvo of epistemological doubt from the enemy camp was to kick existing vines to prove their existence to their supporters. It enumerated the most successful "good direct producers": Seibel no. 1, the Couderc 4401, 201, 503, 3907, the Terras no. 20 – all varieties found widely used, "en grande culture." In replying, Jurie was quick to point out that empiricism was on his side in this case. The "greatest scientific expert on hybridization," Millardet, in collaboration with de Grasset, had created numerous hybrids by crossing *vinifera* and American vines, but about 98 percent of the two hundred vines produced crops whose clusters and fruit were too small for commercial winegrowing.

These vines of Millardet and de Grasset were binary hybrids, whereas the vines referred to in the *RHFA* were hybrids descended from hybrid

---

[10] *RHFA*, no. 8 (1898), pp. 177–9.
[11] "Note sur l'hybridation de la vigne," *RV,* June 18, 1898.

parents. Many *rupestris* vines had hybridized naturally, which meant that some experiments with hybridization already started out with hybrids. Their success was not one of direct marriage between the pure subjects but of a ternary combination, which was infinitely superior in fruit bearing. Jurie did not believe that there was anything wrong with using these vines, which produced robust but not particularly tasty fruit, if they were restricted to areas that produced only *vin commun*. Better a *vin ordinaire* than nothing. Jurie thought that hybrids of true value would eventually win increased support, even from unsympathetic people, although there existed only a small number of wines from hybrids; their supporters should not compromise the wines by promoting them zealously. This advice was ignored.

The *RHFA*'s criticism of Jurie's theoretical point that there could not be a "good Franco-American direct producer" did not rise above the level of an appeal to nature: if wild and cultured vines could not interbreed, it would be quite exceptional, for the animal and plant worlds offered nothing comparable to this mutual repulsion.[12] Jurie's rejoinder to this all-or-nothing bluster was to point out that the explanation of the failure to produce successful hybrids at first try was given by Darwin in his remarkable work *The Effects of Cross and Self Fertilisation in the Vegetable Kingdom,* translated into French in 1877 by Edouard Heckel, then in the faculty of sciences of Grenoble.[13] Whatever the unpopularity of natural selection in the wider world of French biology, Darwin's work in botany at least found an enthusiastic French supporter in the vineyards of Millery. In the preface to his translation, Heckel tells of the fiasco that occurred when he presented to fellow scientists his own research related to the "horror nature has for perpetual self fertilization, whose action is harmful to the development of the species." Heckel's distinguished and skeptical audience was the botanical section of the Association scientifique de Clermont-Ferrand. Hybridizers also missed the point of the work.

Many of Jurie's positions were quite acceptable to Gouy and the *RHFA*. So it is not surprising that he became a regular contributor to the journal. It is also better to coopt your moderate enemies than to turn them into serious ones. Jurie created excitement near the end of the century when he noticed in his collection the return of a group of Couderc 4401 vines to the ancestral maternal form. But Jurie, thinking that this was nothing to get excited about, modestly cited Naudin's famous memoir dealing with the structure of hybrid plants, an authority that could explain away such mysteries. As Jurie admitted, the public – winegrowers – knew very little

[12] *RHFA,* no. 8 (1898), and no. 9 (1898), pp. 197–208.
[13] Charles Darwin, *Des effets de la fécondation croisé directe dans le règne végétal,* published by C. Reinwald (Paris, 1877). The work was translated and annotated by Edouard Heckel, with the authorization of Darwin.

about the structure of hybrids and the mechanism of their variations.[14]
Both the *RHFA* and Jurie could agree on the value of the method of natural hybridization, which he described in the erotic prose of plant sexuality.
Of course nature could be helped if a viticulturalist agent of selection,
using Bouschet's method of hybridization, arranged a meeting of the organs of nearby plants. Simpler still, one could plant a few fathers between
the mother vines and let nature take its course. The mysteries of fertilization might not have been understood very well, but a method did exist for
producing a few useful hybrids.

The arguments had not basically changed much by 1913, though the
role of hybrids in wine production was not negligible. Even in areas with a
reputation to uphold, like the Tonnerrois (Paris Basin, Yonne, and Aube),
winegrowers pushed for the introduction of hybrids in order to have wine
to drink themselves as well as to export. In lower Burgundy a long series
of bad harvests had reduced winegrowers to the state of being able to
drink wine only three or so months of the year. However much their livers
might have rejoiced, the growers were not happy. "Cultiver la vigne alors,
et boire de l'eau!" The four years from 1907 to 1910 had been appallingly
bad. Because of yield-reducing diseases and the cost of fighting them,
winegrowing did not appear to be a profitable activity any more. In the
department of the Yonne between the 1870s and 1900–9 the area devoted
to vines fell from 40,674 to 19,878 hectares and wine production from
970,184 to 527,926 hectoliters. In the period 1910–19 wine production was
less than half this figure, with the Great War causing a strong drop in
production.[15] Growing sturdy direct-production hybrids was the easiest
way to obtain cheap wine in difficult times.

The agricultural society of Tonnerre created an experimental vineyard
for hybrids. A neutral wine without an odd taste would be acceptable for
family consumption. This wine, preferably from direct-producing hybrids,
or in case of absolute necessity grafted hybrids, would also make it possible to sell all of the wine from the traditional vines of Burgundy. It was
assumed that no one would be foolish enough to indulge in the fraud of
blending wine from hybrid vines with good burgundy, thus ruining the
distinctive qualities of *vins de cru*. The experimentally minded were in favor of growing hybrids for another reason: to continue building the "archives of the culture of hybrids in lower Burgundy, archives in which there
were still few documents."[16]

Clearly, the hostile camps so evident in 1898 had not merged into one

---

[14] *RHFA*, no. 22 (1899), pp. 226–9.

[15] Lachiver, *Vins, vignes et vignerons*, annex, pp. 579–625.

[16] J. Mahoux, "Les producteurs directs et les hybrides greffons dans le Tonnerrois," *RV*, no.
39 (1913), pp. 42–6. The direct-producing vines of Laribe at Epineuil had been planted
thirteen years before.

cooperative grouping including the hybrid and the pure by 1913. Roy-Chevrier was still breeding vines for the triumph of hybridization, always just around the corner. He had now evolved into a voice of reason, acceptable to the *Revue de viticulture,* which had itself shifted to a more flexible position on hybrids. At the Congress of Lyon in 1901, Roy-Chevrier had emphasized the limited use of direct-production hybrids. The next year, at the big international congress in Rome, he indulged his talent for astonishing audiences when he went beyond stating the necessity of direct-producing Franco-American hybrids in viticulture to prophesy their triumphant success. More surprising, the introduction to his report was published in the *Revue de viticulture.* The pariah plant was finally winning some acceptance in the official viticultural world. In Toulouse, Louis de Malafosse, long an eloquent advocate of hybrid virtues, rejoiced that his cause now had the support of "this new Saint Paul" to spread the gospel.[17]

After Roy-Chevrier became president of the Société régionale de viticulture de Lyon, his message was perhaps pretty much the same as before: the failure or success of ungrafted vines was determined by their local success. This did not refer to adaptation of the vine to the terrain – a matter to consider for rootstocks – or to its acclimatization, that is, the assumed transformation of the plant through the influence of its environment. Roy-Chevrier was skeptical of this dubious botanical dogma, which, like so many other not very useful concepts, had an illustrious intellectual pedigree. Nor did he hope for the salvation of French viticulture through the rare accidents of acclimatation. Of course certain hybrids would do well in certain soils and poorly in others; perhaps that could be explained in terms of the power of ancestral qualities requiring a certain type of soil to do well. One of his own "Burgundian" hybrids (*oporto-colombeau* × *gradiska*) revealed its qualities fully only when planted in the Gers, the land of Armagnac in the southwest, a long way from his property in Lons-le-Saulnier (Jura), 100 km south of Dijon.[18] Viticulture is based on local knowledge.

Roy-Chevrier had no trouble finding winegrowers to try out his hybrid creations on a commercial basis. Some were happy with the results. M. Perbos of Saint-Etienne-de-Fourgères, près Mondar-d'Agenais, proudly showed a commission from the Dordogne his rich collections of direct-production hybrids. One of his *bayards,* a *couderc 1401* × *gradiska,* proved its pedigree by giving a wine Roy-Chevrier cleverly celebrated as "une encre bordelaise colorée de beaucoup de sympathie." The grape was presented as a derivative of the *lambrusca,* which gave it adequate alcohol and balanced the "sweet water" of the *gradiska.* Its resistance to black rot

[17] L. de Malafosse, "L'inévitable," *RHFA,* no. 68 (1903), pp. 190–2.
[18] J. Roy-Chevrier, "De la localisation des producteurs directs," *RV,* no. 39 (1939), pp. 269–71.

was better than its resistance to phylloxera. In fighting one disease it is sometimes necessary to pay a ransom to another.

By 1900 a double movement seemed to be under way in viticulture. First was the increasing use of new Americo-American and Franco-American rootstocks along with or in place of the old pure American rootstocks; second, the ever-growing favor and rapid expansion of the new direct-production hybrids. In 1906 the *RHFA* published a report on direct producers and rootstocks written by the Italian viticulturalist Clemente Grimaldi, well known for his reconstruction work in Sicily and the Mediterranean littoral.[19] Grimaldi made a survey of French experts known for their hostility to direct producers. He also included Roy-Chevrier, who was reported as then being in favor of sowing some direct producers in vineyards known for fine wine in order to "improve" the wine of grafted vines – that is, to beef up thin greats. Verneuil recommended introducing direct producers to replace local varieties in the Charentes, except for the *grand cru* cognacs. Articles clearly hostile to hybrids no longer appeared in the big viticultural organs – *Progrès agricole, Revue de viticulture*. The hybrid tide seemed irresistible; how high it would rise was not clear.

Yet the enemy camp, including governmental general inspectors of viticulture like Foëx, was not convinced by enumerations of successful hybrids or by praise of the juice squeezed from the fruit. Grimaldi saw some virtues in direct-producer plants and their wines, but they still left much to be desired. He would not make a recommendation until more experiments had been done. His own work on hybrids over a twenty-year period had produced only a few new plants, rootstocks, and direct-production hybrids. Given the well-known genetic difficulties of breeding plants with the desirable characteristics, this was not surprising. The process could be seen as a lottery in which only a few of millions of numbers have any value: the difficulty for the researcher was to arrive at the dreamed-of combination with only one lifetime in which to work.[20]

Perhaps the lifetime failure to produce a good wine had also tempered the abrasive confidence in hybrids that Roy-Chevrier had exuded in his youth. He lamely concluded that, if the pessimists were right, perhaps optimists were not wrong, a paradox resolved by Roy-Chevrier's *pensée maîtresse* of localization – for every vine there is a unique *terroir*. Hybridization, after all, was a reality, producing practical solutions to problems arising from the susceptibility of vines to the American diseases.

---

[19] Grimaldi, "Porte-greffes et producteurs directs," *RHFA*, no. 104 (1906), pp. 747–58.
[20] Durand, "A propos d'hybridation et d'hybrides," *La vigne américaine*, no. 11 (1908), pp. 375–81. Durand, formerly head of the Ecole de viticulture de Beaune, was then the head of the Ecole d'agriculture d'Ecully (Rhône). For more doubts on the virtues of hybrids, see *L'Agriculture nouvelle* (Nov. 30, 1912) and Creusé, "La restauration du vignoble poitevin," *RV* 52 (1920): 127.

Supporters of direct producers took a dim view of the antics of Roy-Chevrier in the *Revue de viticulture*.[21] Gouy subjected the turncoat's articles to an astringent analysis. J. de Bouttes, charmed by the little literary masterpieces couched in a light and bantering style, still noted that Roy-Chevrier had buried hybrids. The Société d'agriculture of Chalon-sur-Saône, less amused by his apostasy, demanded his resignation. The bulletin of the society reprinted the articles of Gouy and de Bouttes, which stated the case for Roy-Chevrier's exclusion from hybrid company. A cynical explanation of the apostasy was put forward by J. B. Poulin. The Société des viticulteurs de France et d'ampélographie had just been created and Roy-Chevrier was interested in joining the world of official viticulture represented by this organization. To become the alter ego of Viala, he had to give up criticizing the Ecole de Montpellier and its followers, with their program promoting the grafted *vinifera* vine.[22] But Roy-Chevrier's tergiversation was the sign of an imminent détente.

Détente: The Congress of Lyon in 1901

In 1901 at the Congrès de l'hybridation de la vigne in Lyon, a great event took place in the history of viticulture: the Ecole de Montpellier was welcomed into the company of believers in hybridization.[23] Emphasis on applied knowledge is not a distinguishing characteristic of the Joycean "bellsybabble" of scientific congresses. Yet a scientific report turned out to be the main event of the congress. Ravaz gave the results of his nearly three-year study of direct-production hybrids in the experimental vineyards of the Ecole d'agriculture de Montpellier. Ravaz was optimistic concerning the resistance of these hybrids to phylloxera – too optimistic for Maurice Dumas, who thought that the period of observation was insufficient to warrant so much optimism. Optimism seemed to be a distinguishing characteristic of the school of agriculture in Montpellier. Producers retained the pessimism of those who deal with nature on a daily basis.

Although the true believers in hybridization could hardly contain their joy over this ostensible apostasy in the ranks of official science and, more important, state viticulture, Ravaz's careful conclusion was essentially only a modest concession to the hybrid camp. Direct-production hybrids should never make up more than one-tenth of a vineyard, whether for ordinary or fine wines. His colleague Auguste Bouffard went further in claiming that these hybrids usually produced strongly colored blending

---

[21] "Contribution à l'étude des hybrides producteurs directs," *RV*, Nov. 17 and 25, Dec. 1, 8, and 15, 1900.

[22] Various articles by Gouy and J. de Bottes in *RHFA*, no. 37 (1901), pp. 9–16; no. 38 (1901), pp. 32–53, including an exchange between Gouy and Roy-Chevrier; and no. 40 (1901), p. 91.

[23] See report by Maurice Dumas in *RHFA*, no. 49 (January 1902).

wines (*gros vins de coupage*), which could be used to supply some back-bone or body and increase the level of alcohol in *vins de pays*. These two professorial concessions, even if viewed as insufficient, were still welcomed by the growers of hybrids, who believed in a fertile future for hybrid wine production: table wine, a light red wine (*vin clairet*), and even white wine – on the far research frontier of hybrid oenology was "the possibility of making white wine."

As the pragmatically minded *RHFA* saw it, Ravaz's sensation-creating report contained two important facts related to the cultivation of vines. The Jacquez vine had sufficient practical resistance for four-fifths of the vineyards of the Mediterranean Midi and nine-tenths of the rest of France's vineyards. As many new direct producers were resistant to phyl-loxera, there was the possibility of cultivating them as non-grafted plants in most soils; some could be grown non-grafted in nearly all soils. For the grower here was official recognition of the practical resistance of many Franco-American rootstocks and non-grafted plants in average soils. This resistance was high enough in certain plants to permit them to be planted everywhere. The resistance of the roots to the insect was to be found in all groups of hybrids.[24] The next congress was organized by the Société des viticulteurs de France et d'ampélographie, which created a section to study direct-production hybrids. But this decision was perhaps a footnote to a basic change in the outlook of the Ecole de Montpellier.

Although the values and moral assumptions of science were not stan-dard items in its repertory, the *RHFA* published an interpretation of the moral results of the congress by its general secretary, A. Jurie, no stranger to the pages of the journal. What impressed Jurie most about the congress was that "this great Ecole nationale de Montpellier" decided to attend. Foëx had come to the first congress, but no more came of the discussions than had of the early attempts by the school to produce hybrids. Once the school dropped its universal hostility to hybrids, the *RHFA* began to praise the school instead of denouncing it – *grande, magnifique,* and *objec-tive* were the adjectives that became applicable to the former enemy of French viticulture. One could even find a place for *V. berlandieri* in the vineyard's genetic pool. The vine's virtues had become more obvious than its old vices, at least as a basis of hybridization. The *RHFA* had already decreed the nonagricultural existence of Foëx, who conveniently died in 1906. Foëx had been used as a symbol of bad science; Ravaz became the symbol of disinterested research whose findings would be of immense use to growers. Science and production were cooperative bedfellows at last, with Roy-Chevrier back in the fold as the angelic propagandist of hybrid-ization.

[24] *RHFA*, no. 78 (June 1904), pp. 136–8.

Mendelism, Heredity, and Hybridization

Some growers at the congress had complained about the boring obscurities of science. To the unimaginative mind, science seems to deal with what Joan Kron called ruffles on reality. Jurie, clever enough to recognize that often we know reality through its ruffles, praised the speeches on the scientific theory of hybridization. Two scientists impressed him: the chemist Armand Gautier, professor at the Paris faculty of medicine, who spoke on theories of the mechanism of hybridization and the formation of races. Following the herd of French scientists, Gautier rejected the role of Darwinian natural selection and the influence of environment in modifying species through slow evolution. Instead, he believed in the new version of the teratological gospel: sudden changes created "monstrosities," which, perpetuated by heredity, produced new races. Then there was Lucien Daniel, by then a botanist at the faculty of sciences in Rennes. A student of Gaston Bonnier's, the first botanist to win the Academy of Sciences's Philippeaux prize in physiology, Daniel argued that grafting produced an exchange of specific characters and that these inherited characters led to new species. This revived eighteenth-century belief in the inheritance of acquired characters served as a warning against the dangers of grafting and a reinforcement of belief in the virtues of hybridization. Some people hoped that sooner or later bad practice would surrender to good theory.

The reason for so much discussion and disagreement on the deceptively simple question of what vine to plant lay in the shallow state of scientific knowledge about hybridization. Learned discussion as well as practical empiricism foundered on the hidden rock of plant genetics, which was not yet part of the scientific baggage of either professors or growers. The extent of ignorance about genetics was widely known and frankly acknowledged. A great deal of knowledge had been painfully acquired through the recorded activities of plant breeders, hybridizers, and even experimenting professors. But little was yet known about how the laws of heredity actually operated to determine the result of a hybridization.[25]

Ernst Mayr has argued that the practical plant breeders, whose "major interest was the crossing of varieties, many of them differing, as we would now say, in merely one or a few Mendelian characters . . . have a far better claim to be considered as direct forerunners of Mendel than the plant hybridizers," who were chiefly scientists believing in the eternal nature of the parental essences of plant species. Perhaps this is true, but in the French vineyard at the end of the nineteenth century we find that the distinction does not hold up: plant breeders were also plant hybridizers, who aimed "to improve the productivity of cultivated plants, to increase

---

[25] Victor Ganzin, "L'hybridation et les hybrides producteurs directs," *RV,* no. 8 (1897), pp. 489–95.

their resistance to disease and to frost, and to produce new varieties."[26] It was a little early to expect much in the way of theoretical results. The work of G. H. Shull and E. M. East on the inbreeding of maize followed by hybridization took place in 1907–10, and these theoretical results were not incorporated into practical breeding programs until the 1920s with the switch over to hybrid corn in Iowa coming after 1933.

Scientists would be working with equal knowledge of the genetics of the vine by the 1930s. The difference was that as knowledge of plant genetics increased, the hybrid vine became increasingly controversial and was even banned in several countries. Good resistance to disease could not be substituted for good wine. Most winegrowers eventually became convinced that even if grafting was not the answer to the viticulturist's prayer, at least it made it possible to make good wines. Eventually a viticulture more solidly anchored in genetics would cause the quarrel to fade away.

What Daniel called specific characters or variations were considered equivalent to mutations as newly defined by de Vries in the three volumes of *Mutationstheorie* (1901–3), his canonical statement of "the first great experimental studies" on the subject.[27] Some contemporaries were unimpressed by this " 'rephrasing of the old idea of sports' or monstrosities."[28] Gautier had seen that the new emphasis was on "the actual origins of new genetic characters . . . the problem of the origin of genetic novelties," rather than on the Mendelian problem of how "already existing factors and characters" are transmitted to offspring.[29] Little matter that "few of the original mutations observed by de Vries in *Oenothera* would now be called mutations. The genetic behavior" of the evening primrose (*Oenothera lamarckiana*) "is unusual."[30] Botanical research and scientific squabbling were redirected to a whole new set of problems, whose solutions would produce modern genetics. Around 1900 all this was a generation away in viticultural research.

The crew of the *RHFA* and their supporters had no doubt that Ravaz would one day deny the faith of his fathers at Montpellier and see the light concerning specific variations in the grafted plant, rare though they were. Ravaz could ignore these variations because his hundreds of experiments showed that in grafting vines the reciprocal specific influence of the subject and the graft was nil – in other words, the scion and the rootstock had no effect on each other. This quarrel had become a matter of logic, an issue of modes of scientific reasoning. Experiment and induction seemed

[26] Ernst Mayr, *The Growth of Biological Thought* (Cambridge, Mass., 1982), p. 649.
[27] Quotation is from Hans Stubbe, *History of Genetics* (Jena, 1965; trans. T. R. W. Walters, 1972), p. 231.
[28] Liberty Hyde Bailey, cited in Richard A. Overfield, *Science with Practice: Charles E. Bessey and the Maturing of American Botany* (Ames, Iowa, 1993), p. 184.
[29] Mayr, *Growth of Biological Thought*, pp. 742–4.
[30] A. H. Sturtevant, *A History of Genetics* (New York, 1965), pp. 62, 63.

powerless to help people choose between Daniel's and Ravaz's research programs. Naturalists could not accept a theory of discontinuous Mendelian genes that did not explain continuous variation. It was not until about 1910 that the Swedish plant breeder H. Nilsson-Ehle "demonstrated experimentally (1908–1911) that quantitative characters, resulting in continuous variation, can be inherited in a strictly Mendelian way."[31]

Our early heroes in the vineyard could only see through a glass darkly. This was not different from biology itself until the acceptance of the dogmas of the evolutionary synthesis by the biological community in the 1940s, thus closing "the gap between the experimental geneticists and the naturalists" by marrying the two different "research traditions."[32] But while the schools quarreled the vines produced. At least the threat of the death of the vineyard had passed, and around 1900 it even seemed that the grafters and hybridists were in the process of arranging an armistice.

Late-nineteenth- and early-twentieth-century biology is a good illustration of the soundness of Eddington's Theory: "The number of different hypotheses erected to explain a given biological phenomenon is inversely proportional to the available knowledge."[33] Of course, an enormous amount of practical biological knowledge was used by plant breeders, who often invoked Darwin and Mendel as their patron saints. (The word *genetics* was coined by William Bateson, inventor "of some of the most important technical terms in the field" and perhaps the rock upon which the science was built.)[34] And after the Mendelian revival of the turn of the century, the great hybridizer was perhaps more talked about than understood; but that put the breeders in good scientific company. At the congress of Lyon in 1901, Mendel was not known or, if known, thought to be irrelevant. Couderc, Ganzin, and Millardet thought hybridity to be governed by the law of fusion or the mosaic hypothesis of Charles Naudin, or some elegant variation of these ingenious ideas. As late as 1929 an article in the *Revue scientifique* noted the imprecision of Mendel's laws, their weak experimental basis, and their inapplicability in normal fertilization.[35]

Ignorance of the way heredity works led Millardet to the wrong conclusion when he crossed *V. vinifera* and the American *V. rotundifolia* in the hope of producing a hybrid that had the same resistance to phylloxera as the American vine. The F1 generation of hybrids had the dominant characters of the *V. vinifera,* and Millardet did not press on to breed several

[31] Mayr, p. 790.
[32] Ibid., pp. 566–70, 790.
[33] Arthur Bloch, *Murphy's Law* (Los Angeles, 1978), p. 48.
[34] Mayr, p. 733.
[35] Pierre Jan, "Les lois de Mendel et le calcul des probabilités," *Revue scientifique,* 67e année (1929), pp. 58–60.

generations (F2, F3, etc.) of these hybrids, including back-crossing to the parents, which would have produced plants with some of the characters of the *V. rotundifolia*. Knowledge of Mendelism might have led Millardet to produce valuable plants in this particular case.[36]

Couderc was one of the rare breeders who took some interest in the variations of plants during their lifetime. The mechanism of hybridization being unknown to these breeders, they could reasonably accept the dogma of Naudin as scientific explanation: "hybrids are a real mosaic." Empirical rules governed plant breeding, with each breeder touting his vine as embodying the best qualities of the *Vitis vinifera*. At the congress of Lyon in 1901 there was even a movement for the creation of a separate realm of nature for vines and their breeders: the hybridization of vines obeyed the laws of *métissage* (crossing between varieties of the same species) more than the laws of hybridization – a unique case in botany. So crossing of very different vines gave offspring with characters that one would only expect from the crossing of varieties much more closely related in the same species. A logical application of the laws of hybridization to the family of *Ampelideae* would not predict fruit-bearing direct producers, but as the nineteen clearly separated botanical species (true Linnaean species) of vines behaved like *métis*, hybridizers like Couderc were able to take advantage of this oddity of nature. According to G. Cuboni, this was exactly the kind of idle speculation that could be avoided by accepting Mendel and de Vries – it seems that few people were careful enough readers to detect disagreements between the two. He saw the theoretical importance of Mendel and de Vries in their transformation of hybridization from an empirical into a scientific practice.[37] Planter-hybridizers had not yet been transformed.

Although A. Bonnet introduced Mendel's laws into viticultural literature in a memoir of 1902, and Viala and Péchoutre advocated a limited application of these laws, the discourse was all on a very abstract level, without any requirement that results be verified statistically or that the evolution of established traits be studied. Mendel's method, the most important part of Mendelism according to Branas, was ignored. An article by Kuhlmann in the journal *Weinbau* in 1914 appeared in French in 1919, but it excluded vines from the scope of Mendel's empire on the ground that his results were obtained from experiments on annuals, not perennials. Kuhlmann was not even converted by the fact that his own experiments on hybrid vines confirmed Mendel's laws. Heady with their practical successes, the old breeders saw no need to try to integrate into their mental

---

[36] This was the conclusion of G. Cuboni, whose articles on the laws of hybridization according to Mendel and de Vries were analyzed by Gouy in the *RHFA*, no. 73 (1904), pp. 31–9.

[37] *RHFA*, no. 49 (1902), pp. 1–2; no. 73 (1904), p. 39.

universe something that they did not understand. Like most other science, unless needed in a disaster, Mendelism could be safely ignored.

Hybridizers could look back with pride to successes in crossing vines since 1824, when Louis Bouschet de Bernard, on his estate of La Calmette in Mauguio (Hérault), had started producing remarkable hybrids based on the male Teinturier and the Aramon grapevines. His son Henri joined him in the family breeding empire. One of their offspring, the Petit-Bouschet, became the hybrid Don Juan of the Midi. The breeders, like the Alsatian technician Kuhlmann, believed that Mendel had systematically studied a law of nature, not created it. Breeders had long used the law, even if in ignorance. Branas lamented that the hybridization of the vine in France remained free from Mendelian influence, but it could not have been otherwise until at least after the First World War.

Not that a knowledge of Mendelism without an appropriate genetic technology of the vine, still in its infancy even at the present time, would have helped much. "Mendelism offered a plausible explanation for the extreme difficulty in obtaining varieties that would breed 'true.' " Considerable cognitive satisfaction could be derived from understanding "the greater variability of new types, . . . the problem of fixing hybrids, . . . why some varieties could apparently not be fixed, despite repeated selection, and why success was so long in coming with others." But Mendelism could not provide the rules by which to obtain specific results in breeding. It can be argued that the "situation would change dramatically with two linked developments: a Mendelian interpretation of the effects of inbreeding (and crossbreeding) and invention of the double-cross method of breeding."[38] True, but "the greatest success story in genetics" turned out to be "the invention of hybrid corn," not the ideal vine, which was not even found by the Germans, busy as beavers though they were in genetic research in viticulture.

Much of the disappointment with hybrids resulted from assumptions based on limited knowledge of the complex genetics of hybridization. It was initially assumed that because the properties of different vines were known, it would be easy to predict the vineyard fate of binary hybrids (from two related parents), because they would inherit modified aptitudes of the parents. But nature had a nasty genetic surprise in store for the innocent winegrower. Heredity simply does not work that way. The hybrid produced by crossing a *vinifera* vine with a *berlandieri* vine was just as resistant to phylloxera as the *berlandieri* – a pleasant surprise. Complex hybrids (their parents are hybrids) made prediction more difficult, even impossible. On the basis of its parents' properties, the direct producer

---

[38] Diane B. Paul and Barbara A. Kimmelman, "Mendel in America: Theory and Practice, 1900–1919," in Ronald Rainger et al., *The American Development of Biology* (Philadelphia, 1988), pp. 295–6.

5813S should have had only a medium resistance to mildew; but it had a nearly total resistance. There seemed to be no simple relationship between the specific composition of hybrids and their growing characteristics. In complex crossings there were an infinite number of possible combinations. Given the existence of 38 chromosomes (19 matched pairs), in 1938 Branas did not expect any rapid progress in the research on using genetics to breed the *Oiseau bleu,* the perfect hybrid. He was right, but the spread of hybrids was not dependent on more than a knowledge of kitchen genetics. The empiricism of the breeder allowed him to rush in where the scientist feared to tread.[39]

Hybridizers depended to a great extent on sowing seed in order to create their vines. The offspring of seeds are not necessarily invariable. Because of genetical instability or mutations, variations do occur, especially in the seeds of hybrids. "Seed therefore presents constant opportunities for selection and improvement; vegetative means of propagation guarantee that the progeny are identical to the parent, for they are of the same tissue" – hence the advantage of propagating by cloning.[40] But before producing genetically identical vines, breeders had to find vines that satisfied numerous criteria of growth and production. This led to an excessive faith in seeds.

Eighteenth-century agronomic literature has some references to growing vines from seed, but the process was of no practical significance. Interest increased in the first half of the nineteenth century, probably less as a result of the various works in agronomy than of the practical successes of men like Bouschet-Bernard and of the congresses of professional seedmen. Obtaining plants from seed became an instrument of scientific farming to improve species and to achieve certain practical aesthetic aims: to avoid plant degeneration, obtain more visually pleasing and early ripening fruit, and produce grapes that would give wine a good color. Phylloxera stimulated new interest in seeds and in hybridization. Some species important in the reconstitution of the vineyards could not be propagated very well from cuttings. Many *berlandieri* vines were grown from seed. The famous hybridizers sowed grape seeds with hope and fervor: from 1881 to 1886 Couderc sowed 20,690 hybrid seeds from 607 combinations, and in 1902 he sowed a record of 333,600 seeds.

The plants for Mendel's experiments had also come from seed: "From several seed dealers a total of 34 more or less distinct varieties of peas were procured and subjected to two years of testing."[41] Branas lamented

[39] J. Branas, "Le point de vue français," *Bulletin de l'organisation internationale du vin* (hereafter *BIV*) (1939), pp. 184–6.
[40] Anthony Huxley, *Plant and Planet* (London, 1974; rev. and enlarged ed., 1987), p. 339.
[41] Gregor Mendel, "Experiments on Plant Hybrids" (1865), in Curt Stern and Eva R. Sherwood, *The Origins of Genetics: A Mendel Source Book* (San Francisco, 1966), p. 4.

that the practical men of the nurseries had not used the statistical method, whose aim would presumably have been to achieve Mendelian certainty. In 1939 the Ecole de Montpellier seemed happily ignorant of the "excessive goodness of fit of Mendel's ratios."[42] Perhaps this ignorance did not make much difference: R. A. Fisher's "belief that Mendel's experiments 'were in reality a [biased] confirmation, or demonstration, of a theory at which he had already arrived' " is quite debatable.[43]

Given our wisdom from historical hindsight, we are surprised that the Ecole d'agriculture de Montpellier did not play much more of a role in hybridization and in viticultural genetics than it did. Branas simplified matters by blaming this lacuna on the separation of hybridization and science during a half-century for putting French viticultural genetics twenty years behind Germany by the end of the 1930s.[44] Montpellier had concentrated on solving perhaps the greatest agricultural problem of the nineteenth century: the production of a phylloxera-resistant vine that would produce quality wine. It was a pre-genetics problem and was solved by pre-genetics research, including work on heredity. Why Montpellier did not later produce a Bateson, or why France produced little research in Mendelian genetics are different questions.[45] The quarrel over hybrids continued in France down to at least the 1950s. While scientists, politicians, bureaucrats, and defenders of the *crus classés* and politicians talked, the believers in hybrids enthusiastically planted them.

By the 1930s, scientists at the Ecole de Montpellier were thinking hard about seed. In an analysis of the use of the mechanism of heredity in breeding plants, Branas began by considering the history of the use of seed, the soul of the plant, its guarantee of eternity. Growth from seed is the natural method of multiplication of the vine, its conservation of characters, its way to propagate and adapt the species. By setting its hereditary mechanism in action, the seed creates new types by new combinations of genetic factors. Modifications of *Vitis vinifera* from the Tertiary period (beginning 65 million years ago) down to varieties of the twentieth century all came from seed. Man the manipulator, as Anthony Huxley calls us, has just arrived in the garden. Traditionally, viticulture used only vegetative multiplication – propagation by cuttings (slips) and its derived techniques – which conserves integrally the existing remarkable types of vines. This fixity of varieties gave rise to a doctrine of viticultural monogenesis: the idea of a common origin of cultivated varieties.

[42] Sewall Wright, "Mendel's Ratios," in Stern and Sherwood, pp. 173–5; R. A. Fisher's famous paper "Has Mendel's Work Been Rediscovered?" is reprinted on pp. 139–72.

[43] Robert Olby, *Origins of Mendelism,* 2d ed. (Chicago, 1985), p. 254.

[44] J. Branas, Bernon, and Levadoux, "La génétique en viticulture," *BIV* (1939), p. 206.

[45] See Richard M. Burian, Jean Gayon, and Doris Zallen, "The Singular Fate of Genetics in the History of French Biology, 1900–1940," *Journal of the History of Biology* 21 (1988): 357–402.

Branas, a polygenist, did not believe in this idea. Although accepted by de Candolle, monogenesis was a speculation more fascinating to historians and litterateurs than to scientists, no friends of the *Urpflanze*. Branas, like J. E. Planchon, believed that cultivated vines were indigenous, descended from wild vines in their regions. The vines in each region have a "family" resemblance derived from an original "hereditary patrimony"; some groups of vines in certain areas have characters in common that give them an indisputable "racial aspect" – for example, the "Kabyles" of North Africa. By the 1950s, polygenesis was in full flower, with numerous different centers (five major ones) designated as having given rise to different varieties of *Vitis vinifera*. It was believed that finding the native vines of a region could lead to the use of their distinctive disease-resistant characters in breeding cultivated vines.[46] Hence the search for these vines as a source of rootstocks on which to graft the French *vinifera* vines, thus ensuring the continued production of quality wine, and perhaps ending the debate over grafting.

Scientific Closure: Staging Montpellier's Victory

The triumph of the Americanists came after a long battle with the defenders of hybrids and the believers in the survival of the old French vines. Protected by large doses of poison, some old French *vinifera* vines lived on well into the twentieth century, but it was a system of survival adopted when the vine-grafting solution was in its scientific infancy, a system no longer scientifically or economically justified by the 1920s, and less so in the new age of plant genetics emerging after the First World War. The victory of the Americanists over the anti-grafters was carefully orchestrated at the national level as well as in local situations. Of course, anything of potential influence on well-known wines, part of the *patrimoine,* was by definition a national event.

The pro-grafting campaign of the Americanists of the Ecole nationale d'agriculture in Montpellier showed that their power depended to some extent on social factors. Their status in the scientific community and their support by the government contributed to the ultimate triumph of their science. The school of agriculture itself was a powerhouse of support for the Americanist cause. Its scientific reputation was firmly based on research. It also served the agricultural community as an experimental and developmental center. The records of the school's experimental vineyard might not have satisfied Daniel, but its flourishing reconstituted vineyard of grafted vines convinced the most skeptical of winegrowers. In the first six months of 1880 the school welcomed 6759 visitors. In 1883 a delega-

---

[46] G. de Lattin, "Allemagne," *BIV,* no. 299 (1956), pp. 12–13.

tion of 120 growers came from the Beaujolais. In 1888 the school received 1400 mail requests for information. After 1879, viticultural meetings, organized by the school and the Société centrale d'agriculture, attracted people from all over the Hérault as well as winegrowers from Italy and Spain. The advice of the school's professors was sought everywhere in France. Viala and Ravaz were consulted in most of the viticultural world. The works of the professors on viticultural topics were bestsellers. The book (80 pages) by Viala and Ferrouillat on mildew came out in an edition of 10,000 copies. The journals *Progrès et agriculture* (1884–) and *Revue de viticulture* (1894–) were extremely successful publications.[47] When coupled with the economic viability of the grafted plant, the Americanist program was impossible to defeat.

In 1907 the Congrès de viticulture met in Angers. It could have been the scene of a national debate over progress in the reconstitution of France's vineyards. By 1907 the hybrid men had enjoyed many triumphs; mini-triumphs is perhaps a more appropriate term, for the non-grafted hybrids did not produce wines of *cru* and the hybrid rootstocks could at best only serve the French *cru*-producing graft. But strong criticisms could be made of the failures of the Americanists; manipulated by a clever fellow, the facts could be a serious embarrassment at a national congress. The men of viticultural power behind the congress – identified by Gouy as Gervais, Pacottet, Ravaz, Capus, Viala, Verneuil, and J. M. Guillon – avoided a public spectacle by making sure that the critics of the Americanist program were not invited.

Among the pariahs were believers in specific variation in the grafted vine: Bonnier, Lemonnier (a botanist at the faculty of sciences in Nancy, he had done experiments "proving" the existence of specific variation), and Daniel (now pointed to by his supporters as a scientist whose research had been recognized by the Académie des sciences). There were hybridizers whose work could be cited in support of this generally unfavored doctrine in scientific circles: Jurie, Castel, Luther Burbank, and Nomblot (secretary of the Société nationale d'horticulture). The ideas and research of these authorities had no place at the congress. Daniel was not far away at Rennes, but like other anti-Americanists, he was not invited. The Americanists could thus concoct a unanimous victory statement in the best tradition of successful science.

The congress took up the three main questions being debated in the wine world and gave clear, certain answers to all three. Gouy was suspicious of this unanimity in a meeting of his fellow Frenchmen. It was particularly surprising for a group to agree on both scientific and practical

---

[47] Jean-Paul Legros and Jean Argeles, *La Gaillarde à Montpellier* (Montpellier, 1986), pp. 142–4.

issues of such importance. The following propositions were approved by the congress: first, that no case of specific variation of the grafted vine had ever been found; second, that grafting being an entirely satisfactory procedure, there was no need to do research for another method of reconstitution; third, the wine of grafted vines was nearly always as good as and often better than the old non-grafted *vinifera* vines. Gouy's explanation of the absence of discussion was that all opponents or critics of the views of Viala and Ravaz were systematically excluded from the meeting.

The hybrid men found it hard to swallow the hubris of the Americanists. Although Gouy admitted that the theoretical criticism which Daniel showered on his opponents was perhaps excessive, he insisted that one had to come back to the basic issue of the condition of the vineyards, sometimes more ominously referred to as the future of the vineyards. In 1905 the most recent statistics indicated a loss of 100,000 ha over two to three years, leaving 1,622,000 ha in production. Of this total, 239,000 ha were old *vinifera* vines, a sharp drop from half a million hectares, or less, a few years before. There were about 1,300,000 ha of American vines, 800,000 ha of which were grafted *riparia* and perhaps 150,000 ha of direct producers. Three-quarters of the grafted French vines were joined to the *riparia*, which lasted only 25 years in soil to which it was well adapted and 10 to 18 years in difficult soils. In 1904 it was estimated that 400,000 ha had to be reconstructed at a cost of about 600 million francs – that is, 1500 francs per hectare. Some people thought that if past programs of reconstruction were any guide, there was reason to be a bit pessimistic about the future. The marquis de Vogüé, of Chambolle-Musigny (Côte de Nuits), president of the Société des agriculteurs de France, was disappointed with the death of *riparia* rootstocks in 2–15 percent calcareous soils. Jean Dufour, of the University of Lausanne and head of the Station agronomique du Canton de Vaud in Switzerland, thought that it was right to campaign against the excessive use of the *Vitis riparia*. He might have included the *rupestris du Lot* in his campaign. An alarming picture.

A similar situation prevailed in the southwest, where phylloxera and mildew had effected a brutal reduction of two-thirds of the Haute-Garonne's 70,000 ha of vineyards by the end of the 1880s. Winegrowers imitated the Languedoc's pursuit of high yields by planting *riparia* vines in the wrong soil and the aramon variety of vine where it would ripen poorly. Louis de Malafosse regretted this, because some cantons used to produce wines of export quality – *la négrette de Fronton* and *le fer,* for example.[48] But citing examples does not produce a scientific argument.

---

[48] Today the production of quality wines is again in evidence with the wines of Fronton, and the southwest generally has begun to emerge from the shadow of Bordeaux: Gaillac, Madiran, Cahors, Irouléguy, and Jurançon are names not unfamiliar to those who look for moderately priced substitutes for the wines of Bordeaux.

And empiricists are incapable of closing a scientific quarrel. The Americanists soon showed their critics how professional scientists win their battles.

At the Congress of Angers (1907), the Americanists confidently issued a bold statement on the scientific certainty of their program. In light of the widespread dissatisfaction and alarm over the dead and dying grafted vineyards, there was a serious need for a bold restatement of confidence in their program. With criticism coming from all sides, the Americanists asserted their belief in the ultimate success of their plan. They also had the advantage of having the only workable general plan, which was being fine-tuned on the basis of an ongoing large-scale research program. The Americanist way was the only way of producing quality wines, except for the old *Vitis vinifera*. This was their final trump card.[49]

The choice of Angers as a meeting place also showed good strategical planning by the Americanists. The old capital of Anjou was a keen promoter of the wines that made up part of its active commerce. The confraternity of Sacavins, founded in 1905, was soon competing in rituals and wine-pushing with comparable if older Burgundian and Bordelais organizations. In the first decade of the century the Maine-et-Loire, unlike most viticultural regions, was undergoing full reconstruction. Gouy noted slyly that because the grafted vines there were still young and healthy, the usual accidents and deaths were not yet noticeable. Viticulturalists of the region could not raise embarrassing practical evidence to contradict the professors. No "krach des *riparias*" yet. The *RHFA*, now renamed *La Revue du vignoble*, was not slow in dishing out standard criticisms, supported by respectable scientific opinion, to its readers, probably nearly exclusively consisting of hybrid-believers. The Americanists avoided contamination from these heresies by reading pro-Americanist publications.

No serious criticism of the Americanists would be complete without yet another denunciation of state viticulture – "l'Administration agricole française, les professeurs départementaux, l'Ecole de Montpellier" – for discouraging the use of direct producers. In spite of his condemnation of "la chimie viticole," with its lifelines of sulfur, tartaric acid, and sugar, Gouy admitted that technical and scientific advances in vinification had led to the production of a number of wines as good as, even better than, those of 30 to 40 years ago. He hastened to add that the wine would have been of even higher quality if the grapes had been from non-grafted *vinifera* vines, or even grafts on Franco-American rather than on American roots. Oenological progress would have taken place anyway and the final

[49] *RFHA*, no. 73 (1904), p. 31; no. 76 (1904), p. 77; no. 83 (1904), pp. 248–52, a summary of the study by L. de Malafosse on "Le vignoble de la Haute-Garonne depuis le phylloxera," in the *Journal d'agriculture pratique de Toulouse*, the organ of the agricultural societies of the Haute-Garonne and the Tarn; and no. 92 (1905), pp. 488–91.

results would have been better. This seems doubtful. Scientists got drawn into the wine business because of the challenge of conquering disease and the encouragement of state and regional organizations. It may be that these scientists were excessively tolerant of chemistry and "viticultural therapies," the tools of their trade, but they would eventually see the dangers of too much of a good thing, especially in the use of sulfur dioxide, and set guidelines limiting doses to much less than the empirical winemakers had done. Much of the criticism leveled against the Americanists was for practices long established in the wine business, having little to do with grafted vines.

A note of desperation is evident in the accusation that the men of science played a role in supplying Paris with artificial wine because they had shown the way by doctoring poor wine to make it potable. Science was thus partly responsible for the unsold natural wine surplus and indirectly responsible for the developing revolt in the Midi. (Poubelle, prefect of police, said that four to five million hectoliters of the wine consumed in Paris, about one-half of consumption, had no identifiable viticultural origin.)[50] The real issue – the production of quality wine – had become lost. It was time to put an end to a useless quarrel.

Clear evidence of closure in the debate over grafting can be seen in the judgment rendered by Clemente Grimaldi in 1906 that nearly all the vineyards of France had been reconstituted by grafting. The situation was no different in Italy and other wine-producing countries.[51] And in spite of financial losses in the first grafted vineyards, loudly advertised by the hybrid men, winegrowing could still be profitable, though subject to the same fickle forces of the marketplace as other types of capitalistic production.

Grafting shortened the ripening time of fruit – of considerable economic importance for more northern vineyards – and increased its quantity, at the price of a shorter life for the plant. But once the difficulties of adaptation of the vine to soil and affinity of the rootstock to the scion were overcome, the lifetime of the vine was long enough to pay for expenses and financing. Some grafted vines were still in production after 30 years in France, and some after 20 years in Italy. True, grafting was held to make vines more susceptible to fungus diseases because of the plant's more delicate tissues and the thinner membranes of leaf and fruit. Extra spraying meant an added expense and more debates over the possible effects of the wine on the drinker from heavily sprayed vines. There was reassurance from the men of science: many drinkers have died of alcoholism, but no

[50] *Revue du vignoble,* no. 118 (1907), pp. 1071–82.
[51] Grimaldi also thought that the quality of fine wines had deteriorated, presumably because there had been an increase in production from planting more vines in old vineyards than had grown there before. See C. Grimaldi, "Portes-greffes et producteurs directs," *RHFA,* no. 4 (1906), pp. 747–58.

one from too much copper sulfate in the Bordeaux spray mixture.[52] Americanists could cite as many facts for their side as the hybridists could for theirs. More important, the grafted vine could produce wine of high quality; the hybrid vine could not.

The Impotence of the Scientific Critique of Grafting

Although viticultural science became overwhelmingly Americanist or pro-grafting, there still survived some criticism of reconstituting vineyards with only grafted vines. The leading critic was Lucien Daniel, whose research became a dangerous nuisance in the view of agricultural bureaucrats and university-based viticultural researchers, who generally agreed on grafting as the only national anti-phylloxera strategy.

In 1899 Daniel brought out a book on variation in the scion (budwood) and the inheritance of acquired characters. Jurie brought it to the attention of readers of the *RHFA* in 1900.[53] At that time Daniel was teaching at the Lycée de Rennes; he would later move to the faculty of sciences. (The scientific dissent of Daniel from the grafting gospel is a good example of how to end a scientific career.) In Rennes, a big agricultural center, apple cider was more important than wine. Daniel's frightening botanical conclusions, irrelevant for his *pays,* were rejected by most of his fellow scientists. His experimental work had been done in various botanical gardens and the laboratory of plant biology run by Gaston Bonnier at Fontainebleau, standard places for doing good botanical research. The viticultural public learned of the work of Daniel, that of Charles Collin, his student, and also that of his *patron* and defender, Bonnier, through serious analyses in the pages of the *RHFA* and *L'Oenophile* of Bordeaux. For a while Daniel also had his niche in the agricultural bureaucracy. The Americanists decided that he had to be refuted and scientifically isolated in order to minimize the damage done by his casting doubt on the basic scientific dogma underlying grafting.

Daniel's work on grafting various haricot beans had provided one of his experimental proofs of the transfer of characters from subject to graft, an argument of considerable interest to hybridizers. In 1894 Couderc had argued that the graft could have a marked effect on the subject, which was of more interest to winegrowers desperate to find rootstocks to withstand phylloxera. Whether the rootstocks produced specific variations on the grafts, as Daniel and Jurie argued, or whether they came from another cause, as Ravaz believed, was not of much importance for the winegrower,

[52] C. Grimaldi, *Viticoltura moderna,* in *RHFA,* no. 104 (1906), p. 74.
[53] Lucien Daniel, *La variation dans la greffe et l'hérédité des caractères acquis* (Masson, 1899); letter by Jurie in *RHFA,* no. 29 (1900), pp. 99–109, followed by an article of Félix Sahut on the subject.

though it was of interest for the hybridizer. The laws of Mendel and de Vries hardly interested the ordinary winegrower. After twenty years of research on Italian vines, Clemente Grimaldi concluded that all the winegrower needed to know was that variations constitute the exception. The key rule for the grower was that, except for a few minor modifications, the grafted vine, like all other ligneous or woody plants, maintains the qualities of the varieties employed as grafts.

Daniel thought that grafting improved some plants and caused others to deteriorate. To the Americanists this was despair. To Daniel it was hope: continued experimental grafting of hybrids might lead to new varieties of ungrafted vines with sufficient resistance to phylloxera. In the state of the art in 1900, Daniel argued that grafting on American roots often changed and modified the qualities of the wine.

At the 1901 congress on hybridization in Lyon, Couderc made the "sensational declaration" that wines from grafted vines were inferior to those from nongrafted vines. At the congress Daniel found himself in the midst of a national debate with international repercussions over whether the American plants were harming the quality of the famous *crus.* Daniel seemed to long for a certainty unlikely to come from science. Viala and his supporters believed that grafting was the best experimental road to follow if good choices were made in selecting partners for grafting. The anti-grafters accepted the dubious idea that the effect of grafting on vines is different from the effect on fruit trees. Winegrowers, men who had first-hand experience in the field and in making wine, were also divided on the issues.

The ministry of agriculture was keenly aware of the quarrels going on in the intersecting worlds of viticultural science and winegrowing. In an excess of prudence in 1903, it assigned Daniel the sensitive mission of studying the effects of grafting in the French vineyard. The Americanists thought that the fox was being sent to investigate conditions in the chicken coop. But at that time it was by no means clear what the long-term effects of grafting would be. During his first year Daniel studied the east of France and part of the Midi. He proceeded carefully – so carefully that the *Revue de viticulture,* a voice of the Americanist camp, carried his "first notes on the reconstitution of the French vineyard by grafting," which started to question the two basic dogmas accepted at the beginning of the reconstitution: total conservation of the resistance to phylloxera by the American and hybrid rootstocks; and the complete conservation of specific characters by the rootstock and the graft, thus assuring the unchangeability of the vines after grafting. Daniel found that his task in Montpellier was frustrated by an "act of scientific vandalism" in the Ecole de Montpellier, which had torn up all its ungrafted American vines after they had served its own research purposes. Enemies of the Americanists immediately suspected a coverup. It was at least convenient for the Ecole that the end of their experiments came just in time to prevent a scientific enemy-colleague from exploiting their data.

While Daniel pursued his study mission, the Americanists accelerated their grafting program and produced an avalanche of scientific studies to show that their plan provided the only salvation for maintaining quality wine production. The negative conclusions of these studies with their more sinister implications, aggravated by their sensational exploitation, often with political repercussions, led many in the ministry to question the wisdom of the mission. Daniel blithely continued his rigorous criticism in conjunction with vocal anti-grafting forces, even in Bordeaux and Burgundy. The *RHFA* continued its hostility to official agriculture, except for the Daniel mission.

Bordeaux was suspicious of the program of salvation through grafting. Bordeaux is suspicious of anything concerning its *crus.* The pages of *L'Oenophile,* one of the leading vine–wine publications, were open to Bonnier and Daniel and other doubters of the virtues of grafting. The Société d'agriculture de la Gironde, after hearing the departmental professor analyze Daniel's book *La question phylloxérique, le greffage et la crise viticole,* approved a favorable report on it. More alarming to the Americanists, perhaps, was the fact that Daniel established friendly relations with some self-advertised successful anti-Americanist winegrowers.

One of the best known anti-grafters was Bellot des Minières, "le roi des vignerons," who acquired Château Haut-Bailly (Léognan, Graves) in 1872 and produced some of the best Bordelais of the late nineteenth century, one of the *premiers crus de France,* the "Margaux of Graves," and, according to Daniel, a rival of Châteaux Margaux and Lafite. Bellot des Minières made a fetish out of conserving his nongrafted *vinifera* vines, some of which were real museum pieces, 100 to 150 years old, it was said; treated with carbon disulfide, all the vines enjoyed splendid health. To kill fungus diseases, Bellot des Minières used sulfur and ammoniocupric sulfate instead of the usual copper sulfate, on which he blamed many oenological crimes. Phylloxera did not frighten him because he believed that it was dying out in the Gironde, where the insect was now laying eggs only twice a year instead of six times. There were even rows of healthy vines that had been left without treatment for four years. This amazing news was spread by Bellot des Minières himself in a book on *La question vinicole* (1902), by Daniel in an article in *L'Oenophile* (October 1907), and by an exuberant Gouy in the *RHFA.*[54]

There was worse news on wine. De Ferrand carried out an experiment

---

[54] No. 79 (1904), pp. 151–8.

in Pauillac comparing wines from grafted and nongrafted vines up to the 15 years of age in the bottle. His conclusion was that the wine from the grafted vines was inferior. A brave man who believed in his own science, de Ferrand ripped out 30 hectares of grafted vines at his Château de Ferrand in Saint-Emilion to return to the direct cultivation of ungrafted *vinifera* vines. He shared the common view that grafted vines have a poorer resistance to fungus diseases. This was a serious defection from the Americanist camp, because for twenty years de Ferrand had been one of the most resolute grafters in the Gironde. Daniel added to his case against the grafted vine the experiment of M. Ricard (Domaine de Chevalier), who had a complete collection of ungrafted French varieties and a collection of French vines grafted on many different varieties of rootstocks. Daniel pointed to obvious changes in the foliage of grafted *vinifera* vines, especially the cabernet sauvignon grafted on the highly touted 41B and the 1202.[55] It was beginning to look as if Daniel would soon be able to cite as many unsuccessful grafting experiments as Viala could cite successful ones. But even Viala admitted that phylloxera could be held at bay with insecticides for a considerable time if conditions were right. The cases cited by Daniel could be viewed as atypical of the general situation in the Gironde. The big question was what would be the long-term solution to the problem of how to control phylloxera?

Another question was what would become of Daniel, now playing the role of Cassandra on his mission for the ministry of agriculture – an anomalous situation, for politicians want scientists to produce cheer, not gloom. On July 25, 1908, Daniel was relieved of his mission. One of the reasons, perhaps the main one, was that Daniel went international with his bad news. The *Times* (London), interested in the eternal and bitter quarrels in France over the viticultural crisis and worried about the wine supply, translated one of Daniel's articles. Official science appeared on the stage when Ravaz wrote a refutation of Daniel's allegations for the *Times.* Many viticultural and agricultural societies protested to the ministry of agriculture. The situation had become difficult, because by 1907 Daniel was granting interviews in which he explained the decline in the quality of the *grands ordinaires* and the *crus classés* as partly due to grafting and partly due to the substitution of more productive varieties of vines for the older vines, which had yielded less generous harvests of fruit but finer wine. The descent of the vine into the plain also increased production in the Midi. This led Daniel to point out the wisdom of the Portuguese in prohibiting planting less than 50 meters above sea level. The Midi was not amused. Nor were growers in other areas happy to hear Daniel's repeated warnings

[55] *RHFA,* no. 79 (1904), p. 155.

about using dangerous products in antifungus or antiparasitic treatments of the vine. Rumor spread of the probability of an *interpellation* of the government in the Senate and the Chamber. Daniel was quickly sacrificed.

Daniel's old professor, France's best-known botanist, Gaston Bonnier, was not reluctant to hold out hope, even in an article entitled "Les caprices de la greffe et la crise viticole."[56] "The word caprice . . . evokes the leaps of the goat, its instability," Georges Duhamel once wrote; so Bonnier had hit on the right metaphor to sum up the attitude of winegrowers. The alarming thing was that the distinctive qualities of grapes and wines were open to question and, worse, the subject of criticism in English newspapers. Bonnier did not think that continuing the established program was the solution to problems. It was all very well to call for new ways of fighting phylloxera or for new ungrafted plants resistant to the disease, but neither research direction seemed very promising. No one ever won the enormous prize of 300,000 francs offered by the ministry of agriculture in 1874 for the miracle cure, as elusive in the case of phylloxera as it is in all attacks of big, rich science on diseases that are inseparable from life. Bonnier's one sensible solution, that of many scientists and well-known viticulturalists, was to improve the plants being grafted, presumably with the aim of making them more compatible with the rootstocks and thus keeping the qualities for which the *vinifera* vine was historically valued.

Another of Bonnier's students, G. Curtel, also tried to produce an improved scion or graft. Not so notorious as Daniel, he had far greater influence in the wine world. As head of the Station oenologique de Dijon, Curtel found himself in the center of the grafting storm spreading over Burgundy. Coming from the apogee of botanical research to the praxis of one of the world's few areas that produced great wines, Curtel was uniquely placed to create certainty out of what his *patron* despairingly identified as all these contradictory opinions. The botanical part of oenology was in an exciting, creative mess, which scientists might have enjoyed more if they had not had to face angry winegrowers whose grafted vines were behaving capriciously or even dying.

Curtel, like Daniel, challenged the ancient dogma that grafting improves fruit, or at least he accepted the argument that what was true for fruit trees was not necessarily true for the vine. After three years of work on the problem, he concluded that the winegrowers were right: grafting lowered the quality of the grape and weakened the constitution of the vine. The fruit of the grafted vine had a reduced tannin content measured in acid salts, which made vinification more difficult. An increase in nitrogenous matter, the preferred food of microbes, resulted in more disease – the

---

[56] *L'Oenophile*, reprinted in *RHFA*, no. 106 (October 1906). As editor of the *Revue générale de botanique*, Bonnier published a paper by de Vries on Mendelism in 1900. See A. H. Sturtevant, *A History of Genetics* (New York, 1965), pp. 26–7.

spread of pasteurization was probably related to this new development. Last, the wine produced from the grafted vine was more subject to "oxidation of its color" and its basic elements, leading to a too rapid aging of the *grands vins*. This criticism seemed all the harsher coming from a strong supporter of grafting, recognized by Curtel as the only way of preserving the vines that could still produce quality wines in Burgundy.

Like the winegrowers, Curtel deduced from this confusion that the reconstitution program had now passed the stage where the major concern was necessarily the resistance of the vine to phylloxera. Factors like quality of fruit and of wine, ease of reproduction by cuttings, and growth of the vine in calcareous soil had become of paramount importance now that the vine was safe, at least for a period long enough to farm profitably. They did not know it, but these scientists and growers were enjoying the luxury of complaining in the middle of a great oenological triumph, the saving of the vineyards. All the rest was experimenting in the paradigm or disciplinary matrix centered on the grafting gospel of Montpellier.

Curtel's work on vine and wine supports this conclusion. First, the vine. In 1898 Jurie had grafted a hybrid (*othello* × *mondeuse rupestris*) on the vine *cordelia rupestris de Grasset*. The hybrid lost its foxy taste and acquired a resistance to phylloxera; ungrafted, it was susceptible. The new vine's properties were transmitted to the cuttings taken successively from the original graft. Curtel and Jurie communicated this research to the Académie des sciences. Real or at least official science was thus created out of an ordinary act of hybridization. Most important, it shows that Curtel believed in the possibility of improving the scions, an advantage of rejecting the dogma of their immutability. Second, the wine. Because of the different nature of the liquid derived from the grapes of grafted vines, Curtel specified that certain practices previously used as precautions in vinification were now essential. In most cases, the wine required an addition of 10 to 15 grams of sulfur dioxide (SO_2) to the vat for every 100 kilograms of grapes. Annual analysis of the wine had to be done to verify the maintenance of a good level of acidity, averaging about 10 to 12 grams measured as sulfuric acid. Precautions were also necessary during fermentation and after to ensure minimum contact of the sensitive must with air, thus avoiding oxidation of the wine. Vine and wine were saved, but the difficult effort was neither cheap nor simple.

The history of winemaking does not give us much evidence to believe that good vinifications in the past were achieved without SO_2, though it was often used without the precise guidance of scientific knowledge. Nor were wines exempt from infections. Pasteur's claim that most wine was infected might have been a bit of promotion for his cure, but sick wine was probably more common than healthy wine. The cult of worshiping and selling "old" wines is an accident of scientific vinification. In the end,

then, it does not seem that anything very new was being done in vinification because of grafted vines; perhaps some fine-tuning. The old oenological paradigm remained basically the same. Making good wine was not
dependent on grafting.[57]

Programs for reconstituting vineyards with grafted vines were hailed as
a success in science and were eventually successful in vineyards producing
quality wines. But the success of the grafted vine was limited and, throughout much of the first half of the twentieth century, was threatened by the
spread of the direct-production hybrid. This is a paradox that needs explaining, as does the rapid shrinkage of hybrid acreage in the second half
of our century. The next chapter examines the rise, decline, and fall of the
direct-production hybrid in French wine production.

[57] G. Curtel, "Les vins de vignes greffées et la vinification," *Bulletin du Syndicat de la Côte
Dijonnaise;* reprinted in *RHFA,* no. 100 (1908), pp. 667–8.

4

The Fall of the Hybrid Empire and the Vinifera *Victory*

Extent of the French Hybrid Empire

The continued spread of direct-production hybrids could not give complete satisfaction to their supporters without the intellectual pleasure of scientific approval of these vines. That approval was withheld, with the result that quarrels over direct producers in French viticulture did not end with the Congress of Lyon (1901) or even with the Congress of Angers (1907). In 1903 Gouy attacked Viala, who had just said bad things about hybrids in the Sicilian review *Vitacoltura moderna:* that no known hybrid had a high resistance to phylloxera and that the quality of hybrid grapes and their wine was low. The editor, Frederico Paulsen, had asked both Viala and Ravaz about growing direct-production hybrids in Sicily. Fortunately for the hybrid cause, the answers given by Ravaz were less negative and could be interpreted by Gouy as contradicting Viala's opinion. Ravaz's stock had risen in the hybrid camp since his apparent apostasy at the Congress of Lyon, much feted in hybrid circles. If scientific approval came at the expense of the Americanist cause, all the better for the hybrid cause.[1]

A set of minor concessions by a leading spokesman of the Ecole de Montpellier did not mean that the official program of reconstitution of France's vineyards was greatly modified. It may be that Gouy finally realized this, for a month after his critique of Viala, he published a definitive attack on *La Gaillarde,* a sort of Krapp's last tape in the warfare that he had been waging for over a decade.[2] A two-pronged attack first recalled the failures of the school in creating direct producers and its weakness in creating hybrid rootstocks. He then arraigned official science before the tribunal of private initiative in order to emphasize the successes of the famous hybridizers. Gouy wondered if this sorry record, as he saw it, explained the slowness of the school in accepting Americo-Americans and the prejudice it still showed toward Francos or so-called French hybrids.

---

[1] *RHFA,* no. 67 (July 1903), pp. 160–4. Paulsen, who established a nursery in Palermo in 1916, did a great deal of work on selecting the right vines for reconstructing Sicilian vineyards.

[2] P. Gouy, "L'Ecole de Montpellier et l'hybridation," *RHFA,* no. 68 (1903), pp. 189–90.

Gouy showed no understanding of the tension that necessarily exists between a purely commercial operation and a research program, even if the
research is oriented in a practical direction. Nor did he deal with the problems faced by institutions responsible for coming up with solutions for all
of France rather than concentrating on the special problems of the Midi.

The growing use of hybrid rootstocks was useful in the general propaganda for direct-production hybrids. Louis de Malafosse made use of this
strategy in 1900 when he boasted that some of the greatest *crus* of France
were grafted on Franco-American rootstocks. He erroneously put both
the vines of Romanée-Conti and Richebourg on the rootstocks of Gamay-
Couderc, Malconsort. He then noted that the *grands crus* – officially, they
are *premiers crus* – of Meursault were grafted on the 1202. More than
fifty famous *crus* of Burgundy and Champagne were producing on hybrid
rootstocks. It was the same in the Touraine and Anjou. After his manager
had investigated winegrowing areas for two years, the duc de Richelieu
spent more than half a million francs to replant his 260 hectares on Couderc hybrid rootstocks. Here was another familiar contrast in the history of
economic activity: bureaucratic bungles and capitalist triumphs. De Malafosse spread his gospel through the *Journal des viticulteurs* and *Le Messager de Toulouse.*[3]

Gouy could not emphasize enough that the work of hybridization was
largely a matter of private initiative. Governmental agencies, having
achieved negligible results, wisely abandoned the field to the increasing
number of private researchers. Gouy saw here a chance to teach a moral
lesson. Winegrowers, like the rest of the French population, were too inclined to believe in the omniscience and omnipotence of the state and its
representatives. This was not a bad generalization; it just happened to
ignore the fact that the viticultural glory of France was reestablished by
grafted vines, not by direct-production hybrids, which did retain a key role
in the program for hybridization in the production of rootstocks.[4]

Vineyards producing fine wines were not replanted with hybrids. The
most serious threat was a lower level of quality in wine production: many
vineyards that produced the *vins de pays* went hybrid. The two great wars
of the twentieth century gave a big boost to the hybrid, which required far
less care and chemicals than the *vinifera* vine. In wartime, materials (copper sulfate and sulfur, especially), and agricultural labor were in short
supply or, more usually, unavailable. After the First World War the resis-

---

[3] *RHFA*, no. 23 (1899), p. 286; no. 30 (1900), pp. 135–9: includes reprints of de Malafosse's
article on "La mode en viticulture" and a letter on "La nouvelle viticulture."
[4] Grudging recognition of the success of grafting was made by Jurie in a talk given in 1904.
See "Une conférence de M. Jurie au Comice du Givors," *RHFA*, no. 85 (January 1905),
pp. 194–7.

tant direct producer was more than ever the cheap vine of the future for the production of a drinkable wine.[5] By 1925 nearly one-half of the wine-growing acreage of the Loiret was planted with hybrids. Even in the Yonne, hybrids occupied one-eighth of the vineyards in 1925. In 1931 the lowly Noah was still poisoning 5000 hectares of the Saône-et-Loire as well as smaller areas of other regions like the Loire-Inférieure. And in the Niè-vre by 1932, hybrids covered nearly 20 percent of the vineyards, or 684 hectares, up from 178 hectares ten years earlier. By the 1920s the spread of hybrids was recognized as a cause for alarm in the Centre, whose wines include Reuilly, Quincy, Menetou-Salon, Sancerre, and Pouilly fumé, wines that make up an essential if secondary part of the viticultural patri-mony. The general fear was that the economic advantages of the hybrid in difficult times might accelerate their spread from the areas of polyculture to exclusively winegrowing areas.

One of the most serious problems in planting hybrids was finding out which variety was best for each area or even microclimate. Few planters of hybrids consulted the professors of agriculture in their departments, for most of them were reputedly hostile to hybrids. This had an unfortunate result. Most people got their information from a nurseryman, whose chief interest was to sell plants. He often had no knowledge of the regions to which he shipped vines. Before 1914, some departmental professors of agriculture had set up small experimental plots to test hybrids, but little came from these shoestring operations. During the governmental war against hybrids, the ministry of agriculture forbade departmental directors of agriculture and professors of agriculture to have anything to do with hy-brids.

In 1920 the regional agricultural office of the ten departments of the Centre region undertook the creation of a center for winegrowing research (Centre de Recherches viticoles) targeted on the special study of direct-production hybrids by large-scale field experimentation. By 1932 about 300 varieties were under scrutiny at the research center in Cours près Cosne-sur-Loire in the Nièvre. The aim of the research was to find varie-ties suitable for the *farmer* of the Centre region. There were very few disease-resistant direct producers suitable for the climate of the Centre that would also yield a decent-tasting wine. Only an institution supported by governmental resources could carry out a long-term extensive project to find such vines and recommend them through an effective system for disseminating information to farmers. Even the Centre's research opera-tion, in spite of its regional usefulness, did not acquire much of a reputa-

---

[5] J. Tibbal, "Les producteurs directs dans le Tarn," *RV,* no. 53 (1920), pp. 343–6; Galet, *Cépages et vignobles de France, vol. 1: Les vignes américaines,* pp. 373–4.

tion beyond its captive audience. The hope was that if experimentation established a list of acceptable hybrids, better control over their diffusion might be achieved through regulation by state and regional agencies.[6]

By 1946, according to an estimate by Paul Marsais of the Institut national agronomique, hybrids occupied about 300,000 hectares in France – Galet gives the same figure for 1939 – nearly one-fifth of the vineyards of metropolitan France. Unlike in Germany, where the ideology of purity included plants as well as people, the interest of the rulers of the Vichy state in promoting the racial purification of the French race does not seem to have had any parallel in the vineyard. Not that Montpellier changed its line on hybrids. In the conclusion to his enquiry of 1942 on direct producers, Professor Branas noted ironically that hybrids were in the process of disappearing administratively. Galet calculates the size of the hybrid empire in 1958 at about 31 percent of a total vine area of 1,302,000 hectares. Hybrids were then producing about 42 percent of the country's table wine. Some departments, like those facing the Atlantic coast, produced up to 90 percent of their table wine from hybrids. The spread of hybrids was encouraged by plagues of mildew in 1948 and 1957, and especially by the great freeze in 1956 that destroyed 100,000 hectares of vines. Even reputable vineyards producing quality wine were raising some hybrids. If quality wine was not selling well, there seemed to be no point in maintaining an exclusive and expensive *vinifera* estate. Few could have foreseen that by the end of the 1980s the space occupied by hybrids in the vineyards would have shrunk to about five percent, as the market and the government, aided by shifts in patterns of farming and university-connected oenology, promoted the return of high-quality varieties of *vinifera* vines.

It is clear that after the First World War the battle over direct-production hybrids entered a new phase in which governmental and bureaucratic measures responded to an ostensibly uncontrolled spread of hybrids, threatening to encroach on the territory of the quality wines. Defenders of hybrids were put on the defensive by a growing movement, evident in a series of laws, to create the system of *appellation contrôlée*, which establishes a legal relation between French *vinifera* vines and their specific *terroir*.

The first law against hybrids in France was the *loi Capus* on *appellations d'origine* of 1919, modified by the law of 1927: wine from direct producers, excluded from legally defined areas of *vinifera* wine production, could not obtain an *appellation d'origine*. The second law affecting hybrids was that of August 4, 1929, which prohibited adding sugar to the musts of hybrids, thus depriving the direct producer of the right to chaptalization, long en-

---

[6] Albert Dupoux, *Les hybrides producteurs-directs dans la région du Centre. Contribution à l'étude des producteurs-directs* (Cosne-sur-Loire, 1935), preface by Chappaz, inspector general of agriculture. pp. 17–22.

joyed by the *vinifera* vine. The so-called *Statut de viticulture* of 1930–1 ignored hybrids because of widespread opposition to a governmental plan to restrict further their spread, but the law of December 1934 specified that any *vinifera* vines torn up could only be replaced by vines chosen from a list of varieties provided by departmental authorities. The Office international du vin (OIV), sensitive to widespread European hostility to hybrids, inspired this legislation, which was as close to the draconian German anti-hybrid laws as French politicians could move. The rapporteur of the law in the Chamber of Deputies was the president of the OIV – a cosy arrangement. Friends of hybrids squealed that this law would dictate the end of research on hybrids. To cry about the death of scientific research is a sign of real desperation. We may conclude from the continued expansion of the hybrid empire that property owners were less impressed.

By the 1930s a formidable international alliance had been forged against the hybrid vine. Romania was a leader in the OIV in getting the congresses, even the one meeting in Paris in 1932, to pass motions hostile to hybrids. Romania had been planted with all imaginable varieties of worthless hybrids when the big estates were divided up after the First World War. To fight against this commercial duping of Romanian winegrowers by unscrupulous interests required international cooperation from countries selling the hybrids.[7] Weimar Germany took rigorous measures to exclude hybrids from its vineyards: in 1929 a governmental decree prohibited the cultivation of direct producers; after 1930 the wine from American direct producers could not be mixed with other wine; indeed, after September 1933 the sale of direct producers and their wine was forbidden. Yugoslavia was also a pioneer in promoting purity: a law of December 9, 1929, forbade the planting of hybrids and selling wine made from their grapes. Winegrowers were compensated for losses and were also helped to replace the hybrids they uprooted. In Italy the direct producer was outlawed after March 1931. In many countries the hybrid was legally tolerated only as a subject for scientific research; reality in peasant vineyards was another matter.

The direct-production hybrid was its own worst enemy. Much of its bad reputation came from the indiscriminate spread of infamous American varieties. The Noah, Othello, Clinton, Jacquez, Herbemont, and Isabelle, outlawed by a decree of January 18, 1935, could no longer be legally planted in France, although they still occupied more acreage than the superior, more recently developed so-called French hybrids. Everyone could agree that out of the thousands of existing hybrids, few varieties could be classified as good or even acceptable. It is not surprising that the viticultural power elite could at once salute the magnificent work of French hy-

---

[7] Details of the legislation enacted by European governments against hybrids were outlined by Largillier-Seibel at the Congrès de la vigne moderne in Toulouse in February 1934; see his "Etude sur l'évolution du vignoble moderne," *BIV,* no. 117 (1938), p. 39.

bridizers while calling for the exclusion of the direct producer from the hallowed ground of the *régions à appellations d'origine*. They held that the only people who should be entitled to cultivate the pariah vine were farmers who grew several crops and wanted to produce enough wine for family consumption without having to take care of labor-intensive and chemical-consuming *vinifera* vines. It was even argued that the direct-production hybrid belonged exclusively in the experimental domain until a variety showed qualities justifying its debut in polyculture.[8]

One of the most interesting things about French legislation against hybrids was that it could be correlated with the existence of three economic categories of viticultural regions. In regions of mixed farming, where the vine was for the production of family plonk, the hybrid could be accepted as a gift of agricultural progress, arousing no debate over the qualities of the *crus,* a concept remote from the peasant mind. Second, in the regions of intensive monoculture devoted to the industrial production of table wine, there was a debate over the possible loss of an important market in the regions of production of the *grands vins.* These regions traditionally imported wine from the mass-producing areas both for drinking and for beefing up their thinner products to cater to consumer demand for color, body, and a higher level of alcohol in their wine. Another effect of the spread of hybrids in the mixed farming areas was its putting an end to the market for imports in these areas, because no French wine could compete with the low-priced hybrid wine produced by the peasants themselves. So far so bad for the hybrid effect.

The most serious opposition to hybrids came from a third area, the viticultural *régions à appellations d'origine,* where the *grands vins* were produced on the slopes. Since the appearance of hybrids the table wine of these areas was produced on the plains in what was dubbed "le vignoble de ceinture des grands crus." This vine belt around the *cru*-producing areas gave rise to controversy because it was in most of the areas of maximum hybrid development – the Bordelais, the Loire valley, Haute and Basse Bourgogne, the areas of Cognac and Armagnac, and Alsace – that hybrids appeared to be a threat to the renowned *crus.* The spread of hybrids seemed economically sound, providing a drinkable wine in place of worse stuff like *piquettes* and *boissons de marc* (wines made from presscake, sugar, and water). The economies of these regions depended heavily on the export of fine wines to the deserving rich in Paris and abroad. This meant that areas producing fine wines also imported much from table-wine areas, a heavy drain on their local fine-wine economies. The hybrid belt solved several problems at once.[9]

The *grand cru* has become an important and profitable part of the na-

---

[8] Georges Chappaz, in Dupoux, *Les hybrides producteurs-directs.*
[9] Address by Largillier-Seibel, invited by the president of the Société de viticulture, at a viticultural meeting in Tonnerre (Yonne) in 1938, *BIV,* no. 117 (1938), pp. 38–46.

tional patrimony. Before the heritage became national, it was provincial: dukes of Burgundy and governors of Guyenne banned less noble vines and decreed, often in vain, that lesser breeds be dug up. In 1395 Phillip the Bold recommended a new snob grape, the Pinot noir, and decreed the death of the "très déloyaux plant nommé gamay" in the vineyards of Burgundy. The eventual banning of the hybrid direct producer would be in the grand tradition of elite decisions to produce quality wines for the establishment. The decisions of modern politicians could be much more effective through the exercise of vast bureaucratic and police powers and, most important, enormous powers of economic persuasion: peasants who cannot be coerced can be bought. But in the 1930s no government was going to arouse a large number of its irascible, depression-hit peasants with an order to rip up one-seventh or nearly 15 percent of French vineyards. (The agricultural enquiry of 1929 had concluded that the domain of the direct producer made up about one-seventh of the *vignoble français.*)

Doubt about the future of the *vinifera* vine caused discussion to center on the taste of wines from American as well as grafted vines. This issue of taste was of great interest to Pasteur, who, chiefly as a result of his experiments on heating wines, had become skillful in the organization of professional wine tasting. Pasteur set up a comparative tasting of eight wines, three of which were from French vines. All had been heated to 60 degrees centigrade by Pasteur before he put them in his cellar. (In the Pasteurian oenophilic aesthetic, heating a wine made it taste better.) The wines were tasted by Apollinaire Bouchardat, well known for his "authoritative oenological knowledge." Pasteur himself drank wine but did not have any particular expertise in tasting. According to the pro-hybrid camp, Bouchardat found two drinkable wines (1872 and 1873) from American vines, mixtures of Jacquez, Lenoir, and Clinton grapes. According to Pasteur's report, only the Clinton was suitable for commercial wine production.

One of the curious things about this quarrel over hybrids was that it had become hopelessly out of date. The old argument against hybrids was that they produced bad wine, distinguished by the American *goût foxé.* By the early twentieth century this was was no longer the case. The foxy taste of *V. lambrusca* and the cooked flavor of the first hybridizations with *V. rupestris* had nearly disappeared by 1907, when Louis de Malafosse reported on the exposition and tasting of 232 bottles of hybrid wines in Toulouse.[10] Eighteen artistic plaques and medals were awarded in a ceremony clearly aping *vinifera* rituals. De Malafosse found that small proprietors were being increasingly successful each year in producing real table wines cheaply.

Using traditional *vinifera* wine arguments based on soil and age of the

---

[10] See *RHFA*, no. 85 (1905), 315–19, for a report by Malafosse on a wine tasting of 282 bottles of hybrid wine in Toulouse. For the tasting of 232 bottles, see *La Revue du Vignoble*, no. 111 (1907), pp. 904–9.

vine, de Malafosse made two general points about a hybrid wine tasting that had taken place in 1905. First, the *terroirs* of the old *crus réputés* had as much influence on the hybrid as on the *vinifera* vines. Second, direct producer-vines produced better wines as they grew older: in seven years their wines were far better than the wines from young grafted vines. Of course, supporters of the grafted vine used similar arguments to much greater advantage. Hybrid vines did not require the chemical fix of the *vinifera* vines, but their wine might need a shot or two. In years of drought or when the grapes matured too rapidly, all red hybrids needed 25–30 grams of dissolved citric acid for each 100 kg of crop fermented. The use of the additive was defended on the grounds that it came from fruit and was close to the malic acid in the wine, it did not foul the casks, and it preserved the wine in a good state – traditional arguments for doctoring wine.

It had not been easy for hybrid wine to reach this point. To get quality wine, the Alsatian Oberlin gave up hybridizing with the *rupestris* vine because it often gave a thin, insipid liquid, low in acid; he switched to *riparia*, and after eighteen years got good results. One of the few hybridizers in the German empire, Oberlin crossed a good many Alsatian and Rhine vines and bred a large number of *vinifera*-American direct-producer vines.

In France the death of many *riparia* rootstocks led to increased popularity of the *rupestris*. The famous hybrid *oiseau bleu* was a multiseed production by Fernand Fournas: *V. auxerrois* × *V. rupestris* fertilized by pollen of a *cot à queue rouge* (malbec). Maureau de Salon used this Blue Bird to produce a wine far superior to the wine of *auxerrois* × *rupestris,* which was regarded as the best hybrid wine by many tasters. Like the wine of most hybrids, it was flat tasting and without a distinctive bouquet, but this was not really a defect for a *vin de coupage* or blending wine, used to give body, dry extract, and color to fine wines. Producers of hybrids were more willing to confess that wines were cut than were the producers of the *vins fins* that were cut. But the hybrid men wanted hybrid wines that could be drunk on their own by people who could not afford the high-priced concoctions of the master-mixers of châteaux and estates. De Malafosse lamented that in spite of the scientific virtues of Couderc, Castel, and the Ecole de Montpellier, there had been produced only the *oiseau bleu,* the *oiseau rouge* (*le Malafosse*), and the *auxerrois* × *rupestris* varieties that could be compared to good table-wine-producing *viniferas.*[11]

It seems that any serious opinion on the quality of hybrid wine had, in the end, to earn the approbation of science. Scientists had to show that the direct producer was capable of giving a good ordinary wine. Wine tasting may be an art, but it is always comforting to have a scientific analysis supporting one's viewpoint or destroying the basis of an opponent's

[11] *RHFA,* no. 73 (1904), pp. 26–7; no. 85 (1905), pp. 315–19.

belief.[12] In 1937, L. Depardon (director of the Station agronomique et oenologique de Blois) and his assistant analyzed 120 samples of wine that they had made in their laboratory from forty different varieties of hybrids cultivated in the Loir-et-Cher and the Indre-et-Loire. They found that the chemical nature of the wine from hybrids was the same as that made from French vines when they were compared on the basis of alcohol, acidity, and dry extract.

What about taste? A mixed commission of wine merchants and agricultural bureaucrats subjected the hybrid wine to a comparative tasting with *vinifera* wines. Scores for the hybrids ranged from 11 to 13.5 out of 20, comparable to the scores for the *viniferas,* which earned an average of 13, although a hybrid rosé sank to a low of 9. Ten years later Depardon was still doing experiments on hybrid wine. In a report to the Académie d'agriculture, he reconfirmed the facts established earlier. This time one of the reds, Landot 244, got a score of 14 and one of the whites, Seibel 11803, got 15, an impressive "note de dégustation." None of the eight reds and six whites in the tasting received a score lower than 10.5. "Certain hybrids are capable of giving ordinary wines of perfectly satisfactory commercial quality." It was also clear that growers had to continue planting the old French varieties in the famous vineyards.[13]

It would have been surprising if a professor at the school of agriculture in Montpellier had found anything good to say about wine made from hybrids, and Professor Branas did not try to hide his low opinion of its gustatory properties. His view was that the wine of only a few hybrids, including the 6905 Seibel, gave an undeniable pleasure in tasting. So their chief use was in commerce as a *vin médecin* when weaker wines needed to be doctored or at least given preventive medicine before being sold to the consumer, who had a certain idea of the type of wine he wanted. Branas indicated that the common consumer is an easily fooled drinker, a bad judge of wine because his taste is rapidly corrupted; he can get used to anything, even the wine of the foxy Noah. But Branas was not sure that it could be argued that one would derive more pleasure from the wine of the Carignan grape than that of hybrids. Everything is relative, he admitted, except that it would be unreasonable to argue that the traditional patrimony of France would profit from the loss of its native vines, the source of quality and delicacy in wine. At least this was a valid argument for those who could afford fine wines.[14]

[12] For a professional and scientific approach to wine tasting, see the classic Maynard A. Amerine and Edward B. Roessler, *Wines: Their Sensory Evaluation,* rev. ed. (New York, 1983).

[13] L. Depardon and P. Buron, "Les vins d'hybrides producteurs directs," *BIV,* no. 121 (1938), pp. 66–8; and *Comptes rendus hebdomadaires des séances de l'Académie d'agriculture de France,* reprinted in *BIV,* no. 184 (June 1946), pp. 81–2.

[14] J. Branas, "Les hybrides producteurs-directs. Conclusions d'une enquête," *BIV,* no. 152 (1942), pp. 81–2; reprinted from *Le progrès agricole et viticole* (Montpellier), June 7, 1942

However, would laboratory results, based on the pressing and fermenting of small quantities of grapes, enable winemakers to predict results for large-scale wine production? Here was another version of the old problematic relationship between laboratory and industry. Paul Marsais, professor of viticulture at the Institut national agronomique, warned in a speech to the Organisation internationale du vin in 1946 that the wines obtained in these laboratory tests were not generally the same as those obtained in commercial winemaking. Scale and volume determine taste. To have any significance for the wine industry, tastings must be based on wine made in a volume that can be measured at least in hectoliters. Blois still had lots of work to do.[15]

Achieving good marks in tastings was not a winning strategy for a producer of hybrid wine hoping to change his outsider status. As late as 1948, in a competition in Mâcon, wines made from Millot and Foch hybrids won first and second places. The next year in the Beaujolais they finished in first and fifth places. Success continued for these two groups in a competition in 1956. Millot and Foch were officially recommended as late as 1964 in five wine-growing districts. Success for a few hybrids out of a bad lot did not prevent the government from prohibiting new plantings of hybrids for commercial winemaking after 1975. Amateur winemakers can still make wines for personal consumption; other drinkers must be content with the wines from vines of grafted purity.[16]

In spite of the atmosphere of euphoria promoted by its publicity machine, private enterprise had managed to create only about a dozen varieties of good hybrids over a sixty-year period, twenty usable varieties if one includes the mediocre ones. Nor did Montpellier's Branas hold out much hope for finding the ideal hybrid, the direct producer absolutely resistant to mildew and oidium that would give a decent wine. Theoretically, such a vine was genetically possible, but it was highly improbable that it would be created because of the infinite number of factorial combinations possible. Two German plant geneticists explained why. According to Bernhard Husfeld, in order to obtain a hybrid *vinifera-rupestris* absolutely resistant to mildew, it would be necessary to study 100,000 plants of the second generation; to get a plant absolutely resistant to both mildew and phylloxera, it would be necessary to sow at least ten billion seeds. Erwin Baur thought that he could find an ideal hybrid if he cultivated about five million F2 vines. Not even the Germans had that much scientific *Sitz-*

issue. *Parker's Wine Buyer's Guide,* p. 763, rightly defends the carignan in California. Wineries like Trentadue and Cline have shown the value of this grape: "A dusty earthiness . . . with its big, rich, black fruit and spicy flavors."

[15] "Exposé du Prof. Paul Marsais," *BIV,* no. 186 (1946), p. 116.

[16] David A. Bailly, "Developing the French-American Crosses," *Wines and Vines,* April 1977, pp. 44–7. Bailly is a Minneapolis lawyer who grows hybrids and defends them skillfully.

fleisch. So the decision of the Ecole of Montpellier to advocate the replanting of vineyards with grafted vines was not only the correct decision for preserving France's famous vines, it was also a scientifically sound decision. Nature, science, and state seemed in rare harmony, except on hybrids.[17]

In 1946, in a version of the historical notion of the two political Frances, Paul Marsais put forward his thesis of the two viticultural Frances, two zones of conflict led by opposing propagandists. The group favorable to hybrids, including the society for the study of and experimentation on hybrids and crossbreeds with its center in Poitiers, saw itself as the spokesman for the future, pregnant with progress. The hybrid part of this diarchy included Bresse, an area of mixed farming, parts of the Aude and the Haute-Garonne, the Vienne, the Vendée, high-altitude vineyards in Savoy and the Alps, and the plains of the Languedoc. The anti-hybrid group, defenders of the sound traditions and practices that gave French wines their preeminence in the world, had created the Comité national des appellations d'origine des vins et eaux-de-vie to defend the *vinifera* vines they so venerated. Their territory included the sacred soil of the great vineyards: la Champagne, la Bourgogne, le Bordelais, and l'Anjou.

Marsais, a firm believer in the virtues of technocracy, advocated looking at the problem from the viewpoint of the technician, who easily saw that these two different viticultures could coexist in France. Although only a small number of privileged people could enjoy the products of the great vineyards, these wines had to be protected. The hybrid empire, "young, ambitious, full of confidence in its destiny," wanted its place in the sun, which it deserved, provided it stayed on the plains and the plateaux, where the soils were too rich to produce fine wine. Neither domain should encroach on the other, which meant that direct producers and grafted hybrids could not invade areas where the *appellation* guaranteed quality. Marais saw this as mainly a matter of the repression of commercial fraud. Thus the great viticultural heritage of France would be safe, with *grands crus* for the privileged and a good sound wine for the average Frenchman.[18]

Supporters of wine from hybrids hoped that science would deliver its salvation; instead, science delivered its commercial death. The scientific blow to hybrids came from the work done by Pascal Ribéreau-Gayon in the 1950s on anthocyanins or red pigments in grapes. Using paper chromatography in a study of the molecular structure of pigments from grapes of different species of *Vitis,* he showed that *V. riparia* and *V. rupestris,* the two American species most used in hybridization, possess the dominant

[17] Branas, "Les hybrides producteurs-directs," *BIV,* no. 152 (1942), pp. 80–1.
[18] "Exposé du Prof. Paul Marsais," *BIV,* no. 186 (1946), pp. 114–15.

property of synthesizing their anthocyanins in the form of diglycosides.[19] (A diglycoside is a compound with two molecules of glucose.) *V. vinifera* does not possess this property, although in some crossings of *vinifera* and hybrid vines, the hybrids may appear without diglycosides. The limitation of the method is minor, because the forbidden fruit in France comes mostly from American rather than Franco-American hybrids. The presence of diglycosides in the grape identifies a hybrid; their absence does not mean that the vine is a *V. vinifera*. In spite of numerous and often commercially inspired attacks on it, the principle held up, finally passing into the gospel of official methods of analysis.[20]

In Germany and France it became too risky to keep up the fraud of mixing hybrid and *vinifera* vines in the production of *appellation* red wines. A buyer who found hybrid wine mixed in with his purchase of *vinifera* wine could refuse to accept delivery on the ground of the illegal presence of hybrid wine, which could be convenient for other reasons – for example, a drop in price between the time of purchase and the time of delivery. That was too big a risk for producers to take, especially after a German buyer did exactly that. Another threat to quality disappeared.

Germany: The Abortive Quest for the Perfect Species Leads to the French Model

By the 1930s, France, Italy, Germany, and Austria supported a considerable research effort in viticultural genetics. In the nineteenth century, interest in vine breeding, long a scientific curiosity, had become an economic necessity. In Italy, as in France, phylloxera was the principal enemy, and it was impossible to avoid replanting the Italian vineyards after 1885. Hundreds of thousands of imported American plants were culled and experiments in hybridization done to obtain rootstocks resistant to phylloxera and to find types suitable for different regions.[21] Italy had nothing to compare with the German and Austrian research programs, which paid close attention to the genetic analysis of progeny. Selection centers in Klosterneuburg (outside Vienna), Neustadt, and Trier, and so on, carried on the type of work begun in 1912 by Rasmusson in Metz (part of the

---

[19] Chromatography is a "technique for analysing or separating mixtures of gases, liquids, or dissolved substances." It was originally used in 1906 by the botanist Mikhail Tsvet to separate plant pigments. See entries in the Oxford *Concise Science Dictionary*.

[20] For a detailed analysis and review of the literature on pigments, see M. A. Amerine and M. A. Joslyn, *Table Wines: The Technology of Their Production*, 2d ed. (Berkeley and Los Angeles, 1970), pp. 255–69. Some researchers have found malvidin (3,5-diglucoside malvin) in some varieties of *V. vinifera*, which has led Amerine and Joslyn to assert that "it is difficult to claim that diglucosides are *always* absent." It is also possible to see science used by Germany in this case as a weapon to keep some American wines containing diglucoside pigment off the German market.

[21] E. Pantanelli, "Le point de vue italien," *BIV* (1939), pp. 205–15.

Reich from 1871 to 1918). The need for nationally coordinated first-class work led in 1928 to the creation of a department of genetics of the vine under Erwin Baur at the Kaiser-Wilhelm Institut für Züchtungsforschung (KWG) in Müncheberg. Bernhard Husfeld, student and collaborator of Baur, could not develop the laws governing selection of the vine until after facilities were provided by the founding of the KWG institute. Husfeld headed the operation for grape breeding in the late 1930s. After the vines were selected for early fruit bearing and fecundity, they were sent to other stations with appropriate experimental vineyards: Naumburg (Hesse) to test for resistance to phylloxera, and Geisenheim, Oppenheim, Wurzburg, and Freiburg im Breisgau to establish levels of fruit production and wine quality.

Collecting the data from the different centers, including Müncheberg, became the responsibility of the *Reichsrebenzüchtung,* a state service for vine selection. Husfeld, the head of this service, was responsible for dividing up work among the other institutes, which supervised the individual selections. Austrian and German research enjoyed (or suffered from) a unity of purpose, though no aspect of genetic and cytological research that could improve the vine was neglected, including hybridization between selected European and American clones to obtain direct producers worthy of replacing existing species. The Third Reich continued the ban on the planting of hybrids, but with the sop to hybrid interests of support for Baur's genetic search for the perfect hybrid.[22] Not even the Germans wanted to reject the possibility of a great hybrid, no matter how much a creature of the *imaginaire.* The idea of a viticulture based on the ungrafted vine dominated a unique integral program.

What struck a French observer about viticulture in Imperial Germany, first of all, was that the universal method of conquering phylloxera through grafting had been rejected, except in the official experimental vineyards, in favor of multimillion-mark chemical warfare, especially the use of carbon disulfide, and the planting of officially approved hybrids. This policy applied to Alsace, part of the Reich between 1871 and 1918. France might have invented the word *dirigisme* (1930), but in viticultural policy France was the land of creative anarchy and Germany the land of extreme *dirigisme.* Viticultural stations had an influence or power their French counterparts could not hope to match. The Germans were especially interested in making some sort of breakthrough in the fight against phylloxera, a grave threat because only one-third of the vineyards had been replanted in grafted vines with roots resistant to phylloxera; hence the large-scale experimentation possible only within the context of what

[22] J.-F. Ravat, "Du rôle néfaste des mauvais hybrides pour la qualité des vins," RV 74 (1931): 58–61. On Baur and his agricultural research programs, see Jonathan Harwood, *Scientific Thought: The German Genetics Community, 1900–1933* (Chicago, 1993).

an official French observer admired as "a disciplined organization of national viticulture."

In 1938, Paul Marsais tried to figure out the historical reasons for the German *Sonderweg* or unique direction in viticultural research, which aimed above all at creating a hybrid of "pure European blood" with immunity to phylloxera and fungus diseases. The basis of the program was Mendelism rather than the disorganized and disorderly work of the French hybridizers. Equally ignored was the work of the masters of Montpellier: Foëx, Ravaz, Viala. The laws of heredity could only be useful in viticulture, however, if large-scale experiments were done. Given the genetic difficulties of using seed, the French had opted for cloning and grafting. Nor were they convinced by German research showing that "impurities" or variation of characteristics of the vine showed up in cloned vines after three generations. The possible and even unknown sources of the problem could be controlled: bad clonal selection; accidental mixing of slips of different clones at times of collection of vine shoots, planting, and harvesting in the nursery; and many other factors related to soil, climate, age of vine, and so on. There was French-type research going on in Imperial Germany. In Barr (Alsace), Dr. Hecker produced several well-known grafted vines after obtaining special permission from the Chancellery to import rootstocks from Teleki in Hungary.[23] Defense of the vine was a matter of highest national priority, but the research direction now seems to have been curiously misguided.[24]

After the Second World War, Paul Marsais, by then converted to the erroneous idea that modern French direct-production hybrids are different from the American ones, hoped that the results of the genetic research done by the enemy would not be lost as a result of the destruction of the laboratories and the death of their personnel, especially in Müncheberg and Dahlem. Much of the knowledge gained in the 1930s had not been transmitted to the vineyards before the Second World War broke out.[25] After the fall of the Hitlerian Reich, the Germans seem to have given up viticultural science fiction to adopt the French gospel of grafting, in practice at least. "Like practically all the wine-producing vines of France, those of Germany are for the most part grafted, and on American roots."[26]

[23] Sigmund Teleki (1854–1910), famous for his selection and confusion of *V. berlandieri*-pure, crossed with *rupestris* and especially with *riparia;* Franz Kober, of the viticultural station of Nussberg in Austria, imposed Teutonic order in classifying the offspring of the ten best selections Teleki sent him in 1902–4. Galet, *Cépages et vignobles de France,* 1:246.

[24] P. Marsais and L. Ségal, "Les principes directeurs choisis pour la reconstitution du vignoble allemand," *RV* 87 (1937): 431–5; and translations of articles by Heinrich Moog (Geisenheim) and Fritz Zweigelt in *Wein und Rebe* and *Das Weinland,* RV 89 (1938): 115–27, 321–9.

[25] Paul Marsais, "Méthodes viticoles d'Allemagne," *Journée vinicole,* Oct. 12, 1946; reprinted in *BIV,* no. 189 (1946), pp. 54–5.

[26] Galet, *Cépages et vignobles de France,* 1:370, notes that the ministry of agriculture followed an absurd recommendation of the IVCC by recognizing naturalized French hybrids in a

The French Victory and Nature's New Challenges

The selection of varieties of vines developed in different directions in different countries. Already in possession of the world's greatest vineyards, French growers did not try to improve them by crossing vines. Rather, through the selection of seeds and hybridization, they created excellent rootstocks resistant to phylloxera and also direct producers of proven, if much debated, value. France became a great exporter of rootstocks to other wine-producing countries. Hungary, Switzerland, Spain, Portugal, and California imported rootstocks from France, tested the French direct producers and finally used clones of quality to obtain the varieties best suited for their microclimates.[27]

Everyone came to agree that the best and surest method of reconstituting a vineyard was to use a graft carrier or rootstock resistant to phylloxera, but this resistance was not simply a matter of the "intrinsic resistance" of a variety of vine. External factors like soil, climate, and the individual plant made a vast difference in how resistant a variety was in practice, in spite of a theoretical absolute genetic resistance. Individual vines could have a "practical resistance" dependent on the harmony of plant and environment. In the absence of a single universal rootstock, plant and soil had to be carefully matched. The natural lack of suitability of a plant's root system for a certain type of soil was only aggravated by grafting. In certain soils (chalky, calcareous) most American vines were useless. Viticulturalists never tired of relating how Viala, after searching for a suitable vine for the phylloxera-devastated Charentes, finally settled on one species in Texas (*Vitis berlandieri,* already well known to French viticultural experts) that would flourish in the very calcareous soils of the Charentes, similar to the vine's native soils in Texas, and the friable calcareous or Jurassic limestone formations of Burgundy. This saga now provided the classic model of success in scientific replanting.

The American vine balanced its vices with substantial virtues, which were recognized by viticulturists and oenologists as the source of French viticultural progress. Viticultural Cassandras might warn that grafting would ruin the great vines of France, source of its glorious *crus.* This was nonsense, though not unreasonable nonsense during the early days of grafting on a large scale. Most viticultural scientists believed that grafting France's native grape varieties onto American rootstocks improved the fruit, speeded up its maturing, advanced the beginning of fruit bearing, and increased yields. The degree of resistance of rootstocks to phylloxera

decree of Feb. 27, 1964. Marsais, professor at the Institut agronomique de Paris, was vice president of FENAVINO. On Germany, see Frank Schoonmaker, *The Wines of Germany,* rev. ed. by Peter M. F. Sichel (London, 1983), p. 36.

[27] Pantanelli, "Le point de vue italien," *BIV* (1939), pp. 211–12. See also Prosper Gervais (secrétaire perpétuel de la Commission international de viticulture), "La question des portes-greffes," *Bulletin international du vin* 133 (1939): 36–43.

was carefully studied, with levels of resistance established according to the latest canons of viticultural science. But in the contingent and tough world of soil life, this proud precision crumbled, leaving only a general guide to what might happen as a result of the combination of soil, climate, and individual rootstock. Some viticulturists were convinced that the solution to the problem of what plant to use lay in hybridization, alone capable of furnishing a range of rootstocks adapted to the complex situations produced by differences in climates, soils, varieties of vines, and cultivation. American types could not satisfy all needs. Without hybrid rootstocks the reconstitution of vineyards would have been much more difficult and, in some soils, perhaps impossible.

Native American vines had provided the raw material with which French breeders created hybrid rootstocks used widely in France and the rest of the winegrowing world. True, each of the three major species used for hybrid graft carriers or rootstocks – *V. riparia, V. rupestria,* and *V. berlandieri* – had their defects. The *rupestris* of the Lot seemed extremely unhappy in the soils of the Midi, which had adopted it widely. Even the *riparia,* object of the learned enthusiasms of the Alsatian viticulturist Charles Oberlin, was criticized for its defects. The *berlandieri* garnered most praise, but Gervais had nearly as high an opinion of its rival, Foëx's 333 EM, a hybrid that had a hard time getting adopted as a rootstock. The opposition to the 333 EM was undermined by the publications of the Station viticole de Cognac (headed by René Lafon and J. L. Vidal) and the Fondation Fougerat. The superior qualities of a vine had to be clearly demonstrated in practice and in print.

Early defenders of hybridization did not have an easy time of it. But as Gervais argued in 1939, forty years of cultivation of the Americans showed that they were resistant to the insect phylloxera and proved their specific virtues of adaptation and affinity or structural resemblance to the French vine. Millardet, Couderc, and Castel enjoyed a posthumous revenge. Work had to go on, Gervais warned, and more research needed to be done on affinity. Grafting might not have changed the characteristics of French "native species" or their essential qualities, but it could not be denied that grafted vines were more susceptible to disease – phylloxera excepted – than ungrafted ones, and, indeed, they even seemed adept at inventing new ones for the viticulturalists to puzzle over and the grower to worry about.

The parasitic infection called fan-leaf (*court-noué*) became alarmingly widespread. The literature of alarm over *pumilus medullae spec. nov.* goes back to at least 1935, when Viala and Marsais published their monograph on it in the *Annales de l'Institut national agronomique*. Rives, L. Pétri, and Branas disagreed over its cause. Rives's hypothesis was that the radicles of the plant were invaded by endophyte mushrooms and endotrope mycorrhizae. Was this contamination of phylloxeral order and origin? Pétri blamed a filterable virus, a speculation taken up by Branas. The frightening specter of large-scale degeneration of the vine made urgent action a priority. The disease appeared to be formidable. Putting his faith in etiology, Gervais believed that finding the cause of the disease would be the source of its cure. The task for the future was a rigorous choice of graft carriers in order to determine their qualities and their faults. Viticulturalist scientists had been thrust into the new area of relations between genetics and disease. Arguments over the cause of fan-leaf lasted into the 1950s, when the viral hypothesis emerged from its cocoon of uncertainty. The biology of diseases of the vine became part of oenological and, in the form of a simplified gospel, even viticultural literature.

It turned out that fan-leaf is "caused by a virus which is carried from vine to vine by a microscopic root-sucking worm known as *Xiphinema index*. . . . Deformed, asymmetric, nettlelike leaves appear, turning a variegated green-yellow. The vine takes on a stunted aspect, its yield falls off, and it dies within ten or fifteen years of plantation."[28] This worm, a parasitic parthenogenetic nematode, was particularly destructive in the Côte de Beaune, Meursault, and Puligny-Montrachet (Puligny's production was reduced by about 20 percent) and was for a time "probably the most serious menace in the vineyards of Bordeaux."[29] It is also a major pest in the vineyards of California.

The problem is being solved according to the general prophecy of Gervais. Tolerant and susceptible species of vines are crossed in order to develop a genetic model fulfilling the grower's dream of clonal selection of vines resistant to the virus. The sex life of captive vines has not been suppressed, for new varieties are still created by hybridization or crossing vines of different species. It has been argued that clonally selected vines may be genetically susceptible to some diseases and may produce a less complex wine than grafted vines. Cloning of existing vines has not obviated the need for genetic selection of the desired qualities of the clones themselves, which can vary greatly, even up to 50 percent in the production of anthocyanins and 20–30 percent in tannins.[30] (The owners of great vineyards like Château Haut-Brion, who planted a group of low-yield clones, are not among the pessimists.)

Fan-leaf was a particularly frightening example of a general problem with vine production since phylloxera. Before the destruction of French

---

[28] Anthony Hanson, *Burgundy* (London, 1982), p. 95.
[29] See David Peppercorn, *Bordeaux* (London, 1982), pp. 32–3.
[30] Hanson, *Burgundy,* p. 95, and especially C. P. Meredith et al., "*Vitis* Species," *American Journal of Enology and Viticulture* 33 (1982): 154–7 (hereafter AJEV). On clones, see these articles in *Wines and Vines:* "Clonal Repository Planned" 56 (1975): 43; C. J. Alley, "An Update on Clone Research in California" 1 (1977): 31, and "Clonal Selection and Evaluation in France" 70 (1989): 46–7.

vines by the plant louse, only a few firms sold vines, chiefly for table grapes. With the replanting of the vineyards, a new, sometimes unscrupulous, often ignorant profession came into existence to provide mother vines of rootstocks. These nurserymen operated without any regulation until after the Second World War. In 1944 the Section de Contrôle des Bois et Plantes de Vigne was created to control plant diseases. From 1954 to 1976 this function was taken over by the ONIVIT (Office national interprofessionnel des vins de table), reborn in 1983 as ONIVINS – the proletarian qualification "de table" was dropped. Now regulation is under the aegis of the European Union, which authorizes 29 varieties of rootstock in France. A national program of selection of rootstocks has been the responsibility of ANTAV (Association nationale technique pour l'amélioration de la viticulture), created in 1963. Certified clean vines are easily available. When one takes into account the chaotic spread of diseases under the old distribution system, the new system with its high prices is cheap. A minor agricultural revolution of a clearly scientific-bureaucratic nature had occurred.[31]

Phylloxera produced a number of striking results in agriculture.[32] In the Hérault and the Gard most of the vines were destroyed. One good consequence of this was the temporary return of cereals, especially oats, to the Hérault and the coming of market gardening to the Gard. Departmental distribution of vines in France was considerably modified, with the relative importance of the Hérault declining in the viticultural Midi from 75 percent in the 1860s to even less than 50 percent of wine production. The big capitalist vineyards of the plains, easily accessible to the railroad, replaced the small vineyards of the hills. This system of mass production of wine necessitated a new *vignoble,* a new means of production based on chemistry and mechanization: fertilizers (guano, manure, and then chemical fertilizer), high yields, grafted high-yield vines, antifungus products, and insecticides. And professors along with their university faculties became as important as chemicals. Sometimes brought in by rail, the products were increasingly produced by the new factories that sprang up near places of production. The Bas-Languedoc-Roussillon established a dominant role in the national production of wine, going from less than 15 percent in the early nineteenth century to well over 40 percent in the late nineteenth and twentieth centuries – an impressive capitalist triumph, even if there was little pleasure for the lover of good wine.

It is probably wise to end a discussion of phylloxera in France with the

---

[31] Galet, *Cépages et vignobles de France,* vol. 1, chap. 4, ("Les portes-greffes").

[32] Concisely summed up by Robert Laurent in "Les quatres âges du vignoble du Bas-Languedoc et du Roussillon," *Economie et société en Languedoc-Roussillon de 1789 à nos jours* (Montpellier, 1978), pp. 18–22; see also Yvette Maurin, "Société et Ecole d'Agriculture de Montpellier devant le Phylloxéra," in ibid., pp. 53–6.

warning that it is easy to exaggerate its impact. One must see the disease in its general viticultural context. René Pijassou has suggested replacing the cliché "phylloxera crisis" with the idea of a long depressive phase in Bordeaux's wine production, phylloxera being only one factor. Vineyards were hit by a series of secondary crises; the conjunction of several distinct diseases produced a powerful cumulative effect. Mildew was considered a more serious problem than phylloxera in the period 1882–92. In our fungicidal age we have to make a special effort to recognize that in the 1880s mildew was a bigger threat to wine production than phylloxera. At Château Latour in 1886, nearly 22 percent of the expenses for fighting vine diseases went into the battle against mildew. A drop in quantity was not necessarily a bad thing in the 1880s, but mildew seriously damaged the quality of the wine as well. With an average drop of 3 percent in alcohol, the wine did not conform to the model of the *grand vin*. Daniel Jouet, the manager of Latour, said of the wine of the mildewed years of 1885, 1889, and 1891 that the only way to make a profit would be to sell it to a pureblooded Parisian, who was incapable of distinguishing between the taste of *piqué* and the taste of *rancio*.[33] Mildew is still a danger in the vineyard, but it no longer takes a toll as it did in the great mildew years of 1910, 1915, 1917, and 1957.

Even the idea of a depressive phase in the production of quality wine in the Bordelais loses much of its meaning within the general context of a big increase in yields during the "phylloxera crisis." Between roughly 1830 and 1910 production at Châteaux Margaux and Latour more than doubled; at Pichon-Longueville it tripled; and at some estates it even increased by 350 percent. Less shocking figures are obtained by comparing the two periods 1854–78 and 1878–1920: increases in production of about 25 percent (12–45 percent) are less scandalous for industries in which low yield is a requirement for high quality. After a long period (1888–1907) of abundance and low prices, the *courtiers* and *négociants,* fed up with low profits, imposed lower yields on the estates they had under contract. Equally eager for better profits, owners agreed to use less fertilizer and to accept less money for wine made from crop yields over certain levels. This survival and even flourishing of quality wine production during the phylloxera crisis should give hope to wine producers in California, now battling the same disease.

Life is fragile because life attacks life. In recent years a supposedly new mutant strain (biotype B) of the phylloxera aphid has started sucking the roots of vines in Napa, Sonoma, and Monterey counties.[34] Astronomic

---

[33] Réné Pijassou, in Charles Higounet, ed., *La seigneurie et le vignoble de Château Latour* (Bordeaux, 1974), 2:471.
[34] On phylloxera in California, see the following articles in *Wines and Vines:* William F. Heintz, "Phylloxera, St. George and De Latour" (September 1975), pp. 42–4; George Barn-

replanting expenses are expected, perhaps 250 million dollars in the next ten years – an eventual total of a billion dollars? As California followed the same program as France in replacing its phylloxera-infested vineyards in the nineteenth century, why should the insect have become a threat again? (The state has an advantage over Europe: it is not afflicted with the "winged version" of the aphid because its summers are not humid enough for its development.) The leading replacement rootstock in California had been a pure selection of *Vitis rupestris,* but it was replaced in recent years by the so-called A × R1, known also as the Ganzin 1, named after its French creator. It is a cross between the *vinifera* aramon and *V. rupestris.* Its popularity, resulting from better and more fruit than the old "*rupestris* St. George (du Lot)" – up to 25 percent higher yield – explains why it covered 65 percent of the vineyards of Napa and Sonoma counties. A × R was not much used in Europe after it succumbed to phylloxera in the dry, shallow soils of Sicily, southern Italy, and part of mediterranean France. All vines are dead or are unproductive eight years after phylloxera is discovered. So viticultural scientists are trying for a new fix, the creation of a new rootstock that will be resistant to phylloxera and to nematodes in the vineyard as well as the laboratory.[35] The debate over responsibility for the new disaster is understandable but shows a naive belief in the infallibility of scientists on the part of captive consumers of scientific knowledge and products.

Life evolves, adopting many disguises and devices to conquer and survive. East of the Rocky Mountains "grape phylloxera and *Vitis* species evolved together" in a life-sustaining symbiosis. Transported to noninfested areas, the insect destroyed its host in a root-sucking orgy in which the vine had no time to be driven by the insect into an evolution toward resistance.[36] Not that it would happen in the vineyard, for vines with a resistance in the wild lose it under the stress of cultivation. All life is better free, it seems, and certainly the vine finds it difficult to cope with both man and insect.

well, "The Big Phylloxera Mixup – A × R1 Is in the Clear" 66 (November 1985): 63; L. W., "Update on So-called Phylloxera Crisis" 66 (July 1985): 17; George A. Barnwell, "A New Rootstock Is in the Pipeline" 66 (March 1985): 27–9; and two items in the issue for February 1991, pp. 27, 32–3: "Could Phylloxera Cost $70,000 per acre?" and "A × R1 Still Popular Despite Phylloxera." See also W. E. Wildman et al., "Monitoring Spread of Grape Phylloxera by Color Infrared Aerial Photography and Ground Investigation," *AJEV* 34, no. 2 (1983): 83–93. Also informative are Anthony Dias Blue, "Bad Bugs," *California,* May 1991, pp. 106–7, and Frank J. Prial's column "Wine Talk," *New York Times,* May 8, 1991, B6; see also Prial in *New York Times,* August 12, 1992, B1–5. *Le Monde,* Sept. 15, 1992, pp. 35–8, reported on "Le 'sida de la vigne' Californienne."

[35] Philip Hiaring, "Harold P. Olmo: A Vineyardist Is Wine Man of the Year," *Wines and Vines* 69 (1988). Olmo has acquired a certain fame for his encouragement of the mating of *vinifera* and *rotundifolia,* which is resistant to nematodes and the fan-leaf virus.

[36] A. J. Wapshere and K. F. Helm, "Phylloxera and *Vitis:* An Experimentally Testable Coevolutionary Hypothesis," *AJEV* 38, no. 3 (1987): 216–22.

What is new about the new phylloxera? What does it mean to talk about phylloxera biotype A and biotype B? If the new biotype is more than an entomological invention, hubristic and ahistorical scientists may be saved some embarrassment for forgetting that Ravaz had warned about the lack of resistance of A × R1 in dry soils. But even people who knew this – Pierre Galet, for example – assumed that the deep, generally irrigated Napa valley soils would provide a safer environment for the vigorous A × R1 than southern Italy and Sicily, where it had unwisely been planted earlier. The new model insect seems to elude linguistic precision. It is different "but not necessarily morphologically," and its "set of characteristics may vary within the biotype itself."[37] This just means that the insect adapts to the nonresistant rootstock. The development of new models of these aphid-like insects may be similar to the development of immunity to pesticides in insects. "If it is a chemical mechanism in resistant rootstocks that rejects phylloxera, the phylloxera in time may develop its own resistance or tolerance to the chemical." New biotypes were announced in German vineyards in the 1920s–1930s, and these biological novelties have not been lacking elsewhere. So an inevitable logic unfolds. New biotype, new rootstock. First, laboratory testing of the plant for resistance, then field trials. History repeats itself, even in science.

It is highly improbable, indeed it would be inconsistent with life as we know it, that scientists will develop "a foolproof resistant rootstock to phylloxera." The best one can hope for is a redefinition of "the relations between living things" in which the vine survives.[38] Even if some viticultural miracle were to cause phylloxera to disappear, there are enough diseases to keep scientists occupied: drying out of stems (a physiological problem since 1970), Pierce's disease (caused by the bacteria *Xyllela fastidiosa*), the fungal dead-arm disease, Eutypa dieback, golden flavesence, and the eternal powdery mildew, for example. The more disease, the more science – the converse is controversial. The phylloxera insect has its own developmental plan to keep alive. Scientist and louse are host and parasite in a mutually supporting system. And so far science has prevented a recurrence of the disaster of the nineteenth century.

Science saved French *vinifera* vines threatened by diseases transferred

---

[37] George A. Barnwell, "A New Rootstock Is in the Pipeline," *Wines and Vines* 66 (1985), which reports on a meeting of grape growers and scientists, including entomologist Jeffrey Granett of the University of California, Davis. See also Lawrence M. Fisher, "A Laboratory to Renew the Vineyards," *New York Times,* September 6, 1992, F9. Phylloxera in California, the "billion-dollar nightmare," was the cover story of *The Wine Spectator* on August 31, 1992, pp. 28–33. Biotype B is reported as capable of reproducing *in the laboratory* up to forty times faster than biotype A and can kill off vineyards in 3–5 years compared to the 10 years required by biotype A.

[38] The quotation in this sentence is from François Delaporte, *The History of Yellow Fever: An Essay on the Birth of Tropical Medicine,* trans. Arthur Goldhammer (Cambridge, Mass., 1991), p. 145.

from North America as a result of the human lust for novelty and experiment. Because these vines produced the greatest wines in the world we are grateful. Science also gave us the chemical lifeline so convenient, if not necessary, for maintaining these vines in production. We are less grateful. Science also shows us a way out of this dangerous pollution of the soil, if politics and profits permit. Having poisoned the soil for over a hundred years, winegrowers, even in Burgundy, are now turning more and more to systems of organic production, including Rudolf Steiner's biodynamic cultivation, which reject chemical fertilizers, herbicides, and pesticides. The general aim is to bring the soil back to life by reestablishing its normally present microorganisms. Presumably the mythic *terroirs* will regain their unique characters, which will be reflected in wine differences originating in distinctive soils, and the vines will be less susceptible to disease once they are taken off their chemical support systems.[39] What the next problem will be is unknown. We reluctantly accept the axiom that in living systems no problem has a scientific solution that does not introduce another problem.

[39] James Lawther, "France's Natural Move to Organic Wine," *The Wine Spectator,* Nov. 15, 1991, pp. 125–31. If you cannot afford Château Margaux or Clos de la Coulée de Serrant, try Château de Beaucastel, Mas de Gourgonnier, or Domaine de Trevallon, among others.

PART II

Laying the Foundations of Oenology

5

Jean-Antoine Chaptal

If one were to write a history of oenology, it would be tempting to begin with Louis Pasteur. Some historians, lured into finding more remote roots of oenological theory and practice, begin with Adamo Fabbroni, author of a prize-winning work (1785) on the art of making wine rationally, which was translated into French in 1801 – a vintage year for oenological work. Pasteur noted Fabbroni's originality in experimentally identifying a cause of fermentation ("la matière végéto-animale," presumably gluten) and saluted him as "the principal promoter of modern ideas on the nature of the fermenting agent." In spite of his favorable if critical reception by French chemists – for example, Fourcroy – and the publication of his ideas in the *Annales de chimie,* Fabbroni has practically disappeared from the history of wine.[1] French scientists merely replaced him by themselves.

Pasteur thought that he himself had replaced just about all scientists who had worked on wine, though he granted previous writers (chiefly Lavoisier, Chaptal, and Gay-Lussac) the status of worthy predecessors. There was one figure who could not be relegated to that limited position: Jean-Antoine Chaptal, one of the power elite to whom Pasteur usually showed the proper deference. Chaptal was a rich man – though he died poor – versed in medicine and science, a respectable chemist, a best-selling author, an industrialist, and a powerful politician of the First Empire. The influence of Chaptal's own book on the technique of making wine was enormous in France and, helped by German and Italian translations, throughout Europe. He cut a brilliant figure in the history of oenology and, to a much lesser extent, in viticulture.

Chaptal's work is often seen as a major turning point in the history of winemaking. It was not only a scientist who spoke when Chaptal pub-

[1] Louis Pasteur, *Etudes sur le vin* (Marseille: Laffitte Reprints, 1873), pp. 4–9. See also Leo A. Loubère, *The Red and the White,* pp. 93, 95–6, and the short entry in the *DSB,* vol. 4, by Mario Gliozzi, who notes that this savant's name is sometimes misspelt as Fabroni.

lished on vine and wine.[2] Chaptal, Napoleon's minister of the interior (1801–4), wrote to the prefects asking for information on progress in departmental vinification procedures. Recommendations for improvement, with the dual authority of science and government, could be disseminated in the same way. This imperial procedure would eventually become part of the program of ruling through bureaucratic science, among other techniques. The solicitude of Chaptal for winegrowers was hammered home by Cadet de Vaux in his *Conseil aux Vignerons:* the minister was the architect, the winegrowers were the workers; following his plan would lead to commercial success.

Whatever one thinks of this authoritarianism, inherent in all schemes of scientific progress, perhaps the reason Cadet de Vaux gave for following Chaptal was a good one. Winemakers should give up the routines of their fathers, makers of bad wine. Renouncing false family gods, the new winemaker would achieve success by following the oenological decalogue of the scientist-ruler. The oenological canonization of Chaptal occurred early, during his lifetime.[3]

Chaptal's Oenological Doctrine

The versatile Chaptal put the weight of science behind the idea of increasing the degree of alcohol in wine by adding sugar to the fermenting must, a process that came to be called *chaptalization* in the well-sugared later decades of the nineteenth century. Chaptal actually preferred adding a syrup made from the grape itself, an idea that has recently been given a great deal of publicity in German vinification. It became legal to used a concentrated must in France after 1931, but sugar was cheaper.[4] Chaptal

---

[2] Writers usually refer to Chaptal's *L'art de faire le vin* (1801; 1807), which simplifies a complex matter of publication. Catalogs also refer to *L'art de faire, gouverner et perfectionner le vin,* 2d rev. ed. (Paris: An X, 1810) "par le citoyen Chaptal." Then there is the *Traité théorique et pratique sur la culture de la vigne, avec l'art de faire le vin, les eaux-de-vie, esprit de vin, vinaigres . . . ,* par le Cen Chaptal, l'abbé Rozier, les Cens Parmentier et Dussieux, 2 vols., 1st and 2d eds., (Paris, 1801); see also versions of 1805 and 1809. But note must be taken of the articles on vine and wine in vol. 10 (1800) of *Le cours complet d'agriculture théorique, pratique . . . ,* 12 vols. (Paris, 1785–1800), edited by the abbé François Rozier (1734–93). The *Traité* was translated into German in 1804, Italian in 1812, and Hungarian in 1813. For more complications, see Michel Péronnet, ed., *Chaptal* (Toulouse, 1988), pp. 309–11.

[3] C. Moreau-Bérillon, *Au pays du champagne* (Reims, 1922), p. 110. Oenologist Michel Flanzy notes that nowadays neither practitioner nor theoretician can subscribe to the Chaptalian program for manipulating wine, which could include sugaring, coloring, perfuming, and adding alcohol. Flanzy, "La science appliquée: Le vin," in Péronnet, ed., *Chaptal,* p. 236.

[4] Chaptal was one of the five authors of an *Introduction sur la fabrication du sucre du raisin* (1810), which was promptly translated into Italian and German. See also Leo Loubère, *The Wine Revolution in France: The Twentieth Century* (Princeton, 1990), pp. 82–3. Pascal Ribéreau-Gayon, *Le vin* (1991), p. 23, points out that concentrated musts have never been able to be used for fine wines because of the lack of quality control in their production.

also noted the suitability of sugar made from beets and cane.[5] In order to reduce the effects of a shortage of cane sugar as a result of the British continental blockade against Napoleonic Europe, Chaptal recommended the manufacture of grape sugar, less successfully pursued than his other related enthusiasms, the manufacture of eaux-de-vie and vinegar and, above all, chaptalization.

Chaptal's *L'art de faire le vin* (1801, 1807, 1839) begins in classic scientific style with a commentary on the mediocrity of preceding writers on the subject. More than a half-century later, Pasteur would say much the same thing as Chaptal, with less justification. Both Olivier de Serres and Chaptal were keen on science; two centuries of scientific progress permitted Chaptal to be somewhat more scientific.[6] The problem, according to him, was that writers, in trying to arrive at a general principle, looked at only one viticultural area and ended up describing a procedure essentially relevant to one locality. A more basic problem was that "science did not yet exist": the theory of fermentation, the analysis of wine, and the influence of climate – sources of the invariable principles that must guide vinification – were not precisely established in a scientific language understood by everyone. Winemaking did not differ from other techniques drawing on "the basic truths of physics." So it was necessary to start with a perfect knowledge of the nature of the very matter used in the operation and go on to calculate precisely the influence exercised on it by the different methods used. Chaptal did not believe that there is a set of rules for a universal vinification process that can be mechanically applied: the fact that different areas produce different grapes has to be taken into account in varying the fermentation process.

One of the basics for Chaptal was the nature of the soil. He believed that ignorance of soil science placed a severe limitation on early oenology. Quite true, but Chaptal's pedological discussion was itself general, if not vague, and was limited to a few pages describing general categories in terms of clay, gravel, limestone, granite, and volcanic soil. In the second edition (1807) he added a paragraph establishing a positive correlation between light, porous, friable soils, like those of Bordeaux and Burgundy, and the production of good wines.

In his discussion of climate Chaptal recognized the virtue of sunshine and the evil of too much rain. Cold and rain prevent the grape from trans-

They can also have other undesirable effects, such as an increase in wine acidity and in iron. See Pierre Sudraud (Directeur central de laboratoire, Service de la répression des fraudes et du contrôle de la qualité, Laboratoire inter-régional, Bordeaux), "Critères de la qualité des vendanges," in Ribéreau-Gayon and Sudraud, eds., *Actualités oenologiques et viticoles* (1981), p. 193; hereafter cited as *AOV.*

[5] Loubère, *The Red and the White,* p. 95.

[6] Olivier de Serres, author of the *Théâtre d'agriculture,* published in 1600, went through 20 editions by 1675. Johnson, *Vintage,* p. 312.

forming its sugar into alcohol and developing a bouquet, resulting in insipid, weak wines with insufficient alcohol to preserve them. A high proportion of extract in these wines causes denaturing changes: *la graisse* (ropiness) and *l'aigreur* (*l'amertume*). (*L'amertume* was probably the fermentation of glycerol by lactic acid bacteria, turning wine sour with a strong bitter taste; this was a curse on the wines of Burgundy in the late nineteenth century.)[7] Chaptal thought that a high malic acid content produced its own specific problem, a nonacetic sourness (*l'aigreur*) that is characteristic of less alcoholic wines. No matter how vague or tautological these explanations may now seem, it is useful to remember that they were once the latest scientific revelations and generally satisfying to demanding minds. Nor did demanding palates seem more dissatisfied with expensive, prestigious wines than they are today. But in the 1830s a chorus of bitter complaints about the bad quality of wines, especially burgundy, may have reached a crescendo in serious oenological literature. It was the prelude to Pasteur's oenological revolution. Chaptal retired from the oenological scene before the wines of Burgundy underwent their great trial, culminating in Pasteurian times, but his work was part of a general scientific concern with wine as a saleable and profitable product. Like Pasteur, Chaptal solved problems whose solutions consumers then saw as new problems.

"The ripe grape rots on the vine." God makes the vine but only man can make wine. Chaptal insisted that wine, unlike water and milk, is not a natural product. For Chaptal wine was a man-made product. So he would have been puzzled by the oxymoronic idea of establishing a natural wine model, something that seems highly desirable to many people today. In fact there are few natural products that humans have introduced into the food supply without changing them: flours, meat, and fruit are all changed through a beginning of fermentation before we eat them. Wine is just the most striking example of this transformation. So man's technique, not nature, makes wine. Chaptal's view of fallen nature is a powerful part of contemporary vinological philosophy. He could easily endorse Pierre Sudraud's view of "chemical corrections of the harvest" as "a way of correcting the irregularities of nature."[8] Different countries may have different ways of getting grapes to the process of fermentation, but Chaptal argued that the operation itself must be based on the principles stated in his *L'art de faire le vin*.

In a discussion of when and how to put the wine into barrels at the end of fermentation, Chaptal noted the unsatisfactory nature of "general methods." As the time of fermentation varies with many factors, a knowledge of the constituent elements of wine is needed in order to make an

---

[7] Emile Peynaud, *Connaissance et travail du vin* (Paris, 1981), p. 241.
[8] Sudraud, "Critères de la qualité des vendanges," in *AOV,* p. 197.

intelligent decision. Chaptal had a low opinion of empirical methods used to determine if fermentation has stopped. Agitating a container of wine to produce froth does not give a convincing proof that the wine has finished fermenting. When discussing the influence of air on fermentation, Chaptal drew on the experiments carried out in Burgundy by D. Gentil demonstrating that a very slow fermentation takes place in a closed vat. Chaptal was also aware of the considerable influence of the volume of the must on the time required for fermentation. But he placed major emphasis on three elements in the grape because they seemed to have the most powerful influence on fermentation: the pleasant sweet element, water, and tartar. Variations in the proportions of these three elements caused the chief difference in fermentations.

In the second edition of his work on winemaking, Chaptal equated "sugar" and "yeast" as phenomena, without any hint of later terminology: "le principe sucré, la matière douceâtre ou la levure." Without the "principe doux et sucré," no fermentation. The more of the sweet element in the liquid, the more alcohol produced by fermentation. Experience confirmed the logical theory. At this point Chaptal introduced a distinction between sugar and the sweet element ("le corps doux"). He knew (and said it in the second edition) that pure sugar alone does not ferment. Of course the sugar exists in the grape and its decomposition produces alcohol, but this sugar is always mixed up with "un corps doux" of varying abundance, actually a true "levain" or "yeast." The precision increased in the second edition, where the phrase describing the "yeast" as "très propre à la fermentation" was replaced by "qui sert de ferment." This distinction between "le principe doux et sucré et le sucre proprement dit" had been established by several people: Deyeux, Proust, and particularly Seguin. The second edition is clear on the distinction between *levain* or *levure* and *sucre*, "les deux élémens de la fermentation vineuse," whereas in the first edition he believed the two elements so closely related that the "principe doux" could in favorable circumstances change into sugar.

Chaptal's name has been so corrupted that it is difficult to believe he really recognized that grapes with little sugar could make a good-tasting wine if the fermentation were stopped before all the sugar "decomposed." His example was some vineyards in Burgundy where the fermentation lasted only 20 to 30 hours. Some grapes that appear to be very sweet make a poor wine of low alcoholic content because in reality they contain very little sugar. A little experience enables the palate to distinguish the very sweet grape of the Midi from the Chasselas of Fontainebleau, nevertheless a pleasant-tasting grape. The perfectly ripe grape gives the best fermentation because the must is neither too watery nor too thick. To avoid the problems of fermenting too watery a must, a characteristic northern problem, the classic technique was to boil off the excess water. Or, Chaptal nonchalantly noted, the water content could be reduced by adding plaster of Paris (calcium sulfate); or the grapes could be partly dried before fermenting, a process best known in the production of the *vins de paille* of the Jura.

The second edition of the essay expanded this concern with fermenting grapes full of water. The *levure* remaining after the decomposition of the sugar would produce a weak wine susceptible to the maladies he called *aigre* and *gras;* probably, lactic acid bacteria produced bitterness and ropiness. One way that defects could be corrected or, better, prevented was to reduce part of the must by boiling and then mixing it with the remainder. But Chaptal's preference was to dissolve sugar, brown sugar, or molasses up to 5–10 percent of the weight of the must (15 to 20 *livres* per *muid,* or 456 liters). The addition of the sugar had the double advantage of considerably increasing the low alcohol content of the wine and preventing the acidic degeneration to which weak wines were subject. Chaptal emphasized that if the grapes were full of sugar, adding sugar to the must was useless, and could even be harmful if there was not enough ferment to decompose it. Grapes full of sugar would need addition of *levure* "to reestablish the exact proportions between sugar and ferment."

Chaptal, among others, identified tartar as an essential ingredient of fermentation that should be boiled up in the cauldron with grapes having a high sugar content in order to produce a high level of alcohol. (He did not refer to its acidity.)[9] Conversely, a must containing excessive tartar could produce lots of alcohol if sugar were added to it. Much good wine has an adequate natural high alcohol content, but the message of Chaptal's gospel is that the level of the alcohol must be raised if it is evident that it will be naturally low.

In the early nineteenth century the justification for chaptalization was chiefly to protect the health of the wine by ensuring that it had a high enough level of alcohol. Now that science and technology have long made this practice unnecessary for that purpose, and chaptalization is recognized as potentially detrimental to some organoleptic properties of good wine, the case for it has to rest on rather flimsy aesthetic grounds. Chaptal's generation did not like to get drunk any more than we do – probably less – but it was nearly always a problem to produce wine with enough alcohol for its own health and stability. Much could go wrong during fermentation. Chaptal blamed carbon dioxide, a well-known product of fermentation, for carrying off a considerable amount of alcohol. Champagne was regarded as a wine in which nearly all the alcohol is dissolved in the gas. On a general level, Chaptal was confirming experiments done by D.

---

[9] Tartaric acid is obtained from deposits (potassium hydrogentartrate) formed in wine casks. Optically active, it was one of Pasteur's early research passions.

Gentil and A. von Humboldt showing that "the carbon dioxide released from wines holds a very considerable portion of alcohol in dissolution." Loss of alcohol was a bad thing.

The transformation of sugar into alcohol nearly mesmerized Chaptal. It is a theme to which he returned again and again because he saw alcohol as the essential characteristic of wine ("L'alkool qui caractérise essentielle-ment le vin"). No doubt this is correct, but alcohol is also characteristic of all the other alcoholic beverages, and drinkers who stick exclusively to wine flatter themselves with the idea that good wine has other qualities that addict them to wine rather than whisky or beer. The idea that the amount of alcohol could be increased by simply adding sugar to the must seemed a powerful economic one to Chaptal. In fact, it was a bit surprising that a systematic application of the idea had not taken place before. And perhaps it would have if someone had done a scientific study comparable to Chaptal's and had been politically powerful enough to spread the gospel.

Chaptal was especially impressed by the experiments carried out by the chemist Pierre-Joseph Macquer in 1776–7 showing that if sugar was added to the must, good wine could be produced from unripe grapes or even from grapes not suitable for anything except producing acidic juice for cooking. From unripe wine grapes Macquer produced a clear, good-quality, very lively, rich, ardent wine, pleasing to the palate. Macquer's generosity with adjectives indicates that the discourse of wine tasting had developed considerably since the early eighteenth century. Several con-noisseurs could not believe that the wine came from unripe grapes with the taste corrected by sugar. This was not the first or last time that experts would be used to show the excellence of a dubious commodity. From the other grapes Macquer obtained a strong wine not totally displeasing to the palate but with no bouquet. It is tempting to believe that in this case Macquer had reinvented a model of *gros rouge*. He attributed the defi-ciencies to the type of grape and the youth of the wine rather than to the procedure or the producer. Chaptal concluded that the lesson to be learned from Macquer's experiments was to follow nature. In Pope's lan-guage, another eighteenth-century case of rules discovered not devised, nature still but nature methodized.

All the chemist was doing in remedying the immaturity of the grapes was to follow nature's own process in putting into the must the quantity of sugar that nature could not give it. Chaptal admired the practicality of the procedure. Not only sugar could be used. Any sweetener would do if it did not produce a disagreeable flavor in the wine: refined and semire-fined sugars, molasses, and honey, which was the old favorite of winemak-ers and recommended by Rozier. In any case, it was a particular applica-tion of a general precept on the "art de gouverner la fermentation."

Chaptal dealt with the science of it all under the rubric of "ethiologie de la fermentation," which was changed to a more impressive "théorie de la fermentation," in the second edition.[10]

Chaptalization became a widespread practice in France, especially after the 1850s, the beginning of the period of modern diseases. Chaptal would probably not have been happy with the quasi-universal prostitution of his curative principle: "One should never forget that fermentation must be regulated according to the nature of the grape and the quality of the wine that one wishes to obtain. The grape of Burgundy cannot be treated like that of the Languedoc: the merit of that of Burgundy is in a bouquet that would be destroyed by a lively and prolonged fermentation; the merit of that of the Languedoc is in the large quantity of alcohol it can develop in a necessarily long and complete fermentation."[11]

In the second edition of L'art de faire le vin Chaptal clearly limited the application of his process, indeed of his whole work. The highly deserved fame of many French wines provided enough justification of the excellence of the methods employed in making them: Bordeaux, Burgundy, Champagne, Hermitage, Côte-Rôtie, Arbois, Lunel, Frontignan, Rivesaltes, Rhône, Narbonne, Roussillon – with an et cetera added to console the regions omitted. Chaptal admitted that he had less intention of working for these rich vineyards, which furnished excellent wines for all Europe, than for the regions where the grapes produced only wine of bad quality. Yet he did not believe that his oenology was useless for the privileged cantons. First, his work was scientific – namely, it transformed the results of a healthy practice into precepts explaining and justifying good methods while also furnishing the means of correcting faults that bad seasons and other factors could produce even in the wines of famous regions. Obviously Chaptal was not going to carry false modesty too far. And it is certain that no ambitious scientist would limit his oenology to bad or even mediocre wines.

Dumas's Side Show: Sugar from Potato Starch

The one major if temporary innovation in chaptalization in the nineteenth century (1820s–1840s) was the use of sugar derived from potato starch. The syrup made from starch was supposed to be identical to that derived from grapes and other fruit. Some writers who kept up on the latest science noted that the chapter on starches in Dumas's treatise on applied

---

[10] Chaptal's ideas on fermentation are dealt with in Chapter 6.
[11] Essai sur le vin, 1st ed., p. 93, in Chaptal et al., Traité théorique et pratique . . . ; the 3d ed., with a description of winemaking apparatuses by L. de Valcourt, appeared in 1839. An Italian translation (in 3 vols.) of the second edition (1807) appeared in 1812–13. On the fifteenth-century disaster, see Camille Rodier, Le Clos de Vougeot (Dijon, 1949; 1980), p. 52.

chemistry stated that potato starch can be transformed into sugar (dextrin) and a small amount of carbonic acid.[12] (Dextrin is a polysaccharide carbohydrate, an intermediate product in the hydrolysis of starch to glucose.) Consulting the *Traité de chimie appliquée aux arts* of Dumas, Jullien discoved that glucose could be derived from potato starch, but he exhibited the good wine merchant's caution in dealing with good news because this glucose was not so well known as the one derived from grape must. It should therefore be used carefully. It was easy enough to leap to the conclusion that here was a substitute for the more expensive cane sugar, highly recommended because in a crystallized form it contains less impurities than any other type of sugar, and indeed is practically pure. The debate over the type of sugar remained alive in the 1880s, when the argument in favor of potato-starch sugar was dealt a fatal blow by Dumas himself.

The postexperimental mind will not be surprised to learn that wide-ranging experiments were able to show the gustatory virtues of starch sugar if not its superiority over cane sugar. Experimentation is a flexible art form. Experiments in Burgundy using the "new sugar" were cited as very satisfactory. The Société royale d'agriculture et des arts de Seine-et-Oise reported on experiments carried out at Rueil, at the request of the starch sugar manufacturers, which showed that the new sugar would give a more active fermentation and a higher degree of alcohol than what was usually obtained from cane or beet sugar. The most convincing experiments were done by Collas at Argenteuil. A harvest was divided into three batches, each producing 320–40 liters of wine. One batch received 20 kilos of starch sugar costing five francs; the second batch received five francs'-worth of cane sugar – the basic issue seems to have been the cost of wine production. The third batch had no sugar added. The first wine (with starch sugar added) was judged to have a flavor and a color superior to the other two wines. Tasters echoed this judgment. Further experiments at Auxerre, Compiègne, and other places confirmed the superiority of starch sugar. And the winemaker got an extra six francs per *pièce* (228 liters) in making the superior wine. Science was finally showing that it was of some real commercial use. The scientific experiment could be used in commerce as an effective tool of salesmanship.[13] Both governments and private organizations eagerly brought scientists into consultation on various thorny issues in agriculture and industry.

Two of the thorniest in winemaking were *vinage* (adding alcohol to wine) and *sucrage* (sugaring or chaptalization). The Société nationale d'agriculture de France, after adopting a report on the modalities of *vinage,* sent it into the parliamentary machinery through the ministry of finances.

[12] J.-B. Dumas, *Traité de chimie appliquée aux arts,* 8 vols. (1828–46), 6:89.
[13] Louis Leclerc, "Le sucre et le vin," *Bulletin de la Société d'oenologie française et étrangère* 3 (1837–8): 214–19.

Although we now distinguish between *vinage* and *sucrage,* the former illegal and the latter legal, the report presented to the national society of agriculture by Dumas in 1882 referred to *sucrage* as a form of *vinage* while also noting that the two are clearly different.[14] *Vinage* might appear to be a simpler method of raising the alcoholic level of wine, but the alcohol added would contain a small amount of what Isidore Pierre and other chemists labeled the toxic elements produced by alcoholic fermentation: amylic alcohol, acetic ether, aldehyde. *Sucrage* had a big advantage because the fermentation of sugar added to the must would produce not only alcohol and carbonic acid but also, as Pasteur showed, two substances having a considerable and favorable effect on the flavor of the wine: succinic acid and glycerine. So Dumas concluded that although *vinage* might be a necessary practice, sugaring was the preferable way of increasing the alcoholic content of wine, "the salutary practice."

The report by Dumas questioned the practice of substituting starch sugar for cane sugar. In spite of the purity advertised by the big Dijonnais manufacturer Mollerat, it had become clear that whatever the good short-term results from the new product, it did not have the same results as cane sugar. Over time the starch-sugar wine lost the delicacy of its flavor and even acquired a considerable bitterness. This did not prevent winemakers in the Dijon region from buying large amounts of Mollerat's syrup, especially in the period 1825–45. Dumas warned that in the case of organic food products needing to withstand the ravages of time it was wise to be careful in making changes in processing. Dumas and the special commission on *sucrage* recommended a return to Chaptal's process, which could use cane or beet sugar and did not change the health, bouquet, or flavor of wine. Science confirmed the wisdom of one of its heroes, the scientist who had risen to the highest level of power in the state and had made a valuable economic contribution to the nation.[15]

The Reception of Chaptal's Doctrine

Chaptalization has been a controversial subject in oenology from its early days. Winemakers just do it. The issue has always been its effect on the

[14] Dumas, famous author of the *Traité de chimie appliquée aux arts,* did not answer the question "What is sugar [saccharide] anyway?" We give the chemical reply: "Any of a group of water-soluble carbohydrates of relatively low molecular weight and typically having a sweet taste" – and so on. Starch is "a polysaccharide consisting of various proportions of two glucose polymers, amylose and amylopectic," found in plants and delivering glucose when digested by animals. Sucrose, derived from cane and beet sugars, is a disaccharide made up of one molecule of glucose (dextrose, grape sugar) bonded to a molecule of fructose (fruit sugar or laevulose, which is a simple sugar, $C_6H_{12}O_6$), stereoisomeric with glucose. Sucrose is laevorotatory, whereas most naturally occurring glucose is dextrorotatory. For definitions, see Oxford's *Concise Science Dictionary.* On sugar and fermentation, see the chapter on Pasteur.

[15] J.-B. Dumas, *Rapport sur le sucrage des vins avec réduction de droits,* presented in the session of May 17, 1882, of the Société nationale d'agriculture de France in the name of a

seductive qualities or organoleptic properties of good wine. In the early 1830s the Société d'oenologie française et étrangère was formed with the dual aim of improving wine and increasing its sales. For a few years it published a serious wine journal (*Bulletin de la Société d'oenologie française et étrangère*), which illustrated the ambivalence of the wine world toward chaptalization or *sucrage* (sugaring). Whereas the baron de Cussy, vice president of the society, denounced the practice, Louis Leclerc, a man of commerce, approved of it along Chaptalian lines, which should have meant that it was wrong to add sugar where none was needed. Leclerc seems not to have really approved of tampering with pure fermented grape juice, but acquiesced in the commercial necessity of *sucrage* in order to satisfy the demand of an undiscriminating public eager for strongly colored alcoholic liquids.

In the various editions (1813, 1822, 1826, etc.) of A. Jullien's *Manuel du sommelier,* chaptalization was an issue of minor importance. This benign attitude toward sugaring was maintained in the revised version published in 1859 by his son.[16] The dedication to Chaptal was dropped in the third edition (1822), but this must be a political salute to the Restoration of the Bourbons rather than an indication that the author had changed his mind about the great man's importance for oenology. Yet, ten years later, in 1832, in a third edition of his *Topolographie,* Jullien was more critical of the practice. He noted that the expansion of vineyards on the hills produced a substantial increase in good burgundy, which compensated for an increase in mediocre wine produced by an expansion of vineyards on the plains or in the low country. Some proprietors had adopted the method of sugaring the must in order to increase the degree of alcohol and improve the body of their wines. This procedure was good only for mediocre wines (*bas crus*), especially in years in which the grapes did not fully ripen. He emphasized that sugaring harms the quality of fine wines because they have to ferment for a longer time. True, they acquire a deeper color, more body, more solidity; but the trade-off is less delicacy: the bouquet and the *sève* (vitality or liveliness); their chief qualities are clearly altered and sometimes destroyed by the secondary fermentations provoked by the addition of sugar.[17] The wines age more slowly.

Jullien speculated that chaptalized wines were also more apt to suffer from *amertume*. As a merchant, he found this a threat to trade in the wine of Haute-Bourgogne because consumers turned to the wines of Bordeaux, which kept better. No idle speculation this: as a merchant he was well placed to document the increasing number of wine consumers switching

---

special commission, a scientific and economic galaxy including Chevreul, Dumas, Léon Say, Gaston Bazille, Bouchardat, and Boussingault.

[16] A. Jullien, *Nouvel manuel complet du Sommelier,* rev. C. E. Jullien *fils,* 1860 (1859). A. Jullien died in 1832.

[17] *Sève* is defined by *Le petit Littré* as a "certaine force qui rend le vin agréable."

to the wines of Bordeaux.[18] One can blame the increasing popularity of chaptalization on the vagaries of climate, the emergence of a new model of taste, the winemakers' desire to save mediocre vintages, the growing lust for more alcoholic beverages, or several other perceived advantages of sugaring. Jullien's warning was increasingly ignored. The growing availability of sugar and the increasingly rigorous control of fermentation, which made it possible to avoid the danger of refermentation from the addition of too much sugar, made chaptalization a standard procedure.

Jullien believed that grape syrup and glucose were *sometimes* used to increase the sweetness of new wines or temper their tartness when the sugar content of grapes was low because there had not been enough sun to ripen the grapes normally. But the wine merchant insisted that this manipulation was practiced only rarely on wines consumed in France; rather, it was done sometimes for wines exported to countries where the consumers, used to drinking wine from southern Europe, appreciated French wines to the degree that they possessed the great sweetness of the "Midi." The Dutch were singled out for their love of sweet white wines, which led merchants to add to the wine a lot of sugar or grape syrup, along with sulfur dioxide to prevent them from refermenting. This unpleasant taste, classified as disagreeably sulfurous by Jullien, pleased the Dutch.

The posthumous seventh edition of *Topolographie* (1859) grudgingly admitted that sugar tends to correct the sourness or bitterness in a wine and to give it body. Part of the reason for lack of interest in the practice during much of the eighteenth and nineteenth centuries may be favorable climatic conditions for wine growing. In the periods 1711–17, 1739–52, 1765–77, 1812–17, and 1850–6, the weather was poor for producing the mature grapes needed for good wine. The periods 1718–37, 1757–63, 1779–81, 1801–11, and 1857–75 were extremely favorable for wine production all over Europe, often even in Germany. "The early wine harvests of the fifteen or twenty years after 1856 are one of the major fluctuations in the history of climate in western Europe."[19] When the sun produced the sugar in the grape, there was no need to improve on nature except to cater to the taste of foreigners.

The Spread of Chaptalization

The adding of sugar was part of the literature on winemaking long before Chaptal developed his axiomatic art. In his *Oenologie* (1770) the Burgundian Béguillet included a note on sugaring in a murky discussion of the

---

[18] A. Jullien, *Topographie de tous les vignobles connus,* 1st ed. (Paris, 1816), p. 78.
[19] Emmanuel Le Roy Ladurie, *Times of Feast, Times of Famine: A History of Climate since the Year 1000* (New York, 1988), p. 61; see also Loubère, *The Red and the White,* p. 140.

fermentation of the "mucous body" of the must.[20] He accepted the ancient practice of adding honey when the natural grape sugar was low, if the honey were first boiled and skimmed to get rid of the wax. His statement is careful, indicating that the practice was not normally essential. "Some add very refined sugar, not only because it is a sweet, extremely wholesome substance, but also because it is analogous with all wine, whose raw material is a sweet and tartarous substance." So syrup or concentrated sugar could be mixed with the wine without any problem, letting the acid and bitter essences in the verjuice turn sweet as they matured. Béguillet was aware of the difficulty of being precise in talking about fermentation because not enough work had been done on the nature of sweet and sour plant saps or juices, their uses, and how to make imitations of them. He indicated no problem in sugaring. Perhaps he had an excessive confidence in his rudimentary science. A little knowledge often leads to great optimism. The issue did not seem of much importance to Béguillet because only *some* winemakers engaged in the practice, and especially because there was no evidence that it worsened the quality of the wine.

According to Chaptal, science confirmed that the addition of sugar was not "contrary to the nature of the wine": fermentation converted the sugar into alcohol; the more sugar, the more alcohol. Supporters of Chaptal who were close to wine producers enlisted nature to support chaptalization. In his work giving advice to winegrowers, Cadet de Vaux pointed out that nature herself indicated how to compensate for the influence of poor weather on the grape. "Remove the pellicule of the grape of the south and you will find crystallized white sugar. In putting her secret under your eyes, isn't nature revealing her secret to you? Is she not saying to you, it is with sugar that I make wine?"[21] Science was a way of reading nature in order to derive rules to imitate natural processes when nature failed to deliver. The good news had then to be communicated to the winemakers. No one was better placed to do that than Chaptal.

The spread of chaptalization has to be understood in its commercial context. Because of the higher level of alcohol in chaptalized wine, it kept better than unchaptalized wine. There was not much point in making subtle wines with a magnificent bouquet if they were going to be destroyed by diseases like *amertume* and *pousse.* So Payen and Herpin, members of the Société d'oenologie, concluded that although chaptalization destroyed *délicatesse* and *bouquet,* the sacrifice was worthwhile if a healthy saleable wine resulted from the operation. Payen believed that the sugar added should most resemble the sugar in the grape itself, but once having stated this seemingly reasonable principle, he recommended the cheaper dextrin, "the

---

[20] See the beginning of the chapter on Burgundy for a fuller discussion of Béguillet.
[21] Quoted in Moreau-Bérillon, *Au pays du champagne,* p. 110.

sugar extracted from potatoes."[22] This deviation from Chaptal's doctrine, however debatable, would eventually produce many poor wines before oenology returned to cane and beet sugars as best for the health of the wine, whatever their gustatory vices.

An evolution of opinion can be seen in Henri Machard, who was so alarmed by the spread of chaptalization by the early 1840s that he wrote a pamphlet (1843) on the dangers of the abuse of sugar in winemaking in Burgundy. This pamphlet was incorporated into the second edition of his treatise on vinification in 1849. Machard noted with disgust that even in 1842 the *négociants* used an enormous amount of sugar in the wine (forty *livres* of sugar for each *pièce* of wine).[23] But by the time of the fourth edition of his *Traité*, in 1865, Machard had changed his mind on chaptalization. He admitted that science and technology had proved him wrong. His past battle against sugar was based on the fact that bad wines, with consequent damage to Burgundy's reputation, resulted from the frequently impure sugars on the market. Good pure sugar was now available, but great care still had to be taken to prevent any problems arising from impurity. When Machard opposed the use of sugar, people had assumed that glucose and the sugar of the grape were identical. This was a continual source of errors, and precision was needed in adding the different sugars in order to control the corresponding changes in the must. Machard came to accept the argument that chaptalization improved wine and produced desirable qualities in it. A complete fusion of the elements supposedly took place as they were perfectly reconstituted. Chaptalization was better than the addition of alcohol, although Machard still accepted the addition of wine alcohol to give a higher degree, a better body, and a longer life to the wine. In science the text follows practice, and Machard's *Traité* conforms to this rule.[24]

Burgundy was not alone in profiting from the science of sweetness. In 1899 the Bordelais magazine *L'Oenophile* carried an article on the abuses of chaptalization, then increasing in all wine areas, including the Gironde, where there was also protest against sugary excesses. Chaptalization was being practiced on a large scale in order to counteract recent smaller harvests, which put pressure on family consumption. In this situation it also became more important to increase the level of alcohol in the wine as a partial protection against the numerous microbial diseases to which wine

---

[22] "De l'amertume et de la pousse des vins," *Bulletin de la société d'oenologie* 2 (1836–7): 257–9.

[23] Henri Machard, *Dangers que présente l'abus du sucre dans la confection des vins de Bourgogne* (Auxonne, 1843), included in H. Machard, *Traité complet de vinification ou guide des propriétaires, négociants, vignerons, etc. dans toutes les opérations qui sont relatives à la meilleure manière de traiter les vins* (1st ed., 1845; 2d ed., 1849).

[24] Henri Machard, *Traité pratique sur les vins,* 4th ed. of *Traité de vinification,* revised and considerably expanded (Besançon, 1865), pp. 52–3.

fell prey. Experts of the ministry of finance calculated that wine from the first pressing of grapes averaged an increase of three degrees in alcohol from chaptalization. Most of the alcohol of *vin de sucre* (a second wine made from the *marc* or presscake, water, and sugar) came from chaptalization. *Vins de sucre* were a great source of wine for family consumption, and also fraud when sold as wine from the first pressing.

Burgundy, the Beaujolais, the Champagne, departments of the Loire, and Alsace-Lorraine have been the big chaptalizers. In Provence, Languedoc, Roussillon, and the Gironde chaptalization was forbidden after 1929. A decree of the ministry of agriculture then exempted Bordeaux from the law. *Vins de sucre* were illegally put on the market by producers as well as merchants, often victims of producers delivering this liquid under the reassuring label of natural wine. The producers of good natural wines, always an expensive commodity, squealed in protest against the illegal competition of the rotgut that drove prices down. By the 1890s sugar was recognized as already having a quite substantial role in the production of real wine.

During the nineteenth century there were good and bad harvests, and fluctuating amounts of sugar were used, with a rise in use in late century. As a result of the post-phylloxera reconstruction of the vineyards and increasingly successful control of disease and fraud, the amount of sugar used fell considerably. After 1900 the average number of degrees of alcohol from the addition of sugar to natural wine dropped to below 2; in 1929 it was down to 1.2 degrees. *Vin de sucre* became a family affair. The wine business returned to a normal state, which now included moderate chaptalization. In 1899, out of a total wine production of nearly 48 million hl, about 4,332,933 hl were given more vim and vigor by the addition of 36.5 million kg of sugar.[25] But this was also the end of a well-sugared period. In the next year, 1900, with a harvest of over 67 million hl, the total sugar used was less than 17 million kg. The figures for the year 1899 show Girondin restraint in using sugar compared to some other departments.[26]

---

[25] *Bulletin de statistique et de législation comparée* 70 (1911): 716–17.

[26] In years of poor harvests, lots of sugar was used: in 1888 a harvest of just over 30 million hl required nearly 39 million kg of sugar to produce the year's wine supply. The following figures (kg of sugar) show that, in spite of a general decline in the use of sugar, it remained an important part of winemaking. The amounts used depended on the quality of the harvest and the type of wine made.

Department	1911	1929
Côte d'Or	345,697	251,988
Gironde	105,610	138,900
Marne	188,879	554,152
Saône-et-Loire	264,014	84,915
Rhône	112,151	134,781

J. Le Bihan, "Les abus du sucrage," *L'Oenophile*, 6e année (1899), pp. 285–6. J. L. Riol,

Department	Harvest (hl)	Sugar (kg)	Kg/hl (per 1000 hl approx.)
Aisne	47,980	288,000	6000
Jura	75,000	403,000	5000
Loiret	120,000	630,000	5000
Gironde	2,355,000	2,217,000	1000

Chaptalization and the Reputation of Burgundy

In his discussion of problems in Burgundy, Louis Leclerc, writing in the late 1830s, identified a variety of factors explaining the production of so much poor wine. He included the planting of bad varieties of vines whose only virtue was high yield and excessive fertilization, but he did not identify *sucrage* as a negative factor in itself, only its abuse. In the same journal, only about a year before, the baron de Cussy had given a quite different analysis of the role of *sucrage* in the decline in quality of burgundy.[27] It is also an analysis in which a drinker has more confidence because it is squarely based on the snobbish concept of the gourmet, so vital to the existence of fine wines. "The art of good drinking depends essentially on the art of good living rather than that of good eating; the rarity of this art explains why you will meet ten *gourmands* before finding one *gourmet*." The gourmet can be distinguished from other mortals by his ability to identify wines by vintage and *cru*. (Up to the 1830s the great year of the century was 1802.) The problem for burgundy becomes clear when we read the baron's lament that fashion, "the pitiless tyrant everyone obeys," then dictated that wines served at a dinner had generally been reduced to three: champagne, Rhine wine, and bordeaux. "Le champagne bien frappé se boit immédiatement après la soupe, et son règne finit au moment où l'on pose le dessert; le vin du Rhin arrive en agréable intermède, et les grands crus de bordeaux ferment honorablement la marche." Madeira as a preface to the meal and *vins de liqueur* as epilegomena had disappeared. And burgundy?

Cussy put the question bluntly: why had the Côte d'Or degenerated to the point of losing the distinguished rank it had formerly occupied? Whatever the causes of previous declines in the quality of burgundy, the cause of low quality in the 1830s was identified by the baron, vice president of the Société d'oenologie, as the procedure of introducing sugar into the

---

Le vignoble de Gaillac depuis ses origines jusqu'à nos jours et l'emploi des ses vins à Bordeaux (1910; 2d ed., 1913), pp. 260–1, gives figures of 136,427, 210,291, and 106,444 kg for the Gironde in 1908, 1909, and 1910. The Côte d'Or dropped from 219,303 kg in 1908 to 62,531 in 1909. Figures are from the *Bulletin de statistique et de législation comparée* for 1909, 1911, and 1929.

[27] Le baron de Cussy, "Des vins de la Côte d'Or. Ce qu'ils étaient autrefois et ce qu'ils sont aujourd'hui," *Bulletin de la Société d'oenologie française et étrangère* 2 (1836–7): 26–9.

wine at the time of vinification. Cussy emphasized that Chaptal had recommended the procedure as a remedy to be used in bad years. But because
of its advantages in production, winemakers of the Côte de Beaune and
the Côte de Nuits soon adopted chaptalization as a standard procedure.
The result was the disappearance of the delicious wines with marvelous
bouquet and flavor, and a perfumed finish that lasted for five minutes after
you drank the wine. But Cussy did not want to attack experimental science. Chaptalization was a piece of science that the baron disapproved of
because it was not the result of experimentation, and thus unlike many
other really useful processes.

If this apparent science was the first reason for the decay of burgundy, it
was not the only one. Cussy added the important role played by a sinister
sociopolitical development. The Revolution of 1789 had destroyed the
monasteries and the structure of property ownership in the wine country,
with the result that viticultural units were "denatured" and fragmented.
The Clos de Vougeot had been created by the quality-conscious monks of
the Abbaye de Cîteaux, whose motto was "Produire peu, valoir beaucoup." By the late eighteenth century, success had made them querulous
and litigious, ripe for depossession in the French Revolution. Some people
believe that they used to harvest the grapes horizontally from three sections of the *clos*. The wine from the top, which was the best, was kept for
themselves and used for strategically placed gifts to royalty. The second
and third regions provided wine for commerce. In a fit of commercial
excess, their successors harvested the *clos* as a unit, cashing in on the reputation of the stuff from the top of the hill. *Monstrum horrendum!* cried the
baron. The inconvenient idea of the trinitarian harvesting of the clos has
been denied by some modern writers, who cite eighteenth-century evidence that the monks made at least ten kinds of wine. Fortunately, Montrachet, "le plus fin des vins blancs que produit notre riche France," had
remained the property of the marquis de La Guiche.[28] Curiously, Cussy
then mentions drinking not Vougeot but some "Romanée Conty" that had
been given in 1782 to de Juigné, who became archbishop of Paris. Both
velvet and satin in a bottle, sighed the baron. Shortly thereafter, sometime
before 1792, the progressive monks of Cîteaux started helping nature by
putting sugar in the wine just after barrelling it. The monks anticipated
Chaptal, because they added sugar only if the wine was low in alcohol.
New owners of the Clos de Vougeot would not be so successful in resisting
the Chaptalian temptation.

The vicomte Alfred de Vergnette-Lamotte explicitly divided the history

---

[28] Cussy would be happy to learn that the estate which Marquis de Laguiche et ses fils still
owns is "the largest slice of Montrachet" (4.75 ha) and that the wine is carefully produced
and distributed by the Maison Drouhin. Anthony Hanson, *Burgundy* (London, 1982), p.
299; Jean-François Bazin, *Le Clos de Vougeot* (Paris, 1987), p. 80.

of the wines of Burgundy into before and after Chaptal, the golden age followed by the dark age. One of the nineteenth century's major scribblers on viticulture and winemaking, his views are doubly interesting and significant. His activity married science and production in a striking and creative way, but it seems to have had little practical or theoretical impact on either winemaking or oenology. A man of Beaune, a noble owner of important vineyards, including Château Meursault, and a winemaker, Vergnette was a distinguished graduate of the Ecole polytechnique and no stranger to scientific experiment. Pasteur, creator of experimental science at the rival Ecole normale, might have asked what one could expect from a graduate of the Polytechnique. That would have been a less rude comment than some others Pasteur made about this nineteenth-century agronomist in a nasty quarrel over who discovered the "pasteurization" of wine.[29]

Vergnette thought that wine alone contains the elements (alcohol, organic acids, including tannin, various salts, and sugar) necessary to protect it from spoiling, except in years when industry and science must replace "what nature refuses to give." Additions to wine could be classified geographically: chiefly but not exclusively tannin in Bordeaux and the Champagne, sugar in Burgundy, and alcohol in urban wine factories. In this conceptual scheme, Chaptal's oenology was reduced to the belief of a man from the Midi who valued the alcoholic strength of wine above all its other qualities. This criterion was the basis of "the sad principles of vinification" still in vogue in mid-nineteenth century. Chaptal's new wine model required that grapes be allowed to mature so that they are as close as possible to the southern grapes of high sugar content. If the weather prevented the development of such grapes, then sugar would be added to the vat or even to the newly made wine when it was drawn off.

The novelty in Vergnette's anti-Chaptalian argument was its insistence on the superiority of a different wine model. Harvested early, before bad fall weather arrived, the grapes were richer in salts, tannin, and acids than grapes left to ripen to an excessive maturity. Vergnette would view today's wine model, based on late harvesting and modest sugaring, the model that makes most wine writers and other winos ecstatic, as the triumph of Midi barbarism. (Most oenologists believe this model to be the results of scientific oenology, a sort of Newtonian triumph tempered by taste.) Believing that "the harvest makes the wine," Vergnette argued that the great *crus* of Burgundy should be made from grapes ripened according to the old system and fermented without chaptalization. These were wines needing

---

[29] Vicomte Alfred Vergnette de Lamotte, *Memoires sur la viticulture* (Paris, 1846), 146–7, and *Le vin,* 2d ed. (Paris, 1868).

nothing nature had not provided: *vins complets* distinguished by "un cachet de finesse remarquable."

If the pre-Chaptalian wines were so good, why did the general practice of chaptalization catch on? Explanations of this apparent paradox either state that the process improved wines or that taste changed – for the worse, it is often implied. Vergnette had no doubt that the chaptalization of wines was simply a commercial convenience; chaptalized wine's patina of pleasure-giving qualities covered a body of vices. The vicomte's science told him that the fermentation of the added sugar produced a hydrogenous body, similar to essential oils, particular to the type of sugar. This body gave wine a bitter, penetrating flavor in the back of the throat instead of the pleasant sensation that came from the first contact of the palate with a mellow wine having good body. Adding more sugar to the must produced more of this unpleasant substance, although it was masked somewhat by the extra alcohol, the *moelleux,* and the deeper color that is in itself a psychological stimulus to expectation of pleasure. So far so good, one might argue, especially if the wines were highly chaptalized (10 kg and more of sugar per *pièce,* which leaves residual sugar in the wine, unlike the modest chaptalizing of 2 kg per *pièce*). Chaptalized wines are pleasant to drink when they are old, but since the first glass so fills you up – is perhaps even cloying – they discourage wine consumption. Or so Vergnette argued. He was one of a few worried wine producers who deplored the meretricious wine of the chaptalizers.

By mid-nineteenth century, perhaps even earlier, the wine business had decided that consumers wanted a more deeply colored wine, much in the style of the claret model invented for the English market in the previous century. Vergnette advised against the consumption of these wines of a rich, velvety color not only on the basis of his own more austere taste model but also on the assumption, still accepted by believers in the natural wine model, that the alcohol fermented from the grape was not harmful to the health of the consumer. Wines with alcohol added were even worse, producing a reaction like that of inorganic poisons invading the mucous membranes of the alimentary canal. Sugared wines must age in order to achieve more *fondu* or mellowness. So fermentation used to be speeded up by heating wine without any concern for the production of small amounts of acetic acid in secondary fermentations carried out at 30–5 degrees centigrade. The request of wine merchants that wines be chaptalized also ruined a lot of wine by secondary fermentations leading to "disease and destruction." Satisfying consumer demand was a risky business.

Vergnette was not convinced of the solidity of the argument that consumers preferred wines more full-bodied than they used to be, thus placing less value on the qualities of delicacy (*finesse*) and *bouquet* that should distinguish the wines of Burgundy. Chaptalization just produced alcoholic, syrupy, *empâtant* (sticky) wines that were no match for their competitors. The wines of Bordeaux, their chief competitors, had none of the characteristics of chaptalized burgundies. Vergnette was inclined to blame the whole problem on the attempt to produce wine from the *pineau* (pinot) grape planted on the plain, where it could produce only weak wine needing chaptalization. Along with poor wines of bad years, these wines of the plain turned the consumer away from burgundies. Thus the consumer of the *grand vin* of Burgundy was lost.

To us this seems a curious explanation, but it was less unpleasant than admitting the decadence of the *crus de Bourgogne,* which was to some degree the result of unscrupulous producers' mixing the unchaptalized "pure *premiers crus*" with chaptalized "second wines." This process disguised the vices of the "second wines" by killing the virtues of the *crus.* The lust for profit replaced quality control and caused Burgundy to lose both quality and profit. Vergnette reminded his fellow producers that Burgundy had already lost out to the Champagne in one sales area. The loss of the market for sparkling burgundy could be partly blamed on the use of poor wine to make this product. Having lost the market, the Burgundians then classified champagne as an industrial product. Even winning back customers for still wine would require basic reforms in wine production.

Vergnette's program was a return to the practices of the old days, when the reputation of the great wines of Burgundy was still untarnished. It is not really clear exactly when this was. A skeptic might argue that Burgundy's reputation was great when taste models were not based on high-quality wine – or at least our model of it – and wines were drunk young. Citing the medically inflicted taste preference of the gluttonous and ailing Louis XIV for Côte de Nuits hardly improves the quality image of a product. If the idea of high-quality wine is a late-seventeenth to early-eighteenth-century creation, the whole argument about a loss of quality may be totally wrongheaded. Quality was not lost, it was not yet achieved. For burgundies as well as for bordeaux and the wines of the Champagne, the eighteenth century was the time when quality control became a serious part of winemaking. (Perhaps somewhat earlier for champagne.)

So Vergnette's mid-nineteenth-century program of a great leap backward to improve the *crus* of Burgundy makes a limited sort of sense. Quality had been achieved in a few cases in Burgundy; it was more a matter of extending techniques for quality control to more producers. In Bordeaux the production of high-quality wines was limited to a few producers, but the greater size of the estates gave it the opportunity to produce a far larger amount of quality wine. It is also true that the Revolution of 1789 had degraded the estates producing fine wine in Burgundy more than comparable estates in Bordeaux. Most political-economic forces drove Bordeaux in the direction of profit through quality and Burgundy in the direction of profit through convenience. Vergnette's program had a charming simplicity about it in one part – the return to the past – and a naive confidence in technology in another part – mild freezing of wines in winter in order to concentrate and preserve them for export.[30] The natural elements in the grape juice were sufficient to preserve wine, according to Vergnette, although he accepted the use of sulfur for preservation and charcoal for filtering. To sum up his program: nitrogenous fertilizers should be avoided, harvesting should be earlier, vats should be classified according to quality, and great care should be taken in fermenting and vinifying. The wines produced would win back customers. Science – namely, chemistry – was limited to an etiological role, identifying the causes of disease and thus indicating more rational methods of culture and vinification. Inspired by Pasteur, scientists in Bordeaux would not be so modest.

The Example of Château Latour and the General Model

Wine production now depends heavily on scientific knowledge and technology. Until well into the nineteenth century, winemakers could do little to change the production of wine once fermentation got under way. Even such simple processes as cooling or heating the must to slow down or speed up fermentation were technically unfeasible, and their scientific explanation was unknown. At Château Latour (Pauillac) the most that might be done was to add some *eau-de-vie* or Armagnac to the vats of the "second wine." Some of the second wine, next in quality to the best wine of the harvest, was added to the best wine of the vintage along with some *vin de presse*. The success of this tricky operation of equalizing (*égalisage*) depended upon a rigorous selection by the manager, who was, in theory and often in practice, an infallible taster.

In 1816 an attempt was made at Château Latour to correct a thin and acid vintage by chaptalization. The manager, Lamothe, hoped that a human technique would compensate for nature's neglect of the harvest. This was a rare event, for nature was in a generous mood from 1775 to 1825, when, according to Pijassou, 80 percent of the fifty wines produced were good to excellent, including the great years of 1795, 1798, 1801, 1802, 1815, 1818, and 1822. Good weather and careful vinification produced good wines. The sugaring experiment of 1816 did not please Lamothe. Although the grapes were in rotten condition, Lamothe concluded that chaptalization would never produce more than a mediocre wine, because technique is incapable of supplying the primary materials of nature that

---

[30] Vergnette's freezing technique is dealt with in Chapter 6.

determine good quality. The issue of making good wine "better," that is, conform to a certain model of taste different from that produced in a year of good weather, had not yet arisen. The mentality and model were still those of Chaptal, no universal chaptalizer. Pijassou admits that "the tradition of high quality in the wines of the Médoc forbade practices like chaptalization."[31]

A new tradition was created early in the twentieth century. By 1908, a humid year of immature grapes, the *négociants* recognized the impossibility of selling mediocre wine in a difficult market. Having the estates under contract to buy their wine, the merchants convinced the producers to chaptalize moderately. Daniel Jouet, manager at Latour from 1883 to 1932, still thought that sugar added to the vat was no more than nature's helper. But when he found the chaptalized wine to have a good bouquet, fruit, and body, and to be clear and mellow, he became friendlier to the practice. Jouet thought that the chaptalized Latour of 1909 (11.6 degrees of alcohol) was superior to the wines of Châteaux Lafite (10.2) and Margaux. A certain minimum degree of alcohol was explicitly recognized as an essential feature of the *grand vin*.

The idea of moderate chaptalization was being incorporated into vinification as a technique that could be used when needed without changing the quality of the *grand vin*. Quality control demanded high-quality cane sugar, supplied by the same firm to Châteaux Latour, Lafite, and Margaux. Along with other estates, Lafite and Margaux followed Latour in employing chaptalization in order to get wines of better color, body, bouquet, and taste. A potentially poor wine could be made into a nearly good one, as in 1912. A wine with certain well-defined qualities of color, body, alcohol, acidity, bouquet, and taste came to require moderate chaptalization in all but exceptional years. Taste determined technique. Unless one assumes that the vineyards, partly replanted with grafted vines and often heavily treated for various diseases, including phylloxera, were producing grapes that gave wines different from the past, it is difficult to understand what had happened to the *grand vin,* the *cru* of previous times. Nor were there any basic *long-term* weather changes – no little ice age, for example – that explain the generalized spread of chaptalization. What many experts had regarded as a vice, partly explaining the existence of so much bad burgundy, now appeared as a Bordelais virtue. Taste changed and so did the wine. People wanted less tannic wines, and chaptalization, through an increase in alcohol, reinforced the new lower-acid wine model.

---

[31] Réné Pijassou, "L'art de produire un grand cru, 1775–1825," in Higounet, ed., *La seigneurie . . . de Château Latour,* 1:275; and "Les temps difficiles, 1880–1920," in ibid., 2:510–3.

Science Reconsiders and Reinforces Chaptalization

One of the most striking features of modern history is the often happy congruence of science and capitalism. Science made the sugaring of wine a universally accepted principle, with its legality enshrined in the wine code, the collection of laws dealing with the legal manipulations that was given hieratic status in the 1930s. One can assume that generally both producer and scientist wanted to make the best possible wine with the highest possible profit. In the late nineteenth century it became traditional to argue that making good wine without chaptalization was not feasible in northern climates. So the realization of a basic aim in winemaking led to a set of practices that are often deplored by people who are inspired by things organic or natural and gourmet but are generally deemed indispensable by scientists and wine producers and accepted, often in happy ignorance, by most consumers.

In contrast to the times of Béguillet and Jullien, when chaptalized wine was said to be for foreigners, by the end of the nineteenth century a new consumer's wine model had developed for French drinkers. Commercial convenience, the thirst of the consumer for more alcohol in the wine, and the scientific *nihil obstat* for the practice all promoted chaptalization. The familiar litany of justification may be summed up with a certain brutal economy. Unlike Burgundy and the Champagne, the Midi could grow grapes with up to 300 grams of sugar per liter, capable of producing highly alcoholic wines. In the north the commercial convenience of picking immature grapes, to avoid possible bad weather during the harvest, condemned the wine to a low degree of alcohol, rarely more than 10 percent, and therefore susceptible to various diseases not usually present in more alcoholic wines.

Some people perversely preferred to take their chances with nature – no matter how it was constructed – to accept poor and lost vintages, and to drink pure wine with less alcohol. This behavior was anathema to commerce and to governments that taxed alcohol. In its desire to reassure drinkers, some producers and merchants sought help from experts, scientists "in the system" – namely, graduates of science faculties and agricultural schools, and heads of agricultural experimental stations – who are the logical people to consult if you are a capitalistic producer worried about the salubrity, appeal, and profitability of your product. Governmental agencies that could compel change in products were scarce and weak and would long have their hands full in keeping people from being poisoned rather than having their aesthetic sensibilities offended.

By the end of the nineteenth century many serious winemakers had come to see the advantages of a scientific approach to wine production, even in Burgundy. But science in Burgundy was like the Enlightenment in the German states – a few beacons beaming out into a night of indifference: Ladrey, the professor; Latour, the engineer-winemaker keen to introduce as much technology and science as was compatible with tradition; and Eugène Rousseaux, a missionary from agricultural science, or rather engineering, an oenologist invading Chablis country. With the phylloxera louse came the men of Montpellier to control it and much of viticulture. Perhaps because of their precarious position between the worlds of science and agriculture, or their fear of despair before the great ignorance to be conquered, or their missionary spirit imbued with their science-based ideology of progress, the scientific servants of the Republic never slackened in trying to sow the seeds of science in the refractory soil of the French countryside.

Agricultural scientists and engineers often exuded an air of fanatical rationalism when they dealt with vinification. Few problems would not succumb to a scientific solution. Young scientists took a particularly hard line on the hidden vices of traditional practices and brooked no criticism of the obvious virtues of scientific oenology. Even where traditional practices were grudgingly admitted not to be bad, they could be improved by the injection of a little science, presumably as much as tradition could take without shattering, though that too would not have been regarded as a tragic event. "Every improvement in the making of wine must rest, not on vague ideas, but on precise and scientific bases."[32]

New scientist spoke with as much authority as old priest, and his cause was equally just. A young agricultural engineer like Eugène Rousseaux, assistant at the Institut national agronomique, just did not believe the argument used against progress by many winegrowers – namely, that their methods gave excellent results, always producing wine of irreproachable quality. The *vignerons* never spoke of the diseases which, commonly resulting from their practices, continued to afflict wine at the end of the nineteenth century. So much for the profound influence of Pasteur in the vineyard. Sick wines have been as common as sick people since antiquity. Every French oenologist knew that the evidence for ancient wine diseases could be found in Pliny. The existence of numerous remedies embalmed in empirical practices also proved that diseases had been serious since antiquity. Of course, not every twentieth-century winemaker suffered from the illusion that history knew only healthy wine. And Rousseaux hammered home his point with an a fortiori argument: every honest person would admit to having known diseased wines, even in the most famous vintages.

In the late nineteenth century it seems that the state became much more conscious of the economic importance of wine than it had been. It could

[32] E. Rousseaux, "Etudes sur la vinification dans le canton de Neuchâtel faites aux vendanges de 1897," *Bulletin du ministère de l'agriculture* 17 (1898): 1435–89; quotation on p. 1435.

be argued that there were more disasters: added to cycles of bad weather and insects were the mildews, phylloxera, and black rot. The peasantry, especially in the Midi, had entered into politics; in the early twentieth century radicalism infected the Champagne. The Second Empire had looked to Pasteur for help against the microbes he had introduced into science. In the Third Republic, governments and desperate or anxious groups looked for help from scientists and from institutions, usually ones it had created for the production of science. The wine business could use lots of help, or so said the people who wanted to clean up the act of the vignerons, often denounced for producing a bad product that got worse with age. The consumer's fertile imagination, fueled by a sensational press that often stumbled onto the truth, did not flag in explaining why wine was so bad, often with a comparison to that of a golden age. Dousing with chemicals against phylloxera, black rot, mildew, and even more exotic diseases; soil exhaustion; insufficient fertilizer – a reproach soon reversed – these were just the more popular complaints. By the 1880s it was possible for an enterprising winemaker to find a local or at least regional organization or authority – agricultural station, departmental professor of agriculture, or even a university professor in places like Montpellier and Bordeaux – to ask the difficult question if vinification could not be modified to produce a good wine of considerable longevity.

In 1897 the Swiss society of winegrowers of Neuchâtel showed its interest in better cultivation and vinification by asking Rousseaux to study their procedures during the winemaking season of that year. The Ecole de commerce put its laboratory at his disposal. Hermann de Pury collaborated with him in observing and experimenting, although the report was done by Rousseaux alone. Rousseaux was able to draw upon the research on vinification that he and A. Müntz had previously published in the *Bulletin du Ministère de l'agriculture.* The fact that Rousseaux was known across the frontier is testimony to the reputations of Müntz, his *maître,* Rousseaux, and French viticultural and oenological research.

The Neuchâtel winegrowers were especially worried about chaptalization. Rousseaux assured them that adding sugar during vinification is not a falsification. Even better, in the case of immature grapes resulting from bad weather, *sucrage* produces a real improvement in the must. This new reality was explained by biochemistry and established by precise practice. One and seven-tenths kg of sugar per hectoliter of wine would raise the alcohol level by one degree; three and a half kg of sugar would increase the alcohol by two degrees. Careful, rigorous procedures had to be followed: the sugar had to be dissolved in a certain quantity of must heated to 60 degrees centigrade and then added to the vat of wine that had started fermenting. A greedy attempt to increase the alcohol content immoderately would result in a bad wine through incomplete fermentation. The sugar could not be added to water, for that would be equivalent to illegal *mouillage.* (Adding sugared water was a dubious privilege extended to Alsace-Lorraine after their reintegration into France in 1919.) Like Chaptal, Rousseaux did not advise chaptalization as an annual routine but only when the grapes could not ripen to a sufficient maturity to produce a wine with a desirable level of alcohol.

What did the men of Neuchâtel gain from Rousseaux's study? Let us assume that they wanted to know what to do when nature forced them to press immature grapes, which have low sugar and high acidity, generally the problem with cold springs and summers (e.g., 1812–17 and 1850–6). The winegrowers got a scientific solution to their problem: the old procedure of chaptalization, now clothed in the latest chemical explanations and technical precision. And if there should be warm springs and hot summers (e.g., 1801–11 and 1857–75)? Certainly the result would be a grape with high sugar, capable of giving a wine of good color and considerable alcohol. But overmaturity, which could happen when ripeness was determined by an inexpert guess, meant too low a level of acid for a good fermentation. Again, science to the rescue. Unripened grapes, never hard to find in northern grape country, could be added to the overripe ones until the level of tartaric acid rose to eight to nine grams per liter for the Chasselas grape and ten to twelve grams for the Pineau (Pinot). Chemistry and instrumentation made this precision possible. Precision was also possible in regulating the level of tannins, now seen as perhaps the most vital group of organic chemicals in good wine. Temperature could be controlled. Vinification was well on its way to acquiring the ability to produce drinkable, technologically correct wines in situations that would have previously been disastrous.

The bottom line of Rousseaux's careful study of the entire winemaking process in Neuchâtel was that the usual problems arising from traditional and empirical practices could be avoided by scrupulous cleanliness, which used to be amazingly rare, the maximum of precision in measurement, including an analysis of the chemical composition of the grape, and careful preparation of any additions to the must. It was a lesson he repeated in his *Notions d'oenologie,* published in Auxerre in 1907, when he was the director of the Station agronomique de l'Yonne.

Chaptal's Doctrine and the New Model Wine

It is wrong if comforting to believe that the *grands vins* have not changed since the eighteenth century. Apart from evanescent evidence based on rare tastings by a favored few, there is solid analytical data, also complemented by the results of professional tasting, of 76 *grands vins* made between 1906 and 1972. In a note (1977) for the Académie d'agriculture de

France, the Bordelais oenologist Pascal Ribéreau-Gayon summed up the evolution of the great reds of the Graves and the Médoc during the twentieth century, developing essentially an evolutionary model of twentieth-century wine.[33] One of the evident trends has been toward more alcoholic wines, partly the result of the use of riper grapes from later harvests, partly the result of the general spread of chaptalization. The use of chaptalization was encouraged by the new ability of winemakers to control alcoholic fermentation, including precision control of alcohol produced by the addition of sugar, which reduces the risks of a stopping of fermentation and of sourness produced by lactic acid bacteria. A moderately higher level of alcohol in modern wine, if it remains below the level at which the hot flavor of alcohol appears, is an advantage, because it brings out the wine's roundness and mellowness (*flaveurs moelleuses*).

Another of the striking changes in twentieth-century wine has been a decline in acidity. The decline in volatile acidity has been relatively recent, a post–World War II phenomenon. The two factors responsible are the greater maturity of the harvest and the general use of malolactic fermentation. The organoleptic properties of contemporary wine remain strong throughout its lifetime, without higher levels of residual sugars and malic acid, serious dangers of even the recent past. At the same time the level of sulfur has not risen much, while being more precisely controlled than it used to be. The oenologist prides himself on quality control through "the mastery of biological phenomena." It is a reasonable boast.

Although most of the changes in the elements analyzed in the 76 wines show the improvement that has taken place, the high – indeed, higher – level of carbon dioxide is a negative factor because its pricking or tingling effect on the tongue reinforces the tannic character of the wine and consequently reduces it suppleness. One of the factors responsible for this high level of carbon dioxide is probably early bottling, which avoids the old harshly tannic style of wine resulting from years in casks, with the attendant dangers to its health and quality. Pascal Ribéreau-Gayon emphasizes that the most distinguishing feature of the evolution of twentieth-century wine has been the reduction of its tannin content. In the period 1953 to 1962, the "tannic equilibrium" of the wines is the best of the five chronological groups (1906–31, 1934–42, 1943–52, 1953–62, 1963–72, see Table 6). The wines of the first three groups are "harsh and astringent" to us, but the wines of the last groups could support more tannin. *Grands crus* must be rich in the good-quality tannin that gives wine body without harshness. The drop in the level of tannin – and extract – has been accompanied by a reduction in the intensity of the color of the wine of the last

---

[33] Pascal Ribèreau-Gayon, "Bilan et perspectives de l'oenologie bordelaise des vins rouges," *Procès-verbal, Académie d'agriculture de France* (January 19, 1977), pp. 120–6.

Table 6. *Analyses of Different Vintages of Red Wines of Crus of the Médoc and the Graves (Analyses Done in January 1976)*

Number of Samples	Period		Degree	Total Acidity (g/l)	Volatile Acidity (g/l)	Total SO$_2$ (mg/l)	Malic Acid (g/l)	Sugar (g/l)	CO$_2$ (mg/l)	Tannin	Extract 100°	Alcohol Extract	Color (intensity)
12	1906 to 1931	Min.	10°0	3.72	0.53	27	0	2.0	110	48	19.2	2.6	0.59
		Max.	12°3	5.59	0.90	83	3.5	4.2	230	86	43.0	4.2	1.45
		Mean	*11°1*	*4.40*	*0.71*	*41*	*0.8*	*2.5*	*170*	*64*	*28.2*	*3.4*	*0.84*
11	1934 to 1942	Min.	9°9	3.92	0.46	27	0	2.0	115	40	20.2	3.0	0.46
		Max.	13°1	5.39	1.10	105	3.5	3.8	335	90	35.0	4.4	1.40
		Mean	*11°2*	*4.32*	*0.75*	*53*	*0.6*	*2.4*	*167*	*59*	*26.5*	*3.6*	*0.77*
17	1943 to 1952	Min.	10°3	3.43	0.45	28	0	2.0	85	44	20.0	2.7	0.35
		Max.	12°7	5.29	1.05	70	2.0	3.8	735	90	35.5	4.45	1.06
		Mean	*11°3*	*4.03*	*0.70*	*45*	*0.2*	*2.3*	*451*	*56*	*26.0*	*3.6*	*0.71*
18	1953 to 1962	Min.	11°1	3.19	0.45	29	0	2.0	95	32	20.0	3.5	0.41
		Max.	13°3	5.29	0.72	80	1.5	2.5	585	100	27.4	4.8	1.24
		Mean	*11°9*	*3.55*	*0.57*	*52*	*0.1*	*2.1*	*305*	*50*	*22.9*	*4.2*	*0.61*
18	1963 to 1972	Min.	11°8	3.09	0.37	51	0	2.0	75	29	19.6	3.7	0.36
		Max.	12°7	3.53	0.50	89	0	2.5	470	60	26.6	4.8	0.64
		Mean	*12°0*	*3.38*	*0.42*	*67*	*0*	*2.0*	*263*	*43*	*22.2*	*4.35*	*0.49*

Source: Reprinted with permission from Pascal Ribéreau-Gayon, "Bilan et perspectives de l'oenologie bordelaise des vins rouges," Académie d'agriculture de France, *Extrait du procès-verbal de la séance du 19 janvier 1977*, p. 6.

group. Viticultural practices result in more or less anthocyanins and tannin in the grape skins. Because vinification can only compensate so far for low levels of tannin in the fruit, botanical knowledge has become important even in the determination of the model of taste desired in the wine itself.

The Logical Corruption of Chaptal's Doctrine

In 1896 the vineyards of the Yonne covered 35,000 hectares in Basse-Bourgogne, including Tonnerre, Auxerre, Avallon, and Joigny. It is now less than one-tenth that size. Rousseaux would be saddened to learn that Chablis is now expensive, rarely great, often the victim of overproduction with excessive chaptalization.[34] There are exceptions, but science is powerless to prevent the prostitution of itself by the very forces of production that brought it into operation. Emergency procedures that produce more profit have a way of becoming routine. And why gamble with the weather when certainty can be achieved through cheap science? Even the northern Côtes-du-Rhône may now be legally chaptalized.

The virtues and vices of chaptalization are nearly the same for burgundy as they are for bordeaux, but the Burgundians receive much more criticism for it.[35] In France, where chaptalization is regarded as an esential part of winemaking, drinkers are less upset by the practice than Americans, especially in California, where the process is illegal. Since 1938 sugaring or chaptalization has been legal for bordeaux, up to the limit of producing two degrees of alcohol. (The fermentation of seventeen grams of sugar produces one degree of alcohol per liter.) Michel Dovaz notes that "All the *crus classés* of Bordeaux add sugar to the musts when necessary, except the great *millésimes* or vintages and the naturally sweet sauternes."[36] Bad examples seem to abound in Burgundy. One winegrower was unlucky enough to have had his Mâcon rouge 1986 found sufficiently chaptalized to raise the content of alcohol by 3.5 percent.[37] Limits are often exceeded by producers who believe that the "the road of excess leads to the palace of wisdom" – or, more likely, of profit.

---

[34] See the *Guide GaultMillau: Le Vin* (1989). Chablis is the most generous flowing white *cru* of Burgundy: 80,211 hl gushed from its vats in 1986, compared to about one-half of that amount for Pouilly Fuissé, its nearest competitor in quantity. Meursault produced 20,168 hl and Puligny Montrachet 11,876 hl of white wine in the same year. For a critical report on high-priced disappointments among these wines, see *Decanter* 16 (May 1991): 62–4. For more information, see *L'Atlas des vins de France* and William Fevre, *Le vrai chablis et les autres* (Colmar, 1978); on p. 27, Fevre notes that Chablis could produce 100,000 hl without the vines leaving the "coteaux kimméridgiens."

[35] See Kermit Lynch, *Adventures on the Wine Route* (New York, 1988), pp. 62–3.

[36] Michel Dovaz, *Encyclopédie des crus classés du Bordelais* (1961).

[37] *Guide Dussert-Gerber des vins de France 1988*, p. 16, citing *Le Canard enchaîné* (Oct. 14, 1987).

Conventional contemporary arguments for the chaptalization of wine fall into two categories: aesthetic and commercial. It is one of the few vices of capitalism that the two do not always perfectly coincide. First, taste. Dovaz takes a friendly attitude toward moderate sugaring, saying that one of its results is an intensification of the feeling of fullness or fatness and rotundity in the mouth – in short, a feeling of harmony. If you like these tastes, without the hint of angularity or the clash of piquancies characteristic of unchaptalized greatness, it seems that you should drink the wines of the *grandes maisons* – many of the *petites maisons* can please you too. However much the conflicting arguments may confuse the consumer, who is kept ignorant of the nature of wine manipulations by the patriarchs of production – and a few matriarchs – it is difficult to deny that wine is "altered by the technique of chaptalization, which completely upsets the original equilibrium of its components."[38] It may not be nice to fool around with Mother Nature, even if making her "soft, lush, silky, full, fleshy, rich, supple" and more alcoholic.

The nineteenth and twentieth centuries saw stupendous growth in the production of many items essential to our civilization: coal, iron, steel, chemicals, electricity, alcohol, food, tobacco, and sugar. In 1800 about a quarter of a million tons of sugar stimulated humanity's sweet tooth. By 1860, world sugar production, including the fast-growing beet sugar supply, had risen to about 1.4 million tons; and by 1890 production was more than six million tons, 500 percent more than in 1860. Sidney Mintz believes that "probably no other food in world history has had a comparable performance."[39]

The corruption of the French palate proceeded at a slower rate than that of many other Western countries. Annual per capita consumption of sugar in 1925 was 20 kg, in 1955 it was 26, and by 1980 it was 37 kg, probably about half of the consumption in Anglo-American societies.[40] The French took some of their sugar in fermented form. After the mid-nineteenth century a series of vine diseases created shortages in the production of wine alcohol, which was soon replaced by beet sugar alcohol. During the phylloxera crisis (late 1870s to the 1890s) enormous amounts of sugar were used to make artificial wine and to raise the alcohol content in wine after the first pressing. The peak year was 1899, when nearly 40 million kg of sugar were used.[41] Another change in metabolism and taste occurred as the French body slowly moved along the Western route to the

[38] Rolande Gadille, *Le vignoble de la côte bourguignonne. Fondements physiques et humains d'une viticulture de haute qualité* (Publications de l'Université de Dijon, vol. 39; Paris: Les Belles Lettres, 1967), p. 306.

[39] Sidney W. Mintz, *Sweetness and Power: The Place of Sugar in Modern History* (New York: Viking, 1985; Penguin, 1986), pp. 73 and 143.

[40] *GaultMillau Magazine*, no. 245 (October 1989), p. 34.

[41] Loubère, *The Red and the White*, p. 166, gives 1888 as the peak year.

sugar fix, as sugar and alcohol became new sources of energy for industrial and agricultural workers.

A closer intimacy seems to exist between language and the commercial justification for chaptalization. Made more robust by chaptalization, wine can stand up to the violence of transport and bad storage conditions often typical of commerce. Dovaz makes it clear that there is a trade-off in getting a solid, stable wine which, for example, can make it off the docks and still be in drinkable condition: "what the wine gains in solidity it loses in delicacy, even in equilibrium." The danger, as Dovaz admits, is that the consumption of these wines outside France will corrupt the taste of the consumer by habituating him to highly alcoholic wines, which will then lead the French producers of fine wines to ignore the quest for better quality while satisfying the tasteless consumer. Or, one can redefine quality in terms of the taste model of the more alcoholic (chaptalized) wine. This is the current oenological and market strategy. "Not only are the wines of Burgundy as good as they were in the past, they are significantly better."[42] The drinker and his guides accept the doctrine and the product. Producers see themselves as adhering to a popular consumer's model of taste, which it was convenient for them to have created.

Once scientists who were part of the agricultural establishment had endorsed chaptalization, as a "pratique bienfaisante" – Dumas's felicitous phrase, so useful in all food technology – governments were not slow in passing laws to regulate the process. Unfortunately for the consumer, the first laws were passed during the height of the phylloxera crisis and were therefore excessively tolerant to the process in order to help an industry in distress. It is difficult to contest Warner's conclusion that the Midi "indulged in . . . an orgy of fraud," producing over twice as much wine in 1902 as officially declared harvested. Northern winemakers were in the vanguard of the new industry. In 1906 the department of the Côte d'Or "used almost four times as much sugar as the four departments of the Midi."[43] A great deal of antifraud legislation, badly enforced, subsequently tried to undo the damage done to the production of real wine. The forces of the marketplace, especially bumper harvests, lower consumption, and a growing demand for quality wines, eventually achieved what legislation had failed to do, ensuring that generally the alcohol produced in wine from sugar added during fermentation is limited in amount and is there legally.

[42] Robert M. Parker, *Burgundy* (New York, 1990), p. 42.
[43] Charles K. Warner, *The Winegrowers of France and the Government since 1875* (New York, 1960; Westport, Conn., 1975), pp. 12–14, 234 (n. 13). Warner's book gives a detailed account of the sugar wars from the 1880s down to the Fourth Republic. See also Louis Ferré, *Traite d'oenologie bourguignonne* (Paris, 1958), pp. 53–64, and Loubère, *The Red and the White*, chap. 12.

The disasters of the second half of the nineteenth century encouraged scientists to be bolder, to try to restore the mythic old excellence of wines or even to claim that it remained to be created. It was not difficult for Pasteur and the Bordelais oenologist Ulysse Gayon to show that burgundy and bordeaux were not beyond improvement. Pasteur had to deal with more serious problems than Chaptal: curing disease rather than improving quality – although that unintended consequence was claimed as a result – and the related difficulty of selling burgundy that easily went bad in an increasingly quality-oriented market.

Pasteur thought it especially unfortunate that, given the economic importance of wine, scientific knowledge of it left much to be desired. Not even its composition was known. Pasteur himself, as he was not reluctant to repeat, had discovered two of its essential principles two years before the *Etudes sur le vin* appeared in 1866. Chemistry called itself modern, but the fact that the best scientific treatise on wine, Chaptal's *L'art de faire le vin,* had appeared more than a half-century before indicated that nearly everything scientific about wine remained unknown. Pasteur saw Chaptal as his oenological John the Baptist. His critical historical appreciation of the oenological work of his predecessors and contemporaries may be open to some question, but his confidence in his own research program seems reasonable. Few of his competitors could boast of having done five years of first-rate research on fermentation. Pasteur had a profound knowledge of the principles on which winemaking rested.

6

Louis Pasteur

The Founder of Scientific Oenology?

In a preface to the last reprinting of Pasteur's *Etudes sur le vin*, Maurice Vallery-Radot declared that this classic work had formed generations of oenologists.[1] Perhaps. But the work was reprinted only once, in 1924, after the second edition of 1873. Perhaps after having achieved paradigm status, Pasteur's book became the classic text of oenologists, at least after they invented themselves in the late nineteenth century. To see the greatness of Pasteur's oenology today is not easy without making a concerted effort to appreciate his revolutionary work in its nineteenth-century context. Of course, even old-fashioned worshipers of the contextless text can appreciate Pasteur's brilliant scientific papers. Much of his writing on wine is as worthy of admiration for its striking novelties, scientific style, cogent reasoning, and epistemological alertness as his great paper on lactic acid fermentation.[2]

In post-malolactic and increasingly post-Pasteurian oenology there may be some doubt that Pasteur deserves *ein Heldenleben*. The panacea of pasteurization for wine diseases perhaps came to be more practiced in Burgundy than in Bordeaux, but it was in Bordeaux that one of Pasteur's most famous students, Ulysse Gayon, founded the oenology of fine wines. And that oenology does not include heating, in spite of Gayon's production of masterpieces of experimental theater showing that oenophiles could not tell the difference between the raw and the cooked. In malolactic fermentation, a specialty of Bordelais oenology, live bacteria assumed a non-Pasteurian beneficial role. As oenology developed, it grew less Pasteurian; but it was rigorously Pasteurian for a couple of generations.

---

[1] Maurice Vallery-Radot, preface to M. L. Pasteur, *Etudes sur le vin. Ses Maladies et les causes qui les provoquent. Procédés nouveaux pour le conserver et pour le vieillir* (1866; 2d ed., revised and enlarged, 1873; reprint by Editions Jeanne Laffitte of Marseille, n.d.); Pasteur's studies had originally appeared in the *Comptes rendus de l'Académie des sciences* in 1863-4. The book was published by the Imprimerie impériale in 1866.

[2] *Mémoire sur la fermentation appelée lactique* (1857), the subject of admiration and semiotic analysis by Bruno Latour in "Pasteur on Lactic Acid Yeast: A Partial Semiotic Analysis," *Configurations* (1992), 1:129-45.

In their classic introduction to wine, Maynard A. Amerine and Vernon L. Singleton show the proper respect for the hero in history. A discipline needs a significant hero to establish its cultural importance and scientific respectability. "The most important contributions of the mid-nineteenth century to the wine industry we owe to Louis Pasteur . . . , [who] represents the application of the scientific revolution to the wine industry"[3] (alternative version: "Pasteur personifies the application of scientific principles to the wine industry"). Amerine's interpretation of Pasteur's oenological work is placed in the logical framework of his brilliant career in alcoholic fermentation. Getting poor yields of alcohol from fermenting sugar-beet molasses can be an unintended consequence of the process of fermentation: the conversion of dilute ethyl alcohol into acetic (ethanoic) acid by an aerobic microorganism of the *Acetobacter* type (organisms producing acetic acid). It turned out to be the same enemy responsible for widespread wine spoilage, caused by the microorganism's oxidization of ethanol. "The normal condition for alcoholic fermentation" being "the absence of air or oxygen," the theoretical prescription was simple: keep the developing wine free from oxygen. The notorious complexity of fermentation was unsuspected by the great trinity of Pasteur, Liebig, and Bernard, and from time to time still mystifies oenologists.

It is possible to doubt whether, as Pasteur claimed, his life in science was guided by an almost inflexible logic, seemingly centered on an obsession with optical activity, from research in crystallography to fermentation and then to disease, but he does seem to have always had a healthy interest in wine.[4] Some writers, addicted to the theory of environmental influence, believe that Pasteur's Jurassian roots gave him a native interest in wine. In the middle of the nineteenth century, the department of the Jura produced about half a million hl of wine annually, although it has since dropped steadily, until a century later production was just over 70,000 hl, and less than 35,000 hl in 1986. (The wine area of the Franche-Comté also includes the departments of Doubs and the Haute-Saône, where production has also declined drastically since the last century.) The wine of Arbois was the best known of the Jura wines; the unique *vin jaune* (13 to 14 degrees of alcohol) – that of Château-Chalon being better known than that of

---

[3] Maynard A. Amerine and Vernon L. Singleton, *Wine: An Introduction,* 2d rev. ed. (Berkeley and Los Angeles, 1976). See also Maynard A. Amerine, "La pasteurisation," in R. Protin, ed., *Apport de Pasteur à l'oenologie moderne* (1968), pp. 22–33. For an economical statement of Pasteur's work in oenology, see Annick Perrot, "Les découvertes de Pasteur," in *La vigne et le vin* (1988), pp. 172–3. Any emphasis on the practical interest of Pasteur's research should not obscure its basic theoretical inspiration. For a defense of the theoretical inspiration of Pasteur's research on fermentation, see Gerald L. Geison, *The Private Science of Louis Pasteur* (Princeton, 1995), chap. 4.

[4] Gerald L. Geison, "Pasteur," *DSB* (*Dictionary of Scientific Biography*), 10:352, with references to René Dubos, *Louis Pasteur: Free Lance of Science* (New York, 1950).

Arbois – and the *vin de paille* (minimum of 18 degrees) give the area a curious distinction and eclipse its other wines.[5]

On the fiftieth anniversary of Pasteur's death, Paul Marsais noted that Pasteur was interested in everything connected with vine and wine: regions, vines, cultivation, varieties of grapes, harvesting, yeasts, fermentation, different methods of vinification, bottling, aging, preservation, and tasting. In dealing with yeasts, fermentation, aging, and preservation, Pasteur laid the foundations of modern oenology – or at least one might be tempted to write this if we still believed in heroes in history and the doctrine of discovery by individual geniuses. Partly for practical reasons, Pasteur was especially interested in analyzing the wines of the east of France: "He had at Arbois, fortunately, some old comrades of his childhood who owned some caves well stocked for home and market purposes, and he easily obtained permission to subject their wines to a microscopic study."[6] In days of small budgets, cheap and even free laboratory supplies were welcomed.

Pasteur possessed a considerable ampelographical baggage as a result of his special interest in the varieties of vines in the Jura: the poulsard, the tressol, the enfarine, and the savagnin, which he compared to the pinot of Burgundy. The poulsard still produces the wines of Bugey and the savagnin blanc rules Château-Chalon, "grand terroir des vins jaunes." Less well known is Pasteur's knowledge of American vines. At the invitation of Maxime Cornu and Léopold Laliman, he examined the wines of Bordeaux made from the first American vines brought to France after oidium had invaded the vineyards. Bouchardat carried out a tasting of the wines in Pasteur's laboratory at the Ecole normale supérieure. Out of the wines from the Jacquez (Lenoir), Clinton, Cunningham, Herbemont, Delaware, and Isabella vines (all used in the post-phylloxera reconstruction of vineyards), only the Clinton did not have the disagreeable taste of the other non-vinifera vines and its wine was even superior to Bordeaux's ordinary wines of the *palus*. Pasteur sent a letter to Dumas, president of the commission on phylloxera, giving the results of the tasting. It is clear that this laboratory knowledge did not trickle down easily to the winegrowers. Even if they *had* known, there is no assurance that the peasants would have acted on the knowledge, for peasants, untouched by bourgeois models of taste, often only planted vines to have cheap wine for their household consumption and sold it to people with similar tastes.

[5] "Le vignoble franc-comtois . . . doté d'une forte personnalité, à maintes égards unique en France. Trois cépages spécifiques y prospèrent: le trousseau (qui donne des vins pourpres et corsés), le poulsard (dont on tire de délicats rosés) et le savagnin ou naturé (cépage blanc)" for the *vin de paille* and the *vin jaune*. Jean Sellier et Fernand Woutaz, *L'Atlas des vins de France* (Paris, 1988), pp. 146–7.

[6] Emile Duclaux, *Pasteur: The History of a Mind,* trans. Erwin F. Smith and Florence Hedges (Philadelphia, 1920), p. 135.

As dean at Lille and later as administrator at the Normale, Pasteur promoted the consumption of the wines of the Jura, an act of considerable oenophilic originality and a useful boost to hometown commerce. Jean Ribéreau-Gayon pointed out that Pasteur's interest had a strictly scientific aspect, for he frequently used wine must in studying alcoholic fermentation to verify the results of his research. Two happy discoveries about wine came from this practice: finding the presence and gustatory importance of succinic acid (an organic acid – $C_4H_6O_4$) and the presence of the trihydric alcohol glycerin or glycerol (0.2 to 1.5 percent of wine). Of more practical importance were a number of Pasteurian principles dealing with the relation between air and life, or oxygen and yeast development. And here Pasteur returned to his key research axis, disease, the pathological as a form of life.

Basic natural processes are notoriously difficult to understand and controversial to talk about in a communal scientific language. It may be even more difficult to use that knowledge to modify traditional practices. In the middle of the nineteenth century something so basic to human existence as fermentation was still a subject of angry debate between big scientists with bigger egos. And when Pasteur's explanation triumphed, it turned out to be only half an explanation, the other half being more complex than anyone could have imagined even at the turn of the century, when the Buchner brothers accidentally came up with a very general statement of the other half of the explanation. Pasteur's power of argument, skill in experiment and demonstration, and expertise in convincing various types of public of the truth of his science make him a good candidate for the Mephistopheles of nineteenth-century science. What was this theory that Pasteur convinced people to accept, at least for a few decades in the second half of the century?

The Debate over Fermentation

Lavoisier's experiments on vinous fermentation led him to write, in his famous *Elements of Chemistry* (1789), of this phenomenon as "one of the most striking and extraordinary of all those that chemistry presents to us." He reduced the process to a simple formula: in the must of grapes the sugar simply breaks up into equal weights and amounts of carbon dioxide and alcohol. In 1810 Gay-Lussac wrote an equation to show that, in the fermentation of one molecule of a simple sugar (a monosaccharide) like glucose (grape sugar), two molecules of ethyl alcohol (ethanol) and two molecules of carbon dioxide were produced. A bit of juggling enabled Gay Lussac to ignore minor discrepancies of yield resulting from some of the sugar's atoms being diverted to other products and uses. Fermentation would be dull stuff if it simply followed Gay Lussac's equation:

$$C_6H_{12}O_6 \rightarrow 2CO_2 + 2C_2H_6O$$

glucose or fructose   ethyl alcohol   carbon dioxide gas

In the interests of simplicity of experiment and calculation, scientists ignored yeast, the agent responsible for fermentation.

Liebig's fertile imagination turned the chemical theory of fermentation into a widely accepted dogma expressed as a subtle piece of chemical metaphysics in which yeast and substances undergoing putrefaction communicated their state of decomposition to other substances. In this scheme, yeast was irrelevant to the process except as a dying substance communicating its state of decomposition; fermentation was not "a vital physiological act of the living yeast cell."[7] An embryonic vitalistic or germ theory of fermentation did exist, a sort of Cinderella to the chemical theory. Charles Cagniard-Latour, Friedrich Kützing, and Theodor Schwann gave living yeast cells the credit for fermentation. The guardians of the chemical faith kept the biological idea among the cinders by the use of power and ridicule, conventional arms of science. The germ or yeast theory of fermentation, according to Wöhler and Liebig, revealed that alembic-shaped animals hatch from eggs in a liquid, eat the sugar, and excrete alcohol through the anus and carbonic acid from a bladder (shaped like a champagne bottle) through the urinary organ.

The first serious blow against chemical exclusivity came from an unexpected source, Pasteur's *Mémoire sur la fermentation lactique* (1857/8), which showed, with Pasteur's typical experimental brilliance, the presence of a special ferment, "lactic yeast," whenever sugar – lactose in milk – becomes lactic acid. Lactic fermentation is responsible for the natural delight of sour milk. Pasteur assumed that the lactic ferment was living. It may be that he was initially drawn into research on fermentation because of his interest in the molecular structure of the products of industrial fermentation – the optical activity of amyl alcohol, for example. His association of asymmetry and optical activity with life may also have "brought him to the view that fermentation depends on the activity of living microorganisms."[8] He went on to study an impressive range of fermentations – tartaric, butyric, acetic, and alcoholic. His memoir on alcoholic fermentation (1860) quickly transformed the biological theory of fermentation from Cinderella to princess, who then forced the chemical theory to live among the cinders. More bad news for the Lavoisier–Gay Lussac–Liebig gang was that fermentation produced more than just carbon dioxide and ethyl alcohol: glycerol and succinic acid, along with trace amounts of other "indeterminate products." Without knowing how ferments did all

---

[7] Robert Kohler, "The Background to Eduard Buchner's Discovery of Cell-Free Fermentation," *Journal of the History of Biology* 4 (1971): 35. For a superb older treatment of "Lactic and Alcoholic Fermentations," see Duclaux, pp. 51–84.

[8] Geison, "Pasteur," *DSB,* p. 361.

this, Pasteur established a line of research that would eventually make it clear how complex wine really is.

In a famous but misleading experiment, Gay Lussac had demonstrated to his own satisfaction, and that of many people who ought to have redone his experiment a few times, that fermentation is started when the must receives a shot of oxygen or air. No air, no fermentation. Then Pasteur discovered that the butyric ferment dies when exposed to air. (Butyric or butanoic acid has a rancid odor; the esters of this weak acid – one finds them in butter and sweat – are useful in making flavorings and perfumes.) Experimenting with other ferments, Pasteur discovered that although yeasts certainly grow in air, they are good at fermenting sugar only in the absence of air or oxygen, functioning anaerobically; so a better definition of fermentation was "life without air."[9] This is the situation when grape juice is fermenting in wine vats: air cannot easily penetrate to the deep liquid layer, while the large amount of carbon dioxide produced by the fermentation provides a protective cloud against oxygen in the air.

But the definition of fermentation as life without air derives from only part of the process going on as yeast cells deprived of oxygen find their energy in transforming sugar. A prolonged fermentation to produce good wine needs the formation of new generations of yeast cells in order for the cells to continue synthesizing the sterols (complex cyclic alcohols) that are an essential part of wine. That process requires traces of oxygen. Yeast ferments sugar to get the energy it needs for the chemical task of synthesizing substances for more yeast cells, which reproduce efficiently by asexual budding. Alcohol and carbon dioxide are by-products of the process. "Yeasts must metabolize more than twelve times as much sugar to produce the same amount of cellular growth when growing anaerobically as they would if oxygen were available to them." So "during alcoholic fermentation the yeasts must 'process' a great deal of the sugar to alcohol with relatively little cell multiplication and therefore relatively little consumption of the sugar and other nutrients to build cells. . . . The fact that the addition of oxygen to fermenting yeasts inhibits the conversion of glucose to ethanol, and gives more cells per unit of glucose consumed, was observed by Pasteur and is termed the Pasteur effect."[10] Nobody knew how to explain it, or what the enzymic mechanism was, until the biochemistry of fermentation was established. By 1940 a dozen enzymic components had been identified in the simplistic explanation put forward in 1897 as Buchner's zymase.[11] A frustrating and understandable ignorance of this

[9] Ibid., p. 364.

[10] Amerine and Singleton, *Wine,* pp. 68–9.

[11] The reader who is looking for a good historical account of the scientific voyage "From Ferments to Enzymes" and a clear account of "Alcoholic Fermentation as Oxidation-

whole amazing process encouraged scientists to take refuge in exclusion-ary dogmatic schemes of explanations, whether it was Berthelot with his modified chemical theory of fermentation or Pasteur with his physiologi-cal theory.

The role of oxygen has been the subject of considerable controversy in the attempts of scientists to explain the complex process of the aging of wine. As Peynaud points out, the problem was misunderstood from the beginning because of a basic confusion, reflected in the opposing views that "oxygen makes the wine" and "oxygen is the enemy of wine." No Bohr appeared in oenology with a theory of complementarity to explain that a single model might not be adequate to explain the observed phe-nomena. The fact is that the assertions apply to different types of wine and their different modes of aging. Aging by oxidation produces certain types of wine, generally made in hot climates: port, oloroso sherry, Ma-deira, and even some naturally sweet French wines, fortified with alcohol and exposed to long contact with air. But most wine ages without much access to oxygen, and fine wines are carefully protected from exposure to air. Pasteur contributed to the nineteenth-century conflation of categories. As in the case of the complex relations between yeast and oxygen, the difficult details of scientific explanation of the maturing and aging of wine were not revealed before much of the twentieth century had passed.[12]

Pasteur relentlessly pursued working out the life–death cycle as a chemi-cal decomposition carried out by two types of microorganisms: the anaer-obes, which indulge in a putrefactive feast of nitrogenous substances in liquids, a fermentation that produces simpler if still complex substances; and the aerobes, which work at the surface of a liquid to produce the "simple binary combinations" like water and carbonic acid. Many books tell us that one of Pasteur's striking successes in the early 1860s was to show that vinegar, a dilute solution of acetic acid, is the creation of the microorganism *Mycoderma aceti,* the "mother of vinegar" that operates on the surface of the liquid, wine in the case of the method used in the vinegar capital of France, Orléans. True, Pasteur gave this fancy Latin name to acetic acid bacteria or vinegar ferments, but in spite of its place of honor in many texts for a long time, it is, as Peynaud points out, incor-rect: the bacteria belong to the genera *Gluconobacter* and *Acetobacter.*[13]

The nonbiological theory, in which "acetic fermentation was widely viewed as a chemical, catalytic process . . . was more in accord with the German method of manufacturing vinegar." A "fermenting medium con-

Reduction" should read Joseph S. Fruton, *Molecules and Life: Historical Essays on the Interplay of Chemistry and Biology* (New York, 1972).

[12] Emile Peynaud, *Connaissance et travail du vin,* 2d ed. (Paris, 1981), pp. 228–35.

[13] Ibid., pp. 236–9.

sisted of a dilute alcohol solution, a trace of acetic acid, and . . . sharp wine or acid beer" – unstable organic matter, which, according to Liebig, began fermentation with beechwood shavings in casks facilitating the process. Pasteur found that *Mycoderma aceti* was responsible for acetic fermentation in the German as well as French processes. The inconvenience of his own theory that fermentation is "life without air" was temporarily shelved. The successful scientist cannot hope for total coherence and clarity, at least at the beginning of important novelties. Hence the importance of style. And ambiguity is a necessary ingredient of style.[14] Nevertheless, Pasteur's research on the manufacture of acetic acid or vinegar by molds had important industrial applications. He took out a patent on the process.[15]

When Liebig finally got around to criticizing Pasteur, it was too late to stop the growing acceptance of the Pasteurian explanation of alcoholic and acetic fermentations. Liebig died in 1873 without accepting Pasteur's typical challenge to settle the issues by public debate. The chemical theory of fermentation, born in France in the late eighteenth century, was buried with Liebig in Germany. The biological theory, largely German in origin, triumphed in France with Pasteur. Continuing debates in France forced Pasteur to clarify and modify his position, especially by recognizing at least an initial role for oxygen in starting fermentation and the dual nature of yeast as an aerobic and anaerobic organism. Pasteur eliminated from his theory any fermentations not conforming to a model described as a proper fermentation ("fermentations proprement dites") which meant that his theory was, as Gerry Geison says, a virtual tautology, but obviously the type of circularity that leads to fruitful research.[16] Bread, wine, urine, sour milk were proper fermentations, the work of a living organism; processes like diastatic fermentation, in which starch is converted to sugar, were not. Had Pasteur been a Scot, whose whisky depends on diastase (the enzyme that converts the starch of malt to sugar), his research might have taken a different direction.

The debate between Liebig and Pasteur fizzled into a reluctant admission by both that neither the biological nor the chemical theory of fermentation provided a total explanation of the process. It was already known that yeast possesses the soluble ferment we call invertase, an enzyme that converts sucrose into simple sugars. The chemical theory had a powerful

[14] "Avec des certitudes, point de style," E. M. Cioran, *Syllogismes de l'amertume* (Paris, 1952), p. 11.
[15] Geison, "Pasteur," *DSB*, pp. 365–6.
[16] For a destructive analysis of this idea, see Denis Temple, "Pasteur's Theory of Fermentation: A 'Virtual Tautology'?" *Studies in the History and Philosophy of Science* 17 (1986): 119ff. Bruno Latour, "Pasteur on Lactic Acid Yeast: A Partial Semiotic Analysis," *Confrontations* (1992), 1:140–1, gives the basic bibliography on the squabble over Pasteur's possible circularity of argument.

supporter in France, Marcelin Berthelot, who had isolated the *ferment/ glycosique* (invertase) from yeast. A sharp clash of these two great men of science occurred in 1878, when Berthelot got in a surprise blow to Pasteur by publishing the results of some secret experiments that Claude Bernard had done in the Beaujolais before he died. A secret critic of Pasteur's theory, Bernard claimed to have isolated a soluble ferment capable of producing, without live yeast, an alcoholic fermentation in grape juice. Pasteur refuted Bernard in a blaze of scientific rhetoric.[17] Embarrassed scientists say both men were right, and delighted historians add that both were also wrong.

Scientists came and went but, as Pasteur recognized, the issue of soluble ferments remained. Hoping that time would vindicate his theory, he had to move on to other conquests, the solutions to problems that looked solvable and which the state wanted solved. Pasteur himself tried to find this phantom soluble alcoholic ferment, given the absurd name of alcoholase, in yeast cells. Why did he fail? One might suggest an unconscious resistance to ruining his own theory of fermentation, even if it is the living cell that makes the enzyme. It may not be irrelevant that the spokesman for the chemical theory was Berthelot, whose support for any theory might be enough to arouse Pasteur's hostility. Alexander Lebedev gave a more plausible explanation. The yeast used by Pasteur was useless for this type of experiment. Yeast used to be terribly fickle, varying in potential from day to day even when it was from the same brewery. But Pasteur's experiment was doomed for another reason: the yeast called "Parisienne" did not produce an activating liquid when pressed in a mechanical fashion, or even when macerated in Lebedev's simpler procedure. The Munich yeast used in Buchner's experiments nearly always gave an activating extract. "The isolation of zymase can serve as a classic example of the role sometimes played by chance in scientific discoveries."[18] This was a case where chance did not favor the prepared mind.

Research on Yeasts: Pasteur and After

Pasteur's work on wine yeasts was the beginning of an important tradition in oenological research. In the famous hothouse experiment of 1878 in Arbois, he refuted the Bernardian idea of an internal origin of the yeast capable of fermenting the must. Bernardians can now take comfort in the

---

[17] See Mirko D. Grmek, "Louis Pasteur, Claude Bernard et la méthode expérimentale," pp. 22–5, and Jean-Paul Gaudillière, "Catalyse enzymatique et oxydations cellulaires – l'oeuvre de Gabriel Bertrand et son héritage," pp. 118–19, in *L'Institut Pasteur – Contributions à son histoire,* ed. Michel Morange (Paris, 1991), pp. 22–5.

[18] Alexandre Lebedeff, "Extraction de la zymase par simple macération," *Annales de l'Institut Pasteur (AIP)* 26 (1912): 8–37. See Jacques Duclaux, *La chimie de la matière vivante* (Paris, 1910), p. 82, and the opening pages of chap. 6.

fact that the biochemical process called the Krebs cycle (a series of enzymic reactions basic to the metabolism of aerobic organisms) also occurs in grapes and produces most of the acids found in wine. Pasteur found that there are a great many different alcoholic yeasts, which give rise to different flavors and other qualities in fermented liquids. Ripe grapes have enough of these yeasts to produce fermentation in normal circumstances. So, unlike in the case of brewing beer, the must does not need to be inoculated.

Nevertheless, adding yeast is a general practice in warmer wine-producing areas like Australia, California, and South Africa, and in France. Indeed, if one were to judge on the basis of professional oenological literature and the publicity for yeasts, the practice would be universal, and it is certainly widespread. Perhaps some European resistance to the idea of "routine inoculation with pure cultures of selected yeasts" is partly based on the belief in the unique environment of each vineyard, or at least area (*terroir*), with its "complex flora of microorganisms." The Californian (Davis) oenological gospel of Amerine and Singleton recognizes these small but significant differences but despairs of the difficulty of controlling their actions, which "often tend to be undesirable."[19] The difficulty, according to Peynaud's Bordelais oenological gospel, is that selected pure yeasts are not really *selected*. "They are cultivated yeasts, isolated locally, and named after their origin, which is no guarantee of their value." The best use of added yeast would be in cases where the wild yeasts had been eliminated, something very difficult to do in making red wine, although quite feasible in the case of the more homogeneous and manipulated must for white wine.[20]

The dream of the late nineteenth century that wine could be made in a scientifically controlled fashion similar to brewing beer is not attainable or desirable, at least for good red wine. In 1883 Emil Christian Hansen succeeded in producing bottom-fermented beer on a large scale for the first time, using "the so-called pure culture yeast."[21] This made it possible to prevent beer from being spoiled by the action of wild yeasts in the vat. After Hansen's revolution, which led to the mastery of the industrial fermentation of beer, research on beer yeasts was less interesting scientifically than research on wine yeasts. Emile Duclaux, the heir to much of Pasteur's scientific empire, claimed that his laboratory first drew attention to the role of yeasts in producing different wines. The isolation of a yeast of the wine of the Champagne led to an elaboration of its distinctive characteristics. Duclaux had shown how to solve the old problem of finding

---

[19] Amerine and Singleton, *Wine,* pp. 54–5.
[20] Peynaud, *Connaissance et travail du vin,* pp. 93–4.
[21] Mikulás Teich, "Fermentation Theory and Practice: The Beginnings of Pure Yeast Cultivation and English Brewing, 1883–1913," *History of Technology* 8 (1893): 117–33.

"for each variety of grape or each must the yeast that makes the best wine."[22] This eventually led to the industrial production of yeasts that could be advertised with intimations of greatness in their names.

The bouquet of good wine has long been a subject of speculation and inquiry by science and industry. Around 1900 serious attempts were made to test the controversial opinion that the bouquet and taste of wine come from the action of yeasts upon the grapes. If this proposition were true, it could be argued that the yeasts of different regions contribute to the diversification of the tastes and bouquets of wines. The use of so-called selected yeasts became an increasingly important topic among winemakers after Pasteur did research on their key importance in the fermentation of wine. Pasteur's aesthetic judgment still carried weight in the 1890s: "the taste, the qualities of wine, depend certainly, for a great part, on the specific nature of the yeasts that develop during the fermentation of the grape crop."

But results were still less sure and less happy than in the brewing of beer, where a certain simplicity made possible in the creation of taste a precision unattainable in winemaking.[23] The many substances existing in feeble proportions in wine make it difficult to predict its taste. No yeast can turn a Neuchâtel into an intriguing version of a Chablis *grand cru.* Unlike some historians and sociologists of science, who produce only prose, the winemaker and their oenologists had to make a clear distinction between science and magic, at least in the vat. The balance of factors producing a great wine could not be duplicated by a manipulation of its main components.

The poor results obtained in numerous experiments done on the use of pure yeasts in vinification probably explain why this became a neglected area of research in French oenology until the 1950s.[24] The cultural prejudice in favor of the natural model of wine might also have been a factor. It was finally recognized that the experimental failures resulted from the fact that the natural yeasts on the grape simply overwhelm the pure yeasts introduced into the vat by the optimistic vintner. Even when the pure yeasts were activated by introducing them first into a small quantity of the must – the method called "pieds de cuve" – the results could be even worse

---

[22] On Emil Christian Hansen (1842–1909), see the article by E. Snorrason in *DSB,* 6:99–101; on Duclaux's isolation of the wine yeast of the Champagne, see his *Traité de microbiologie* 3:442–8.

[23] "L'emploi de levures pures en viticulture a donné des résultats souvent fort contradictoires; tantôt on a constaté une amélioration, tantôt l'effet a été nul." M. E. Kayser, "Rapport sur les expériences de vinification en 1892," *Bulletin du ministère de l'agriculture* 13 (1894): 67–79.

[24] Yeast has risen from a lowly status in research to become "biology's favorite tool," at least in genetic research on cancer; see Natalie Angier, *Natural Obsessions: Striving to Unlock the Deepest Secrets of the Cancer Cell* (New York, 1988), pp. 174, 271.

than leaving the must to ferment with its own yeasts. Why? Perhaps, it was suggested, there is an antagonism between the two groups of yeasts; so pure yeasts would work well only if the natural or wild yeasts in must and vat had been killed by sterilization. Duclaux noted that there was the risk that heat sterilization of the crop before adding the "artificial yeasts" risked ruining the organoleptic properties of the wine, but, as Kayser and Barba showed, it could be done. Indeed, Duclaux accepted the view of many tasters that the process gave a better wine than the control wine in the experiment: it was more colored, higher in alcohol, and disease-free.

The caprices of yeast being of considerable interest to the general scientific community, the latest research, especially that of commercial interest, was quickly reported in the scientific press. In 1897, W. Kühn's experiments were widely discussed.[25] At harvesttime he concocted three musts crushed from aramon and terret bouschet grapes, used for mass production of wine in the Midi. One must fermented on its own, producing ordinary white Midi wine. Another had pure yeast added but was not noticeably improved. The third must was subjected to the rigors of the Kühn sterilizer and inoculated with pure yeasts of the Champagne. The rumor spread – meaning that numerous tasters gave their weighty opinions – that the third batch of wine inoculated with the yeast of the Champagne had been "absolutely metamorphosized and was clearly the type of white wines of the Champenoise region." This result was hailed as being of the highest economic importance for areas like Algeria and Tunisia, noted for their uncompleted fermentations caused by the hot weather and often defective yeasts on the grapes. The new technology would allow sterilized musts to be stored until the temperature fell to a point necessary for a good fermentation, which could not be started by inoculation with pure yeasts. Vinification would be a scientific industry like brewing.

Although the French Empire did not start competing with the great champagne houses, the practice of adding yeast cultures is now regarded as a sound one for countries with hot climates. The Davis gospel puts it optimistically: "wines of the highest quality are produced (and more consistently so) in countries where the most modern and scientific technology is used, usually including routine inoculations with pure cultures of selected strains of wine yeasts."[26]

Experimenting is like writing poetry: one imitates others in order to be original, as François Coppée put it. In 1900, A. Hébert reported on continuing experiments on yeasts as if they were the latest novelty.[27] A.

---

[25] X. Rocques (former head chemist with the municipal laboratory of Paris), "Vinification rationnelle par l'emploi des levures pures, après stérilisation des moûts de raisin," *Revue générale des sciences pures et apppliquées* 8 (1897): 87–8.

[26] Amerine and Singleton, *Wine*, p. 55.

[27] M. A. Rosenstiehl, "Sur le bouquet des vins obtenus par la fermentation des moûts de raisin stérile," reported by A. Hébert in *Annales agronomiques,* vol. 28 (1902).

Rosenstiehl tried to establish experimentally whether yeasts really determine the tastes of wine. His research was a logical continuation of his earlier work showing how to obtain sterile musts. During the harvests of 1901–2, Rosenstiehl put yeasts of very different origins in several portions of the same grape must. At the same time, in another set of experiments, he used one single yeast on the musts of very different varieties of grapes. Tastings were then done by "experts of indisputable competence." The procedure was a judicious if necessary marriage of science and art, like wine itself. Most of the yeasts not only made the grape sugar ferment but also acted on an unisolated substance, present only in the classic varieties of grape (*cépages nobles*). This mysterious substance existed in grapes ripened in both good and bad locations. It also had a certain resistance to mold. The conclusion of this piece of science was that what characterized the regions of the *grands crus* was less the quality of the grapes than the yeast that grew spontaneously on them. Yeasts were recognized again as being site-specific. This was a pronouncement on the uniqueness of microclimates, including the soil: the magic of the *terroir.*

One of the anomalies of French oenological research was that, in spite of the large amount of work done on wine yeasts in the first half of the twentieth century, the French contribution was comparatively weak. In Italy a group of systematic studies of yeasts was done for the whole country. Beginning in the 1930s, oenologists, especially T. Castelli and his collaborators, studied Italian microclimates based on the two factors of latitude and height above sea level. At the Station oenologique de Bordeaux, Suzanne Lafourcade and Madeleine Lafon did a thorough study of various properties of wine yeast, including their metabolism. It was in the 1950s that Simone Domercq did a classification of the wine yeasts of the Gironde.[28] In five years Domercq left little uncovered about 28 different species of yeast, including their capabilities and behavior in the fermentation of red and white wines. Scientific facts of great importance about this tiny ubiquitous organism get discovered by seemingly obscure research.

The alcohol-making power of yeasts varies greatly from one species to another: *Kloeckera apiculata* makes three to six degrees of alcohol, compared with eight to seventeen degrees produced by *Saccharomyces ellipsoideus.* Generally, the yeasts of the Libournais make more alcohol than those of the Médoc. *Botrytis cinerea,* of Sauternes fame, secretes an "anti-

---

[28] Simone Domercq, *Etude et classification des levures de vin de la Gironde,* a thesis for the degree of Ingénieur-Docteur, faculty of sciences, University of Bordeaux, 1956. A survey of French work on yeasts is given by Alexandre Guilliermond, *Les levures* (1912), with a preface by E. Roux. The bibliography is heavily German. Among the people who did research on yeasts were G. Jacquemin, Louis Marx, E. Dubourg, A. Fernbach, and especially Edmond Kayser. Kayser was the head of the Laboratoire des fermentations at the Institut national agronomique. Kayser collaborated with E. Manceau, head of the Station oenologique Moët et Chandon, in a work on *Les ferments de la graisse des vins* (Epernay, 1909).

yeast antibiotic." The alcohol-making power of the yeasts of the white wine regions seems higher than that of red wine regions. Different yeasts operate at different times in the must and wine, with some, especially *Sacch. oviformis,* surviving even in old wines and presenting the danger of refermentation; its presence is desirable in wines that should have no sugar but must be eliminated in the wines requiring a residual sugar. Pasteur's view is still valuable: that the more the winemaker knows about yeasts, the better the wine he can make.

An Anomaly in Pasteur's Theory

One of the wines that Pasteur found fascinating was the *vin jaune* of the Jura. It is made only from late-harvested savagnin grapes, which frequently suffer from a mild infection of the noble rot *Botrytis cinerea* (the unique mold that penetrates the grape without breaking the skin and reduces its water content, thus increasing the proportion of sugar in the grape). In the cask, this wine develops a film of yeast on its surface, something that would be death to ordinary wines. The yeast (*Candida mycoderma*) is a heavy breather that loves, not sugar, but organic acids and alcohol, which it oxidizes into acetaldehyde.

Pasteur had to explain this anomaly in his theory. The resulting masterpiece of scientific scholasticism was not entirely wrong and, given the state of knowledge on a very difficult subject, not a bad explanation by any means. Pasteur compared the film of vegetation on the surface of the wine to his blanket designation of *Mycoderma vini,* which in some cases kept the bad company of the ill-named *Mycoderma aceti,* although in a normally evolving wine, the *aceti* fungus gave way to the *vini* fungus. (The Pasteurian terminology *Mycoderma vini,* still beloved of writers on Pasteur, refers to yeasts belonging chiefly to the species *Candida mycoderma,* although other species belonging to the genera *Pichia, Hansenula,* and *Brettanomyces* can also vegetate on the surface of wine.)

Another problem for Pasteur was to explain how this Jurassian wine, kept in casks for six to ten years without ullage, was protected from the ravages of oxidation? First, Pasteur used what we can now call the type of tautological explanation popular in scientific circles since antiquity. What takes place favors a limited oxidation of the wine giving it its special qualities. Air yes, but not too much, that is, not enough to destroy Pasteur's general theory about oxygen as the aging factor in wine: the aerobic growth of yeast formed a protective screen for the elements in the wine that would otherwise be destroyed by oxygen in the air. Obviously a subject that should have kept researchers busy for a while, but it didn't, probably because these fascinating wines of the Jura had little commercial importance in the French economy and far less significance than Sauternes in the gastronomic weltanschauung.

We can be amused by the irony that wines from Pasteur's own *pays* knocked a big hole in his oenology, but more important is the fact that the fine sherry wines of western Andalusia develop in a way somewhat similar to the *vin jaune* of the Jura. A French researcher might have politely concluded in 1941 that it seemed opportune to expand, if not revise, French oenological conclusions on the yeasts that sometimes form surface films on wines.[29] The Spanish oenologist Juan Marcilla was less reticent in stating the significance of the *flor* or *fleur* or flower film of yeast. A variety of *Sacch. ellipsoideus,* it can live in a wine containing a higher degree of alcohol than is possible for the ordinary *ellipsoideus.* The whole process, according to Marcilla, differs strongly from the rules of classical oenology. Those rules had been derived from an oenological science excessively general in its facts and norms, a science too inclined to doze in the sun of the discoveries of Pasteur. Marcilla, well known for his work on the wines of Montilla and Jerez, isolated the special yeast *Sacch. Beticus* – the Roman name for southern Spain was *Baetica* – responsible for the production of the relatively high level of aldehydes in *flor* sherry. The yeast is now reclassified as *Sacch. fermentati.*

Flor likes to grow mainly in Spain and the Jura. Even if it can be enticed to grow – reluctantly – on the surface of wines in other countries, it does not produce anything comparable to Spanish sherry.[30] Although the oenologist William Cruess brought *flor* yeast back to California, a great deal of that state's sherry is produced as a baked fortified wine, more like Madeira. The drinker who wants organically produced aldehydes in his sherry had better ask his wine merchant if it was properly infected by the *flor* created by *Sacch. fermentati.*

The search for the mechanism by which yeast transforms sugar into alcohol came to an end different from what Pasteur had hoped for, but his dual interest in fermentation and disease as related processes was a popular idea that continued in the biochemical dispensation. *Zymase* is a word known only to biochemists and avid scrabble players. Antoine Béchamp, a longtime critic of Pasteur's work, used the word in 1864 for what came to be called *invertase.* Both Béchamp and zymase dropped out of sight; the word was revived by Eduard Buchner. In biochemistry the extraction of zymase was an exciting project at the turn of the century, when scientists knew about two dozen soluble ferments, lumped under the rubric *diastase* by the French. The important technical achievement of the Buchner brothers in extracting the fermentation-causing cell juice from macerated yeast "has also been long recognized as one of the most important

---

[29] See Pierre Laurent (chef de travaux à l'Institut national agronomique de Paris), "Les vins jaunes du Jura et les levures en voile," *BIV* 145 (1941): 59–62.

[30] Julian Jeffs, *Sherry,* 4th ed. (London, 1992), p. 183. Jeffs points out that there are four *flor* yeasts, "all of the genus *Saccharomyces: S. beticus, S. montuliensis, S. cheresiensis,* and *S. rouxii.* The first is the strongest and the most important, especially on the younger wines."

foundations of the new science of biochemistry." That science's "central dogma" was "the belief that all the physiological functions of the living cell would turn out to be mediated by enzymes." Kohler points out that "For nearly fifty years the germ theory of disease and the cell theory of fermentation ran a parallel course." The theory of cell-free fermentation went hand in hand with Hans Buchner's new chemical physiology of disease infection.[31] Understanding how the normal and the pathological differed, how disease and health were related – in a way the process of life–death – drove research in both areas along their typically parallel paths, which sometimes converged in research as well as in the scientific rhetoric itself.

Pasteur not only reestablished the germ theory of fermentation on a solid experimental basis but went on to apply this theory to wine diseases, relating different diseases to specific organisms. He saw disease everywhere, and no crevice or discharge was exempt from scrutiny. During the debates over puerperal fever, Pasteur was the first person to examine the *lochia* of women in labor to show that those of healthy women are germ-free while those of sick women are heavily infected.[32] Because the nineteenth century was, like most other centuries, a century of disease, it is not at all surprising that Pasteur would move in this direction. And wine was nearly as subject to disease as people were.

The Studies on Wine: Origins, Aims, and Significance

Pasteur's remark that wine can be a safe, healthful drink is well known; the context of his carefully phrased remark is usually ignored. The more Pasteur thought about wine diseases, the more he became convinced that empirically based winemaking procedures had their raison d'être in the conditions of life and ways of acting of the parasites that spoil wine; so devising a simple operation to prevent wine from spoiling would undoubtedly establish a new art of making wine. This sounds suspiciously like the annunciation of a new oenology. Nor would this scientific coup be without economic importance. The new vinification process would be much less expensive than the traditional one and, above all, much more effective in avoiding the losses that result from wine diseases. The process would therefore be quite appropriate for the expanding commerce of this foodstuff. It was at this point in his reflections that Pasteur justified his aim of developing a new process to make wine by pointing out that wine can be

[31] Robert Kohler, "The Background to Eduard Buchner's Discovery of Cell-Free Fermentation," *Journal of the History of Biology* 4 (1971): 35, 41. See also René Dubos, *Louis Pasteur: Free Lance of Science* (New York, 1960), pp. 237–40.
[32] I. Straus et D. Sanchez-Toledo, "Recherches microbiologiques sur l'utérus après la parturition physiologique," *Annales de l'Institut Pasteur* 2 (1888): 426.

considered a safe drink. "Le vin peut être à bon droit considéré comme la plus saine, la plus hygiénique des boissons" – an early if partial and vague statement of "The French [health] paradox."

Pasteur's basic point was that wine is a food. He meant for the working class. Pasteur thought that wine has two distinctive virtues: it is a stimulant, and it is a food. The bourgoisie may drink wine as a stimulant for its jaded palate; the working class needs wine as both stimulant and food. Gladstone, who as chancellor of the exchequer was responsible for getting duties lowered on French wines imported into the United Kingdom, was in basic agreement with this point of view: the "great gift of Providence to man" might tempt the people of England, if they could afford it.[33]

But what about wine for the tea- and beer-drinking working class? Pasteur suggested that an attempt be made to put a low-priced natural French wine on the table of the English working class. "Le vin de France aliment, c'est-à-dire le vin naturel dont Dieu a largement gratifié le beau pays de France." It would have to be a wine comparable to that supposedly drunk by the French working class: a cheap, young wine lower in alcohol than *les vins de cru*. But the only wines that could be safely exported to England were those of high alcoholic content, the *vins vinés*, wines with the alcoholic content raised to 18 percent or more. This was drink for "la table du lord d'Angleterre," and even he should drink only a glass or two per meal. The problem was how to transport *vin ordinaire* across the Channel for the gastronomic seduction of the English proletariat. Pasteur's solution was to heat the wine in order to kill the disease-causing microorganisms that spoiled cheap wine before it could be drunk. No *ferments* or *filaments*, no standard disease. And the enormous English market would open up, according to this good Second Empire thinking. He confidently concluded that until a contrary proof was provided, he would believe that he had solved this important problem of the wine-exporting business.[34]

Pasteur's studies on wine were published in 1863 and 1864, then collected in a volume entitled *Etudes sur le vin. Ses maladies – Causes qui les provoquent – Procédés nouveaux pour le conserver et pour le vieillir.* The first edition (1866) of three thousand copies – the same number as the second printing of Darwin's *Origin of Species* – sold out quickly, and the work saw another edition in 1873 and a reprinting in 1924.[35] Of course, once a winemaker accepted the Pasteurian technique for killing bacteria, the problem was not scientific but technical – namely, to find an apparatus that would heat the wine to the proper germ-killing temperature without causing a deterioration in its organoleptic qualities. Pasteurization did not become a widespread practice immediately after the appearance of the

[33] Asa Briggs, *Wine for Sale* (Chicago, 1983), p. 33.
[34] Pasteur, *Etudes sur le vin: Oeuvres de Pasteur,* 3:152, 356–7.
[35] See n. 1 above.

Etudes sur le vin, but not because the work was contested by chemists who, even if they followed the public in reading the work, ignored it more than they discussed it. A short report by Dumas in 1867 was interred in the *Annales* of the Indre-et-Loire. The ministry of agriculture stimulated interest in the wine industry by holding a competition for the best apparatuses. Gayon and two of his collaborators, in a study of wine-pasteurizing equipment (*Appareils à pasteuriser les vins* . . . , 1897), hailed Pasteur's work as a great contribution to the "future advances of oenology and to the growth of public wealth." The pasteurization of wine was "the sovereign method for the preservation of wines." This seems like a quite reasonable mimesis of Pasteurian discourse, which was by now part of a Republican mentality geared up for glory.

Pasteur's *Etudes sur le vin* had become part of the national patrimony in 1866, when the volume was published by the Imprimerie impériale at the request of the emperor, although Pasteur paid the printer's bill. A steady stream of information on the benefits of heating wines appeared in viticultural publications. Nearly everyone seemed eager to make a contribution to Pasteurian science. Wines of the Loire, including the best wines of Anjou, also underwent a great improvement, according to a piece written in 1897 by Abbé J. Voleau. But large numbers of consumers who believed that "heat kills wine" had to be convinced that they were wrong. Nor were the advocates of the new technique able to conquer the great Bordelais châteaux, for pasteurization never became part of the process of vinification of the fine wines of Bordeaux, in spite of the convincing and dramatic experiments of Pasteur and Gayon. Jean Ribéreau-Gayon points out that pasteurization does affect the first stage of the aging process of fine wines, when the bacteria often deacidify and soften new wines. After this first stage, Pasteur and Gayon may be right.

In his analysis of the Pasteurian "theater of proof," Bruno Latour points out that in the 1870s there was no relation between an infectious disease and a laboratory. The laboratory scientist soon overcame this problem by shifting his activities to the locale of the problem, but he still maintained a connection with an institutional laboratory.[36] This argument might apply equally well to the period of Pasteur's research on wine in the 1860s, when he went into the wine-producing areas of the Jura and Burgundy "in order to study the processes of fermentation and especially the microscopic plant that is the only cause of this great and mysterious phenomenon." Not that Pasteur had any illusions about immediate industrial applications of his work, notoriously slow in coming. He carefully informed his govern-

---

[36] Bruno Latour, "Le théâtre de la preuve," in Claire Salomon-Bayet, *Pasteur et la révolution Pastorienne* (Paris, 1986), pp. 335–84.

mental supporters that his aim was more modest. "I want to get to know better the cryptogamic plant that is the sole cause of the fermentation of grape juice."[37]

Pasteur's *Etudes* conformed to the good nineteenth-century custom of beginning with a short historical sketch of previous scientific work on the subject. After a brisk five and one-half pages on the "old opinion on the cause of wine diseases," Pasteur turned to his "new opinion," the correct one. Standard oenological works, such as that by Ladrey, received little or no reference, because they had, after all, not subjected wine to the scrutiny of experimental science based on the germ theory of fermentation. Alfred de Vergnette-Lamotte received all too much critical attention in a polemic over priority in the invention of "pasteurization."[38] There is no doubt that because of his rigorous laboratory experiments on wine, all carried out within the framework of his biological theory of fermentation, Pasteur believed his *Etudes sur le vin* to be the only serious and original oenological work since Chaptal. This infusion of the latest biological science put Pasteur's work in a very different category from that of *savants* and others who still held on to the ambiguous "principe végéto-animal" of fermentation.

In 1865 Vergnette and Ladrey, as well as Pasteur, presented communications to the Académie des sciences on how heating preserves and improves wines. Ladrey merely reviewed some of Pasteur's work on wine, commented on the usefulness of cooling in keeping wine from spoiling, and endorsed heating in bottles as the best way of curing disease and of preserving wine. At that time, heating, done mostly in secret, was used in only a few vineyards of the Côte-d'Or. The process ruined quite a bit of wine because of poor temperature control in primitive equipment. In his communication to the Academy, Pasteur emphasized, in a friendly fashion, the independence of his work, which had been communicated to the Academy eighteen months before that of Vergnette, who had sent him some wine on which to experiment. In a later note Pasteur was less friendly. After reading Appert's *Traité des conserves,* he concluded that Vergnette had done nothing new on the subject. It was also clear to Pasteur that the heating of wine was of no further research interest because the scientific problem had been completely and satisfactorily solved. It was up to vineyard owners to learn "how to profit from the results of science." The communica-

---

[37] *Correspondance de Pasteur,* 4 vols. (Paris, 1940–51), 2:88.

[38] C. Ladrey, *Chimie appliquée à la viticulture et à l'oenologie* (1857); A. de Vergnette-Lamotte, *Le vin,* 2d ed. (Paris, 1868). An initial friendly correspondence between Pasteur and Vergnette degenerated into an acrimonious exchange of letters over priority in inventing the process for the preservation of wine through heating. As usual, Pasteur won. See the letters in the *Correspondance de Pasteur,* vol. 2 (1951), collected and annotated by Pasteur Vallery-Radot.

tions of Ladrey and Vergnette were presented to the Academy under the rubric of of domestic economy, whereas that of Pasteur was given higher status as applied chemistry.[39]

Because of the improbability of immediate applications of his research on wine, Pasteur declined an offer of support from the court made by the emperor's aide-de-camp, Colonel Favé. He also turned down the offer of the mayor of Arbois, le comte de Broissia, and the municipal council to create a laboratory for him in Arbois. Pasteur believed that his accepting such an offer could compromise the independence of his research. He was desperately and vainly trying to escape the role of scientific miracle worker that politicians were eager to thrust upon him. To ensure the cooperation of local politicians and bureaucrats, Pasteur obtained a letter from Favé stating the emperor's full support for this scientific foray into wine country. There was also more tangible support from the government. The minister of education, Victor Duruy, granted Pasteur's request for 2500 francs to cover the traveling and living expenses of four assistants. It was a Normalien research team (Le Chartier, Raulin, Gernez, and Duclaux) that went to the Jura and Burgundy in 1863. (Pasteur promised Duruy that he would spread the good news that the emperor was constantly thinking about how to develop the country's agricultural riches. The pursuit of objective science did not preclude a scientist's touting the imperial ideology.)

A later trip in 1863, undertaken because Pasteur thought he had insufficient experimental data on some important issues, was granted another 1500 francs through the less formal offices of Colonel Favé. Pasteur's letter to Favé promised that the research results would be of use to producers in preventing undesirable alterations in their wines. It seems that on-site research changed Pasteur's mind about the immediate applicability of his work. It is also true that the research on wine was his most expensive work up to that time because of expenses for travel, his team, and matériel. Governmental support of science was not yet Pactolian, but the well-connected Pasteur never had to stop research for lack of funds.

In the preface to the first edition of the *Etudes sur le vin* (1866), Pasteur disarmed and captured the reader through two powerful, connected rhetorical strategies of basic self-promotion. As a result of his studies on fermentation, he had seen the possibility of writing a useful work on wine diseases. The economic and social importance of wine ensured that the study of its diseases would result in a research program of national sig-

---

[39] *Comptes rendus hebdomadaires des séances de l'Académie des sciences* (1865), 60 (1re semestre), 895–9; 60 (2e semestre), 976–80; 61 (2e semestre), 274–8, 865–6: Vergnette-Lamotte, "Des effets de la chaleur pour la conservation et l'amélioration des vins" (read by Boussingault); Pasteur, "Nouvelles observations au sujet de la conservation des vins"; and C. Ladrey, "Etudes sur les procédés employés pour l'amélioration et la conservation des vins." Pasteur wrote a brochure, *Sur la conservation des vins,* which he presented to the Academy with the note.

nificance. In 1866, Pasteur admitted that there was much more to be said about the topic than he had said in his "observations," but the need to speed up the industrial applications resulting from them had justified their publication. Of course, they would, he concluded ritualistically, serve as the basis for more profound future studies. The big scientists of the nineteenth century believed that sooner or later their research would be of great economic significance, in addition to demonstrating its clear importance on the theoretical or cognitive level.

The second part of Pasteur's rhetorical strategy tied in the economic significance of his research with the highest level of politics. The Cobden-Chevalier commercial treaty between Great Britain and France was signed in January 1860, but freer trade with Great Britain would not be of any advantage if wines leaving France in ostensibly good condition ended up diseased and unsold in British shops. In a draft of a letter (dated August 1, 1861, but not sent) to the minister of education, Pasteur pointed out that the value of one of France's great agricultural riches had increased because of the treaty. So everywhere in wine areas there was concern with improving the quality and increasing the quantity of wine that could be profitably exported.[40] In July 1863 the emperor had encouraged him to do research on wine diseases. Politicians were beginning to look to the state-funded scientist for the quick fix. For the scientist, the ultimate satisfaction then lay in serving his country well by doing good science.

Pasteur kept insisting on his primary interest in pure science, perhaps because he was keenly aware of the technological problems that would have to be faced by the introduction into industry of even scientifically simple practices like pasteurization. Yet all the time he allowed himself to be drawn into the agro-industrial world of the winemaker. Pasteur himself sometimes seemed to recognize that the separation of the scientific from the industrial was more an academic conceit than a reality of the world in which the scientist had to work. From the beginning of his oenological research, Pasteur recognized and even boasted of the importance of his work for France's agro-industry. The first and second editions of the *Etudes sur le vin* were awarded prizes by the Comité central agricole de Sologne, which was suitably convinced of the importance of Pasteur's claim to have put certain processes for the improvement and preservation of wine on a solid scientific basis. In research in both Paris and wine areas, Pasteur had to work closely with wine producers.

Even in the published text of the *Etudes* it was clear, especially in the section on pasteurization, how closely the research was linked to production and selling. His "Cahiers d'expériences sur les vins" give the material details of a working relationship between researchers and winegrowers in

---

[40] Pasteur, *Correspondance,* 2:87–8.

Arbois, Gevrey-Chambertin, Romanée, Volnay, Saint Georges, and so on. In May 1865, Pasteur also talked to MM. Pol Roger of Epernay. The wine was not always free: Château Chalon 1842 cost five francs a bottle. Pasteur soon built up an experimental cellar of hundreds of bottles of wine, divided into heated and unheated lots of the same wines. One of the "Cahiers d'expériences" shows that Pasteur was also interested in a wide range of agronomic research, including plant assimilation of nitrogen from the air, fertilizer, and seeds. Such were the frivolities of "Vacances Orléans 1867." No doubt Pasteur's research on wine had its pure moments and dealt with some theoretical issues. It was also firmly anchored in industrial science, a novelty of which Pasteur was one of the leading prophets and practitioners, perhaps even one of the pioneers. This was clearly recognized by Jean Ribéreau-Gayon, himself a researcher who lived in the intersecting scientific and commercial worlds of modern oenology.[41]

The year of the second edition of Pasteur's *Etudes sur le vin* (1873) being too early in the history of the Third Republic for an icon of Marianne-France to have replaced Napoleon III, the preface dealt only with science, and especially technology. Important additions were made to the new edition: the description of numerous wine-pasteurizing apparatuses conceived by industry to assure the preservation of wine; the proof of results obtained by pasteurizing wine and by some other procedures; and the influence of heating on quality improvement in wines. In this edition Pasteur also defended, with his customary vigor, his priority in inventing the commercial pasteurization of wines. Not for him the convenient sociological doctrine of multiple independent discovery. Vergnette was also a tenacious defender of his own claim to priority. Pasteur conceded that the public would eventually decide who was right, but in the meantime it was vital to protect one's own turf officially and publicly.

Being invited to the imperial court at Compiègne forced Pasteur to think hard about the industrial significance of his oenological research. The empress remarked to Pasteur that it would be absurd not to derive a practical advantage from such works, and Napoleon III agreed. All that Pasteur could think of replying at the time was that French scientists believed that such action caused a scientist to fall from grace. But he later wrote that he was sure of their majesties' appreciation of the real reason for a scientist's resisting the call of industry. By getting involved in industrial applications, the scientist stops being a man of pure science, as his life and the usual order of his thoughts are encumbered by preoccupations that paralyze his spirit of invention for the future. Pasteur declared that if he had followed the industrial path leading from the practical results ob-

---

[41] Bibliothèque Nationale, Manuscrits, *Nouvelles acquisitions françaises, Louis Pasteur, Papiers,* nos. 17945–50 and 17951.[1–5]

tained by his research on the manufacture of vinegar, he would not have escaped the consequent industrial problems for years, if ever.

To follow his practical results through to industrial success in winemaking would, Pasteur believed, be a full-time job for a long time. Instead, he followed, with scientific curiosity, the practical use others made of his research results, while keeping his mind free of technological problems. In principle, the scientist thus avoids the technological trap into which he could fall because of his usual ignorance of technology and its relations to both industry and science. After two years of work on wine diseases, Pasteur was ready to heed the call of the government again and take on silkworm diseases during leave from his responsibilities in higher education. Meanwhile he took out a patent on wine-pasteurizing equipment. He hoped thereby to obtain money for his daughters' dowries and be able to leave a bequest to the charitable Société des Amis des Sciences.[42]

Science and technology were becoming increasingly relevant to winemaking at the levels of production and marketing. The economic crises of the 1840s and the revolutions of 1848 depressed the wine trade. The drop in the sale of luxury wines on the Continent caused some imaginative producers to think about finding other markets, America and warm countries in particular. That would require technological and even scientific solutions to certain problems. Vergnette experimented for years on how to condition wines to withstand the fatigue and heat of long-distance transportation without being mixed with hermitage, chaptalized, or *vinés* (alcohol added), all processes that destroyed the delicate bouquet for which burgundies were famous. He was not happy with the result of heating wines to 70 degrees Celsius according to Appert's method, for the wines retained a cooked taste similar to that of untreated wines shipped to warm countries. One exception: white wines, which were the subject of Vergnette's experimenting on wines shipped to the Antilles. To ensure the quality of wines exported, Vergnette came up with a revised version of an old practice, freezing or rather cooling the wine to the point where some ice formed and was removed. A simple apparatus produced a concentrated wine that could be reduced up to 50 percent in volume at an expense of two and a half francs per hl. Vergnette was also keen on shipping wine in insulation to protect it from high temperatures.[43]

Not many people were convinced of the usefulness of the cooling procedure. P. Bastilliat, a well-known pharmacist and the author of a treatise on French wines, pointed out that white wines were left out to freeze in the Centre, and some people liked them and some didn't. Bastilliat's nos-

---

[42] Letter to General Favé, Dec. 19, 1865; letter to the Directeur du *Messager agricole,* with a copy of the *Etudes,* Sept. 24, 1866, *Correspondance,* vol. 2.

[43] Alfred de Vergnette-Lamotte, "De l'exportation des vins de Bourgogne dans les pays chauds," communication to the Société nationale et centrale d'agriculture, 1851.

trum for keeping burgundies or any other wine from getting travel sickness was to add tartaric acid, one gram to each liter of red wine and one-quarter gram to white. He too had his successful experiment. In 1838 he sent his doctored burgundy on a round-trip to Santo Domingo; after five months of abuse, it was perfectly preserved, even after freezing at Calais.[44] Wine has always been much abused in the name of experiment, although the consumer is often less easily convinced than the experimenter that the wine has been improved.

Pasteurization: Debates, Experiments, and Diffusion of the Technology

Technology is not easy to define. If one takes an oenological viewpoint, which is concerned with the use of technology rather than its definition, Pasteur appears as a significant figure in the history of technology because his work gives the appearance of being unified by a method. Jean Ribér-eau-Gayon believed that Pasteur's work shows the virtuous efficacy of method in being especially unified by his constant attempt to answer questions, whether theoretical or practical. This seems to be a version of the familiar view of science as problem solving. Although Pasteur's means of arriving at solutions might be more scientific or more empirical, depending on the problem, his fruitful solutions usually came from an interaction of the two. Ribéreau-Gayon saw this approach as a prefiguring of the present-day technological method, which approaches problems from all angles.[45] Not that any one man can play Pasteur nowadays. Having become too complex to be mastered by any one mind, science and techniques can cohabit in technological union only in institutes. This fact was generally recognized by the late nineteenth century, when the French provincial university invented the institute of applied science. When Pasteur needed a description of the "Industrial apparatuses for the heating of wine," he called upon Jules Raulin, his former student at the Normale and frequent collaborator, whose technical study of pasteurizing equipment and its functioning occupies an important place in the second edition of the *Etudes sur le vin*.[46]

Once the ministry of agriculture and some scientists in wine areas had decided that pasteurization was a good thing, a serious campaign was

---

[44] P. Bastilliat, *Traité sur les vins de la France* (1846), p. 147.

[45] "Ce qui fait l'unité de la méthode de PASTEUR, c'est surtout sans doute qu'il cherchait toujours à repondre à des questions: elles étaient d'ordre théorique ou d'ordre pratique; les voies de leur solution étaient surtout scientifiques ou surtout empiriques. Mais les uns éclairaient les autres et les réponses toutes profitables. Pour une part, cette méthode préfigure celle de la technologie d'aujourd'hui qui aborde les problèmes par tout leurs aspects." Jean Ribéreau-Gayon, *Problèmes de la recherche scientifique et technologique* (Paris, 1972), p. 27.

[46] *Oeuvres de Pasteur,* 3:266–310.

undertaken to promote it as a standard part of wine making. This professorial-bureaucratic sector had a formidable arsenal of propagandistic weapons at its disposal. A special competition instituted by the government was reported on by three specialists: Gayon, then departmental professor of agriculture for the Gironde, who was responsible for the practical tests; Charvet, professor of agricultural engineering in the Ecole d'agriculture in Rennes, who was responsible for judging mechanical construction; and F. Vassillière, university professor and director of the Station agronomique, who was responsible for laboratory tests and tastings. A curious problem existed. After the appearance of Pasteur's *Etudes sur le vin* in 1866, although the advantages of heating wine were not seriously contested, the technique did not spread rapidly. Pro-pasteurizing professors denounced prejudices, mercenary interests, and egoism in the wine world, but they also admitted that heating wine was a chancy business fraught with frequent imperfections of execution. People in the wine business were uncertain of its efficacy, and vineyard owners hesitated even more than *négociants*.

Nonetheless, the potential profits of selling this panacea touted by science lured entrepreneurs to gamble on manufacturing equipment that could prevent disaster for the winemaker. Improvements came quickly because professors encouraged technicians to produce machinery worthy of great scientific ideas, capitalists worked for the profits of perfection, and the government encouraged competition by rewarding technological prowess. Nature helped most of all by producing invasions of "microscopic mushrooms" (mildew and especially black rot) in the vineyards, making the wines susceptible to alteration. The wine dealer became more ready to listen to the men with a means of producing a wine that would not spoil, even if he was suspicious of the detrimental effects of heating on the taste and bouquet of wine.

Small-scale tests might satisfy the scientifically curious, but the business world and big organizations required large-scale experiments before spending money on the process. Fortunately, Pasteur had one very interested client, the navy, which had considerable difficulty with spoiled wines in its ships sailing to the warm parts of the French empire. A naval commission, headed by the director of naval construction at the Ministry of Marine, de Lapparent, was eager to find a way of preserving wine on long voyages so that, as Pasteur put it, the colonial crews would not have to drink vinegar. A test was carried out on 500 liters of wine by dividing it into two equal amounts of heated (60 °C) and unheated wine and leaving it in two casks on the *Jean Bart* for ten months. Although both wines survived, in a state of advanced aging, the unheated wine had to be drunk immediately. Then, in 1868, the director decided to carry out, under Pasteur's supervision, a mega-experiment, which took on the mythic attributes of an *experimentum crucis:* 650 hectoliters of heated wine and 50 hectoliters of the same wine unheated were sent from Toulon to Gabon on the frigate *Sibylle*. The heating, done in two days, cost only about five centimes per liter; so the navy was immensely pleased when the experiment was declared a success. Several merchants in Narbonne had also successfully carried out large-scale heating of wine. The procedure was clearly shown to be quite successful before October 1868, when a stroke incapacitated Pasteur for several months.

Between 1865 and 1872 Pasteur's experiments on the wines of Burgundy, the east of France, and the Midi showed that wine was not changed in flavor and could even be improved by pasteurization. Later studies by A. Bouffard and by Müller-Thurgau demonstrated that heating wine to 60 degrees centigrade – generally 55 degrees was not hot enough – prevented microbes from altering the wine (*tourne, pousse, amertume, graisse, piqûre, fermentation manitique,* etc.) and also prevented *la casse* by fixing the coloring matter. Would pasteurization also work for the fine wines of the Gironde? All scientific doubt was eliminated by Gayon in a set of experiments carried out on 42 samples of three groups of wines (one healthy white wine and two reds, one sick and one healthy). The experiment was also designed to allay the fears of suspicious gastronomes. A commission of the chief wine tasters of the Gironde did tastings of the wines at four different ages – three and one-half months, one year, two years, and six years after pasteurization. Some wines were heated to 60 degrees centigrade and some to 55 degrees, but the latter group showed evidence of incomplete sterilization. The quite elaborate results showed no appreciable differences in the taste ratings of the heated red wines and their unheated counterparts. The wines did not lose the development that comes with aging. Heated white wines did even better than the reds in the competition. Among the wines tasted were a *premier cru* Médoc 1878, a *deuxième cru* Margaux 1874, a *premier cru* Pomerol 1870, and a *premier cru* Sauternes 1878. What was good for burgundies was good for bordeaux.[47]

With success came difficulty and failure. In Gayon's opinion, the fault was not in the idea but in the "material imperfections of execution." Better pasteurizers had to be built. The idea of a competition was promoted by the ministry of agriculture, a general inspector of agriculture, and the Chamber of Commerce of Bordeaux, which was generous with financial support. The gold medal for heating in bottles went to the Société du Filtre Gasquet of Bordeaux; silver medals went to the two Paris manufacturers Besnard and Houdart. In the category of heating big quantities, the gold medal *grand module* went to Périllot and a gold medal to Nabouleix,

[47] Review of U. Gayon, *Expériences sur la pasteurisation des vins de la Gironde* in *RV,* no. 2 (1894), pp. 136–40.

both of Bordeaux, while a silver medal *grand module* went to Tamarelle of Libourne. Nothing was left to chance. Three series of tests over four months laid bare the structure of the wine through chemical and microscopic analysis – alcohol, extract, acidity, sugar. (Nowadays, science easily detects changes produced in wine by heating. Two examples: the dehydration of fructose can produce a harmful quantity of hydroxymethylfurfural; and wine colloids can change.[48] It is unknown how this knowledge has altered the perception of wine tasters.)

Then came the moment of truth, a tasting by a jury of eleven members, presided over by the prefect of the Gironde and including Gayon and Jouet, the manager of Château Latour. It would not be unreasonable to conclude that manufacturers of wine-heating equipment in Bordeaux had acquired some expertise from having been in the business for quite a while. Buyers of the equipment managed to keep their names secret. Pasteurization could even be useful for great wines, although experts no longer found the vintage pasteurized to be great. Lafite was pasteurized in 1928, "and the same is said to have been done with Haut-Brion," because a high level of heat during fermentation raised the bacteria count to a danger point.[49] As late as 1945, Château Cheval Blanc pasteurized the wine from half its crop because of the threat from *Acetobacter,* bacteria that produce acetic acid. For a *grand vin,* this exception was rare enough to be news.[50]

Because of the recent flap over Louis Latour's pasteurization of his great reds, the process today is more likely to be associated with burgundy than bordeaux. (The Italians seem to escape controversy, perhaps because what they pasteurize is not part of *haute* oenophilic culture.) Although Pasteur had done experiments on famous burgundies, on an official level Bordeaux seems to have been ahead in pushing the practice. On being authorized by the ministry of agriculture to undertake experiments on pasteurization, L. Mathieu, head of the Station oenologique de Bourgogne, first made a tour of other stations, especially that in Bordeaux, the center of Gayon's campaign for the adoption of Pasteur's panacea. Propasteurization propaganda was carried by *L'oenophile,* a leading viticultural and vinicultural publication in Bordeaux. Converted to the use of the process, Mathieu praised its virtues in *L'oenophile* in 1899.

Mathieu's argument shifted the ground from the antibacterial strategy, common to its advocates up to this time, to the growing concern of government and public with adulteration of food and drink along with their preservation through the use of chemicals dangerous to health. The Brousse law (July 12, 1891) had taken a step in the right direction in pro-

[48] M. A. Amerine and M. A. Joslyn, *Table Wines,* 2d ed. (Berkeley and Los Angeles, 1970), pp. 563, 652, 793.
[49] Edmund Penning-Rowsell, *The Wines of Bordeaux* (New York, 1979), p. 577.
[50] David Peppercorn, *Bordeaux* (London, 1982), pp. 48, 128.

tecting wine drinkers by prohibiting the use of boric and salicyclic acids and their analogues as preservatives, but fluorides, formaldehyde, and other dangerous substances were not named. Mathieu asked why any product from the chemical arsenal should be used to preserve wine when, even for fine wines, there was a perfectly legal process, a great discovery of a great scientist – pasteurization? Perfectly functioning apparatuses existed, even "a large commercial model made by the German firm of Boldt and Vogel."[51] Industrial firms would pasteurize wine sent to them or do it on the owner's premises for the reasonable price of a half to one franc per hectoliter.

In 1899 *L'oenophile* published a series of plates illustrating the elements in wine that people then found fascinating. Bacteria and yeasts were of the highest interest. The plates, of a high technical quality, were taken from a manual by Frantz Malvezin on the pasteurization of wines and the treatment of their diseases. The existence of such practical works indicates a widespread use of the process in wine production. A. M. Desmoulins wrote a series of articles in the *Moniteur vinicole* casting the usual doubts on the wisdom of heating fine wines. The pro-pasteurization forces were not long in replying. An a fortiori argument was at stake here: if the *grands crus* can be pasteurized without damaging their qualities, the lesser breeds can be heated without fear. Malvezin believed that the pasteurization of wines was a preventative treatment comparable to the vaccination of people against smallpox. His enthusiasm for heating wines was regarded suspiciously by winemakers because of his financial interest in the manufacture of pasteurizing equipment. Like Gayon, he argued that the red wines of the Gironde should be pasteurized *en primeur,* young; if good equipment were used, the process was always useful, never harmful.

What about white wine? (In 1852 about 26 percent of French production was white; in the 1930s it was 20–22 percent; the Graves region of Bordeaux was famous for its white wines, with Châteaux Carbonnieux and Olivier being the biggest producers.) To pasteurize or not to pasteurize was an important economic question. Malvezin declared that it was not necessary to pasteurize white wines. They had never suffered from bacterial diseases, partly because they were more sulfured than red wines, but Malvezin saw no problem if a winemaker wanted to indulge in this superfluous technology. The first part of his statement still stands as good advice on fine white wines.[52]

The drums of science kept up a steady beat in Bordeaux. Microbic

[51] Loubère, *The Red and the White*, p. 195; see pp. 193–7 for an excellent short treatment of pasteurization.

[52] Anthony Hanson, *Burgundy* (London, 1982), p. 264, on Latour; p. 226, on Moillard-Grivot. Frantz Malvezin, "La pasteurisation des vins fins," *L'oenophile*, 6e année, no. 7, (July 1899), replying to Desmoulins in the *Moniteur vinicole,* 13, 16, 20, 27, 30 June, 1899.

civilization – according to scientific bulletins, an organized, powerful enemy – made it necessary for serious winemakers to introduce the microscope into their daily work. Science had become a necessity not a luxury, indispensable to the wine trade. Some wine merchants did not know this, but they would pay a heavy price for their ignorance and soon join the scientific throng. One advantage of pasteurization was that it allowed even new wines to be sold immediately. The Bordelais merchants liked this, the wine producer liked it even more, and both told the wine drinkers that they liked it too.[53]

Pasteurization was recommended by some of the leading publications of the winegrowing world. In 1901 *Le petit viticulteur bourguignon* gave extensive coverage to the bacterial alteration of wine in bottles including a general recommendation to pasteurize. In February of 1901 the Institut oenologique de Bourgogne organized a giant set of tests in a big hall rented for this spectacle of bringing science to the people. The institute supervised the tests. The Syndicat des vins de la Côte d'Or, the Syndicat viticole de la Côte dijonnaise, and the Union des syndicats agricoles et viticoles de Bourgogne et de Franche-Comté supported the event. Producers and merchants flocked to the fair to find out if the latest gift of science was really as useful as the propagandists said and, more important, if they could afford it, collectively if not individually. One constructor, smelling the profits of victory in this sci-tech event, speeded up the manufacture of equipment. The results of the tests were perfectly predictable by this time: another victory for the pasteurizers. Although equipment was not yet at the desired level of perfection, the machines of the Société du filtre Gasquet of Bordeaux seemed to inspire confidence, even in Burgundy.

A list of procedures to follow in this manipulation of the wine made it clear that the operation was a fairly complex one in spite of the simplicity of the basic idea. Wines with more than nine percent alcohol did not require so high a temperature as wines with less than nine percent. Wines with incipient diseases required different temperatures, according to the diseases. Scientists could warn people that pasteurization did not make a bad wine good, but the producers saw the technique as making it possible to sell wines that would never otherwise arrive on the market in drinkable condition. The time at which to pasteurize was specified as the point at which the sugar had been transformed so that the glucose content of the wine measured only between 0.5 and 1.0 parts per 1000. This meant from February to June for normal wines, even earlier for lower grade or common wines. If it was certain that the wine was clear, the heating could be done in bottles at a temperature of 62°C; if the wine was the least bit

---

[53] D. Cazemave, "La civilisation chez les microbes," and J. Le Bihan, "Un peu plus de science," *L'oenophile* (1899), pp. 371–3.

cloudy, filtering before heating was recommended to prevent any particles from dissolving in the hot liquid and altering the *finesse* and natural bouquet of the wine. The best results were obtained with a wine that was naturally clear after racking or a traditional fining. For the process to be absolutely safe, a preliminary test was recommended in which a bottle of the wine was heated to 62°C to verify that there was no danger of albuminoid precipitation. The recommendation to filter was from Gayon, but the idea of the need to preserve the distinctive aesthetic of the *cru* was as strong in Burgundy as in Bordeaux.

In 1913 the *Revue de viticulture* praised both pasteurization and a specific apparatus built by Depaty, a company in Cognac.[54] This praise probably indicates not only that the Ecole de Montpellier had given its *nihil obstat* to the doctrine of pasteurization but that it had also integrated it into its oenological program as an important piece of winemaking technology. In addition to being a good pasteurizer, the Depaty machine possessed the practical qualities required of a good industrial machine able to treat from 10 to 20 hectoliters of wine without being cleaned. The temperature of the water used to heat the wine could be regulated immediately and precisely, even above 100°C; nor was there any danger of the wine's acquiring a cooked taste. An ingenious arrangement of pipes within pipes let the hot wine in the pipe heated by the water heat the cold wine in the interior pipe. This large-volume machine was safe, cost-efficient, and easily modified for pasteurizing varying amounts of wine.

Pasteurizing did not permit the winemaker to dispense with a scrupulous cleaning of all containers used to hold the wine. Where steam cleaning was impractical, sulfur dioxide, the winemaker's eternal companion, could carry out its "priestlike task of pure ablution" at a cost of one to two centimes per hectoliter, depending on the price of fuel. The Depaty gospel recommended that new full-bodied wines be pasteurized at 75–8°C and light tart wines at 68–70°. A red wine already sick or a wine likely to suffer from *la casse* should also get a dose of sulfur dioxide of 30–40 mg per liter before being pasteurized. The combination was "extremely efficacious." After being sulfured with 300 to 350 mg per liter, white wines could be heated to 80° and even higher for *vins moelleux* or sweet whites. Pasteurizing white wines kept them from refermenting (malolactic fermentation), as well as preserving them in good condition. These two arguments would not prevail when a more complete knowledge of fermentation and a better technology of vinification became available to

[54] René Mallet, "Le pasteurisateur de Depaty," *RV* 39 (1913): 356–61. W. Kuhn's earlier innovation of sterilization of liquids under pressure seems to have made more of a splash in the beer business. See W. Kuhn, *Notice sur le procédé W. Kuhn: Stérilisation pastorienne de la bière et autres liquides fermentescibles appliquée à l'expédition en fûts* (1892), and *Sur un nouveau procédé de stérilisation par la chaleur sous pression* (1897).

winemakers. In 1913 the standard attitude still was that it is better to be safe with the sulfured and cooked than sorry with the sulfured and raw.

With the right machine, pasteurization could be made a fairly straight-forward procedure, but in technology things rarely stay that way for long. The *Revue de viticulture* introduced a distinction not usually made in the literature: heating and pasteurizing wine are two different processes. In heating wine, the temperature was not raised to the point at which the liquid would be pasteurized, that is, sterilized. To treat wine by heating meant to raise its temperature to 65 degrees centigrade, or at least this was the temperature recommended by *négociants*. Heating of wines was widely practiced in France in the red wine and wine liqueur trade. No positive desired effect of heating was observed on white wine, which therefore usually escaped this manipulation.

Wine dealers were keen on heating wine for a number of sound commercial reasons, which were not devoid of a certain aesthetic appeal at the time. Heating took off the wine's rough edges, gave it a seductive *rondeur* – much like the high alcohol does in the *grands vins* today – and speeded up the aging process, or at least gave the wine the characteristics of an older wine. It seems that consumers could not tell the difference between the reality of age and the mask of age in wine, or if they could, they didn't care. The technique of ersatz aging, or speeding up the process, as defenders would say, consisted of sending a shot of air into the container when the wine was at its maximum temperature. The degree of alcohol was not significantly lowered because the operation occurred under pressure. If oxygen were used instead of air, a new wine would appear to be two or three years old or, passing through a biochemical time warp, it really was three years old although its chronological age had not changed.

The heating process could also be a marvelous tool for facilitating some traditional manipulations of wine. Heating is particularly good for making blended wines quickly drinkable. The mélange of heated wines becomes homogeneous and softer. Because there is always a danger of introducing germs and ferments when adding new wine to older wine, the heating is useful in killing the intruders before they infect the mixture or even re-ferment it; although it is not completely without danger when used for the clarification of wine, because it coagulates pectic and albuminoid matters, which otherwise produce a persistent murky quality and make some wines hard to filter. The Depaty company recommended a clarification procedure that consisted of adding albumin to the wine, heating it to the coagulation point, and then pasteurizing the concoction, which was finally filtered to give an extremely brilliant wine.

Wine was made potentially sick to be cured, was deprived of its natural clarity, which was not clear enough, to become artificially clear – that is, brilliant – a key word in the taste model recommended to the consumer

by the manufacturer of the equipment that did the job. The Depaty procedure also produced a slight aging of the wine, but more important was the
fact that, in years when wine was subject to *la casse,* it removed deposits
of coloring matter already oxidized. It is not surprising that any process
capable of solving so many of the problems afflicting the winemaker since
antiquity would be widely accepted in the wine world.

The oenologists of the Ecole de Montpellier, like those of Bordeaux,
devoted as they were to the cause of better wine through science, could
hardly not be enthusiastic in recommending pasteurization. It seemed like
a panacea for so many problems in the production and selling of wine:
preventing disease by killing germs; avoiding excessive volatile acidity;
guaranteeing the conservation of wines duing their exportation; using the
coagulation of albuminoid matter to produce a clarification depriving the
yeasts of the nitrogenous matter that was their chief source of food; and
instant aging that obviated the expense of tying up stock for several years
while nature took its quirky course. The word from Montpellier on pasteurizers was that no winery should be without one. A kind of *futurismo*
had arrived in the wine business: "in our epoch . . . one must move quickly
and . . . there is not always enough time to let the wine age before shipping it."[55]

More important than quick aging in wine preservation by pasteurization was standardization. The consumer was perceived as wanting above
all an unchanging type of wine that would always develop and age in same
way. The policy "the best surprise is no surprise" is not without considerable commercial advantage in dealing with conservative wine drinkers. It
was argued that, thanks to a rational use of heating and pasteurization,
certain regions of France were able to achieve the perfection and finish in
the presentation of their wines that helped them conquer first place in the
world market. And less favored wine regions would improve the commercial value of their wine by emulating the great ones. This was the crux of
the message from Montpellier.

There was little oenological interest at Montpellier in the effect of heating on fine wines because the wine produced in the Midi was mostly table
wine. A quite different situation existed in Burgundy, as proud of its *crus*
as Bordeaux. We find at least one striking difference between the scientific
mandarins of Bordeaux and Burgundy in the immediate post-Pasteurian
period. Georges Curtel, clad in the armor of university science – doctorate
and *agrégation* – and the head of the oenological institute in Dijon, expressed serious reservations about filtering and pasteurizing really great
wines. Both procedures were judged to be excellent for most wines; Curtel
himself carried out laboratory experiments to develop small apparatuses.

[55] Mallet, *RV* 39 (1913): 361.

But the *finesse* and bouquet of "les très grands vins" tolerated these manipulations rather badly. Given the recognized need to improve equipment, this was not a surprising judgment. (The technique of flash pasteurization introduced at the Institut Pasteur before the First World War did not spread until later. It made a triumphant reentry into France under the American name for it. J. Gorgerat, a chemical engineer, was not surprised: the French typically praise to the heavens whatever comes from abroad. Or denounce it, he might have added. Perhaps "pasteurisation rapide en couche mince" was too much of a mouthful.)

There was a similar reaction from Germany. In the 1870s the new *Annalen der Oenologie* (Heidelberg) devoted considerable attention to a critical analysis of Pasteur's experiments on wine. Adolph Blankenhorn, a founder of the *Annalen* and head of the oenological institute at Karlsruhe, concluded that heating would destroy the bouquet of delicate German wines. Experiments by Pasteur on French wines denied any such effect. Students of the cultural context of science may suspect the role played by national culture in the German case, given the state of Franco-German relations at this time and Pasteur's anti-Germanic rage, but the question of the influence of heating on the taste of wine became a much debated issue.[56]

Curtel's results directly contradicted Pasteur's magisterial judgment on the appropriateness of pasteurization for the great wines of Burgundy, whose frequent inability to withstand export had inspired his "great discovery." Perhaps Curtel was a stronger believer in the objectivity of taste. Pasteur was forced to recognize the strong influence of imagination after he saw his *commission d'expertise* completely reverse itself on the same wine in the course of a few days. Most serious tasters have had this experience, and the real experts declare that the wine has changed.[57] We should also remember that at this time science could provide no evidence for changes in the wine that were induced by heating. So the taster had no scientific evidence to back up his professional judgment on changes in quality.

Pasteurization opened up a good opportunity for scientists to extend their empire into the making of wine, hitherto pretty much exempt from their activities. Phylloxera gave them a strong foothold in the vineyard. It was also an occasion for demonstrating the usefulness of oenological science, an important matter when the funds for growth came mostly from departmental and municipal sources. And professors seemed eager enough to appear in publications read by producers. Chavastelon, chemist

---

[56] Blankenhorn's magnificent oenological library ended up in the Staatliches Weinbauinstitut in Freiburg im Breisgau.

[57] "Expériences de pasteurisations organisées par l'Institut oenologique de Bourgogne," *Le petit viticulteur bourguignon,* 1re année, no. 2, pp. 1–3.

and professor of oenology at the faculty of sciences in Clermont-Ferrand, appeared in the *Revue du Vignoble* in 1907 to recommend both filtering and pasteurization for the preservation of wines. And if the wines were to travel long distances or be guaranteed to be in drinkable condition, these processes were indispensable.

Not that the Bourbonnais and Auvergne were big exporters of wine to foreign parts, but in the period 1890–9, the Puy-de-Dôme averaged an annual production of more than a million hl, big enough to be important in the local economy. In the thirteenth and fourteenth centuries, the wines of Saint-Pourçain and the Côtes d'Auvergne (Châteaugay, Chanturgue, Corent, and Boudes) enjoyed royal and noble favor. Although changing fashions of consumption and shifting economies replaced these wines with those of other regions, the near deathblow came with phylloxera. A modest restoration has taken place recently with the help of viticultural and oenological science and a small market for the wines.

Like Curtel in Dijon, Chavastelon did experiments in his laboratory on filtering and heating wine to help bring both pieces of technology to the small producers. As expensive equipment, special knowledge, and skill were needed in these operations, he recommended collective action rather than individual prowess. At that time, only the great shippers of Burgundy and Bordeaux were known to be pasteurizing their wines before shipping them long distances. Some experiments had been done in Saumur, but pasteurization had not become standard practice for the historically famous if now mediocre red Saumur-Champigny.[58] Perhaps Chavastelon hoped that his scientific experiments would lead producers to filter and pasteurize in both the Saumur and his own region. He agreed with the argument that heat perceptibly modified wine by removing its astringence: the wine appeared to have aged, which would satisfy the consumers whose model of taste prescribed suitably aged wine for conspicuous consumption and would fool the people who said that heated wine did not age.[59]

The decision at the end of the nineteenth century by the fourth Latour winemaker to introduce pasteurization into his famous Beaune winery has finally turned a rather academic question about the difference between earlier and contemporary wines into a more passionate debate. A quarrel had existed over the relative virtues of pre-phylloxera and post-phylloxera wines, but it died rather quickly, surviving only as an esoteric topic for eccentric Englishmen. Maison Latour recently defended its "gentle option" of pasteurization, really too crude a term for a delicate modern tech-

---

[58] See the *Guide Dussert-Gerber des vins de France 93,* which blames a poor vinification of the wine for losing the virtues of the specific taste of the cabernet franc grape. Sparkling Saumur is a better bet. *Parker's Wine Buyer's Guide* (3d ed.) agrees with Dussert-Gerber.

[59] *La Revue du Vignoble,* no. 114 (1907), pp. 981–3; see also R. Chavastelon, *Conseils pratiques aux viticulteurs d'Auvergne* (Clermont, 1905).

nique that is less "tiring" for the wine than fining and "sterile" filtration, which remove color and flavor from the wine along with yeasts and bacteria. The wine is heated to 72 degrees centigrade for three seconds while it "passes continuously through a modern stainless-steel heat exchanger where it comes into contact with sheets heated by a separate circuit of hot water." The wine comes out "of the machine through a cooler which returns . . . [it] to ambient temnperature before its final gentle filtration and bottling."[60]

Anthony Hanson, author of a standard guide to the wines of Burgundy, is not pleased with the result: the red wines are "early-maturing, high in alcohol, very round – not my idea of fine Burgundy." He rates Latour's whites as "often splendid." But Hanson gives different numbers (70 degrees for one minute) from those in Latour's house journal for the heating process. Matt Kramer just reports that some drinkers are disturbed by the practice of flash pasteurization.[61] But no one seems interested enough to conduct rigorous tastings, like Pasteur and Gayon did, to test the difference between "the raw and the cooked." Thermo-unified wines, those made from grapes heated to 70°C and more before fermentation, have an initially different character from traditional vintage wines but soon become very similar to them. Oenologists have little interest in the matter, or disapprove of the practice.[62]

Louis Latour, a bit puzzled by the recent excited discussion over the fact that his red burgundies are pasteurized, wonders why there is so much controversy over applying to wine a fairly innocuous procedure that is widely accepted for other foods. Latour believes that the oenophile's weltschmerz is based on the mythical power of French belief in the importance of staying close, whenever possible, to one's peasant roots and a Barrèsian belief in the essential purity of what springs from the soil before it is corrupted by man – Rousseauism applied to crops. Latour's answer is rooted in the mythological connection of food, especially wine, with the soil, implying that products should not be tampered with by modern technologi-

---

[60] Maison Louis Latour, *The House Journal,* no. 2 (Spring 1986), p. 3.
[61] Hanson, *Burgundy,* p. 117; Matt Kramer, *Making Sense of Burgundy* (New York, 1990), p. 290.
[62] Some articles in Pascal Ribéreau-Gayon et Pierre Sudraud, eds., *Actualités oenologiques et viticoles (AOV),* published in 1981, show that it was still a topic of some importance in the 1970s (see, e.g., pp. 283–7) but was not of much interest a decade later; or at least it was of no significance at the *4e Symposium international d'oenologie:* see the *Comptes rendus* in *Actualités oenologiques 89* (1990), ed. Pascal Ribéreau-Gayon et Aline Lonvaud. But heating the must during fermentation, either throughout the process (thermovinification) or at the end ("macération finale à chaud") has certain advantages. See Ribéreau-Gayon, "Conduite de la macération," and P. Martinière, "Thermovinification et vinification par macération carbonique en Bordelais," in *AOV,* pp. 298–310. See also Ribéreau-Gayon, "Les phénomènes oxydatifs dans les moûts et les vins," *AOV,* pp. 205–12, on the use of heat to kill dangerous enzymes in the prefermented must. For a balanced treatment of pasteurization and heating, see Amerine and Joslyn, *Table Wines.*

cal means. Having moved so far from our peasant roots in nature – we ignore the fact that the peasant's nature is as artificial a construct as that of urban man – we insist on as little modification in organic products as is consistent with sanitary demands.

Although by the 1970s heating wines to kill harmful bacteria was quite popular, it has been pointed out by Jacques Puisais and other oenologists that heating destroys the enzymatic system of wine, with the result that the wine is no longer a living liquid. Of course, as Pascal Ribéreau-Gayon has shown, there are enzymes and enzymes, and some can ruin wine. True, it was possible to use less sulfur dioxide in vinification, which is good news for health buffs, but the heated wine, like pasteurized beer, is harder to digest. In the account book of life processes there seems to be no credit without a corresponding debit.

Pasteur as Oenological Icon

About ten years after giving a lecture on "oenology yesterday and today" (1954), which was fairly critical of Pasteur's oenology, Jean Ribéreau-Gayon presented a minihistory of the viticultural and oenological sciences in Bordeaux in which he seriously upgraded Pasteur's position in the oeno-logical pantheon.[63] "Pasteur is truly the creator of scientific oenology (1861–6). He discovered the role of the yeasts that make wine and that of the 'microscropic parasites' or lactic 'ferments' that alter it; he studied aging; he created new methods of analysis; he discovered the secondary products of fermentation in wine." Pasteur would have been smugly satis-fied to learn that Berthelot, subtle supporter of the so-called chemical theory of fermentation, earned only two paragraphs for his work on wine, although his *Chimie végétale et agricole* (1899) contains two hundred pages on it.

Ribéreau-Gayon's promotion of Pasteur to king of the oenological gods was done in the context of presenting a royal succession of oenologists who ruled in Bordeaux. Obviously this sort of genealogical begetting to trace institutional origins back to one of the greatest scientific minds of the nineteenth century and the creator of the discipline itself does not allow for a criticism or catalog of deficiencies in research that would have been appropriate for other occasions. Bordeaux seems from the beginning to have been destined for oenological greatness by both geography and the *esprit oenologique*. The copy of the second edition of Pasteur's *Etudes sur le vin* at the Institut d'oenologie in Bordeaux carries a dedication from

---

[63] Jean Ribéreau-Gayon, "L'oenologie d'hier et d'aujourd'hui," in *Les sciences de la vigne et du vin à l'Université de Bordeaux* (Bordeaux, 1965).

Pasteur to Gayon: "it bears witness to the bonds uniting Bordelais oenology to the Pasteurian school."[64]

Nor does Ribéreau-Gayon forget to mention that Gayon was the heir apparent to the Pasteurian throne in Paris. In 1904 Gayon could have succeeded Duclaux in the chair of biological chemistry at the Sorbonne and become the director of the Institut Pasteur. (Duclaux, notorious for having imposed a four-volume *Traité de microbiologie* [1898] on his readers, had a respectable reputation among oenologists for the long memoirs he published in the *Annales de chimie et de physique* on the higher alcohols of wine and the transformations of color in wine as it ages.) Acting on a variation of the Miltonic principle that it is better to rule in hell than serve in heaven, Gayon decided to continue building his empire in Bordeaux. His provincial virtue was rewarded by the addition of the first chair of physiological chemistry to the faculty, given to E. Dubourg, whose course devoted proper attention to industries of fermentation. In the *Naturwissenschaften* few Bordelais professors escaped the happy burden of oenological relevance.

It seems difficult for the French to subject Pasteur to critical scrutiny. He has icon status. Not that it is difficult for his admirers to admit that some of his ideas have been shown to be incomplete or even wrong. Two obvious cases are the Pasteurian view of fermentation as an exclusively biological or vital process, excluding the chemical action of the enzymes produced by the yeast, and his simplistic hypothesis that the aging of wine is essentially the combining of oxygen with the wine. "Mistakes" of this sort are made by any scientist; the more creative and productive the scientist, the more mistakes he is likely to make in trying to get one big thing "right." The really serious part of the Bordelais critique of Pasteur is of a more general and systemic nature, indicating a fundamental flaw in his oenology – namely, the infection of his work by the Baconian disease, empiricism.

Jean Ribéreau-Gayon sees Pasteur's work as only an intermediary step in the development of scientific oenology because the Pasteurian explanations of alcoholic fermentation and of the alteration of wine retain a large dose of empiricism.[65] What does this mean? Not much that Pasteur could have done anything about. He could not identify the mechanisms and modalities of the intervention of the yeasts, the bacteria, and oxygen in the winemaking process. Ribéreau-Gayon regarded the method by which Pasteur worked as a typical stage between empiricism and the effort of

[64] J. Ribéreau-Gayon, "L'actualité de Pasteur," *Revue des questions scientifiques* 25 (June 1964): 338; text of the Conférence d'ouverture au symposium international d'oenologie, Bordeaux, June 1963, in ibid., pp. 337–53.
[65] Ribéreau-Gayon, "L'oenologie d'hier et d'aujourd'hui," pp. 18–24.

scientific research. Pasteur appears in this history of science as a part of the evolution of science as it heads toward a higher destination, the scientific oenology of Bordeaux.

In the end, after the triumph of the new oenology, including the conquest of university oenology by Jean Ribéreau-Gayon and Emile Peynaud, Pasteur's ideas supposedly remained nearly as important in many respects as Pasteur and his disciples thought they were. In making use of Pasteur's iconic status, Peynaud gives Pasteur his due. "If, on the one hand, contrary to the ideas of Pasteur, certain bacteria can play a favorable role in the elaboration of wines, on the other hand, the harmful action of bacteria that cause spoilage is much more frequent than is generally supposed. Also, when malolactic fermentation is finished, Pasteurian principles regain all their value."[66]

In the preface to the first edition of his immensely successful *Connaissance et travail du vin,* Peynaud gives a statement of an oenological philosophy with which Pasteur could probably agree: "Oenology is not an abstract science; it was born from the search for solutions to practical problems. But if facts are observed at the level of work in the wine cellar or warehouse, explanations cannot be given, laws cannot be established; progress can only be born at the higher level of the study of phenomena. Thus oenology has its deep roots in physical chemistry, biochemistry, and microbiology."

Going beyond Peynaud, the historian friendly to Pasteur might concoct a naive and satisfactory syllogism about the origins of scientific oenology: scientific oenology began with microbiology; Pasteur was the first to apply microbiology to wine; therefore Pasteur invented oenology. The making, vinification, aging, and spoiling or conservation of wine are, for the most part, applied microbiology. As microbiology changed, so did oenology, from the effects of microorganisms – disease and decay and life and growth – to the physiology and biochemistry of microorganisms. The study of wine became part of the study of life processes.

Oenology now seems firmly established in a post-Pasteurian paradigm, especially as a result of "the transformation of winemaking" through the understanding and mastery of malolactic fermentation, "the great cultural revolution of the *grands vins.*"[67] In an interview in *Decanter,* Peynaud gave a proper scientific reaction to Pasteur's work: it was basic but is now obsolete. "Oenology had made little progress after Pasteur. Although his work was fundamental, it has little value today." An unintended consequence of this judgment is to make Peynaud, and particularly Jean Ribéreau-Gayon, "the father of modern oenology" – as important as Pasteur in oenology

---

[66] Peynaud, *Connaissance et travail du vin,* p. 117.
[67] Peynaud, *Decanter,* 15 (March 1990): 38.

but without assuming that they too will pass into scientific limbo. Scientists who talk about dead scientists find the discourse of paradigms to be congenial but slip into the discourse of cumulative processes when they talk about their own work. "When we [Peynaud and Jean Ribéreau-Gayon] started no one would attack Pasteur, 'the god of wine,' but we did!"[68] Whether an icon to worship or to smash, Pasteur is a major figure in the history of oenology – the founder of oenology, he himself would say, and many would agree with him.

[68] Ibid. Peynaud indulges in a bit of historical invention: "We were helped by the fact that Ribéreau-Gayon's grandfather, who invented the Bordeaux Mixture spray against mildew [he should have added 'along with Alexis Millardet'] and founded the Bordeaux Institut d'Oenologie in 1870, was a pupil of Pasteur" – a trifle early even for the founding of the Station agronomique.

PART III

Oenology in Champagne, Burgundy, and Languedoc

7

Champagne: The Science of Bubbles

The champagne industry is one of the world's most successful and profitable enterprises, striking evidence of the marketability of distinction and carbon dioxide in drink.[1] Annual production is about 200 million bottles, barely enough to satisfy a greedy elite, at least in times of prosperity. The rise of the champagne industry partly compensated for the economic and demographic decline of the Champagne in the late nineteenth and early twentieth centuries. With the fall in prices for wheat and beet sugar, land was left fallow; in 1914 agricultural production was half what it had been around 1850. The area covered by vines dropped to 38,000 hectares because of phylloxera and competition from the Midi. In 1835 vines covered 18,495 ha of the Marne; by 1900–9 the figure was 14,860; and in the period 1920–49 the figure was always less than 9000 ha. By 1980–7, vines covered 19,214 ha. The amount of land covered by vines in the Marne has just more than doubled in the twentieth century, with production of wine having increased about fourfold.[2]

The champagne trade came to be controlled by the *négociants,* whose power had its origins in the eighteenth century but grew strikingly in the nineteenth century with the rise in sales. One of the distinguishing features of the industry was the presence of the champagne widows (Bollinger, Clicquot, Laurent-Perrier, Pommery, and Roederer). In recent years financial groups have replaced many of the famous family houses, the "rois du Champagne." The market in the eighteenth century was a few hundred thousand bottles a year, minor compared to the boom under way by 1850, when eight million bottles were sold. Sales of champagne have steadily increased since the nineteenth century: 13 million bottles in 1866, 17.5 million in 1870, 28 million in 1900, 38 million six hundred thousand in

---

[1] Bubble (*la bulle*) power in French is *la mousse* (froth, scum, foam, lather; effervescence). *La mousse:* "mot employé en Champagne, pour désigner non pas l'écume, mais la faculté de la produire." Maumené, *Indications théoriques et pratiques sur le travail des vins, et en particulier sur celui des vins mousseux* (1858), p. 400, n. 1.

[2] Irène Peiffert-Henriot, *Histoire des champenois* (1980), p. 365; Marcel Lachiver, *Vins, vignes et vignerons,* "Annexes," in "Le vignoble français (Paris, 1788–1987)," pp. 579–625, a magnificent collection of data on land, production, and export.

1910, and 44 million in 1913. In the second half of the twentieth century this luxury product became somewhat democratized, with sales approaching two hundred million bottles in societies enjoying, according to a classic exaggeration in Rostow's model of the "airborne economy," their age of mass consumption.

Champagne making was a capitalistic joy because of the possibility of large-scale production of a product of high quality, something difficult to achieve in other areas of winemaking. Production was labor-intensive, the economist might pontificate. In 1827 the operation of bottling *vin mousseux* required five workers (an atelier). Production from Epernay and Reims had expanded to over four million bottles. Moët alone sometimes had ten ateliers bottling at the same time.[3] The solution of scientific problems unique to champagne making and the introduction of new technology permitted the development of a giant industry. Moët et Chandon (Epernay), number one in production, corked 25 million bottles in a recent year. The financial octopus of the business is the conglomerate Moët-Hennessy-Louis Vuitton (LVMH).[4] Moët et Chandon always dominated the *négoce* champenois; LVMH now controls 24 percent of the champagne market.

Vins de Champagne and Champagne Mousseux

The white wines of northern areas, like the Champagne, have a tendency to keep part of their sugar after the first fermentation, which means that if bottled at the end of winter the wine ferments again with the coming of warm weather in the spring. Champagne may have been first produced by sloth, for the Champenois winegrower spent his time on his red wines, letting the white ones go their own delinquent way. Their destiny was to produce a weak *mousse* or froth, a process best described as *crêmer* rather than *pétiller*, a subtle but important difference for the philosopher of *mousse*. Once it was decided that drinking this curiosity was a desirable experience, various steps could be taken to improve the product, or at least to keep more of the gas in the wine. The use of cork toward the end

---

[3] J.-A. Cavoleau, *Oenologie française* (Paris, 1827), p. 210.
[4] LVMH, the booze and baggage group, in the lingo of *The Economist,* in recent years controlled eight famous labels (Moët et Chandon, Veuve Clicquot, Mercier, Canard-Duchêne, Ruinart, Henriot, Pommery, and Lanson). Pommery and Lanson, with their 500 hectares of vines and 50-million-bottle stock, were recently acquired for 3.1 billion francs, which is 39 times the earnings of the two companies. A few years ago Georges Duboeuf enjoyed a comparable share (15%) of the Beaujolais market. (See *Le Monde,* Dec. 9–10, 1990, p. 17, and *The Economist,* Dec. 15, 1990, p. 71.) For a short historical treatment, see Louis Bergeron, "Permanences et renouvellement du Patronat," in Yves Lequin, *Histoire des Français XIXe–XXe siècles,* vol. 2: *La Société* (1983), pp. 242–9.

of the seventeenth century made it possible to keep the carbon dioxide in the wine and, more often, break the bottle.

Obsessed with the search for novelty in historical explanation, we run the risk of ignoring the high reputation that champenois wines had before the new *vin mousseux* usurped their place. It is certain that the new wine was able to exploit that reputation, which gave it an even more advantageous position than that from which the new clarets benefited in coming out of Bordeaux. The Champagne enjoyed a reputation as one of France's best wine-producing areas long before "the night they invented champagne." From the fourteenth to the sixteenth century, the Flemish in the Burgundian wine trade would also buy wine in the Champagne (Châlons-sur-Marne, Epernay, Reims), especially during years of small production in Burgundy. The wines of the Marne valley, particularly Ay (used as a designation for the wines of all La Rivière de Marne in the sixteenth century), had the best reputation. Farther north, the red wines of the Montagne de Reims, where Verzenay had an excellent reputation, inspired the abbé de Boisrobert, a playwright as well as secretary to Richelieu, to a eulogy in 1646 of the "grand vin du mont de Reims." Meanwhile, the wine of Orléans, Coucy-le-Château, and the "wines of France" (around Paris) were degenerating and going out of fashion. For the "vins de Champagne" the mood was upbeat in the second half of the seventeenth century. At the end of the century the still or nonsparkling wines of the Champagne were three times more expensive than the wines of Burgundy.[5] The switch of the Montagne de Reims from red to white wine took place between roughly 1825 and 1850, as *vin mousseux* became a more marketable commodity than the red wine of the area.[6]

The term "vins de Champagne" came into use in a general sense at the end of the sixteenth century but was unusual before the eighteenth. The wines of Ay and Reims might have been sold earlier as "vin français" – probably the language of governmental regulation – or "vin de France," a phrase having a certain geopolitical-historical justification. After the late seventeenth century some ordinary white wine was baptized "tisane de Champagne" (now a term for light champagne). The wines of the Montagne de Reims improved in the seventeenth century; yet it was inevitable that their producers would envy Burgundy, because its wines had a deeper red color. The Champenois sometimes established a pernicious equality by adding elderberries or more vile coloring agents like the "teinte de Fismes," a liqueur recommended by some scientists because of its medici-

---

[5] Michel Dovaz, *L'encyclopédie des vins de Champagne* (1983), p. 14.
[6] Loubère, *The Red and the White*, p. 109, points out "the leading position of red wine" in the Marne as late as 1832. Out of a total wine production of 480,000 hl, only 50,000 hl were "white for sparkling" and 310,000 hl "ordinary red for local use."

nal effect on wine as well as its coloring power. Unaffected by the color craze, purists denounced this mélange of nature's products as a type of fraud to change an unsaleable Champenois wine into saleable red burgundy.

The squabble over color was probably less important than the economic downturn of the 1720s in causing sales to the former high-flying consumers to drop by about 50 percent. There was still a market for stuff changed to "petit bourgogne." An understanding intendant justified the practice as better than permitting the ruin of the wine merchants. In bad years, the Burgundians used the wines of the Champagne to make burgundy and save their market from competition. This so-called fraud, widespread after the sixteenth century, is often taken to indicate the inferiority of the red wine of the Champagne to that of Burgundy. This is a bit of a simplification, for the reds of Bouzy – it should be in the bluffer's guide to wine – and even of Vertus, Damery, and Cumières had little to fear from most of the wines of Burgundy, which were also often beefed up with wines from the south to enable them to travel better and conform to market demands. The vinification of these Champenois wines was difficult because they were low in alcohol and high in acidity. Adding good cognac was one good way of getting the level of alcohol up to 12 degrees. Like the wine of Burgundy, that of the pinot of the Champagne was also subject to the diseases *amertume* and *tourne*. In the nineteenth century, some winemakers in the Marne mixed 10 liters of good bordeaux in a *pièce* of 200 liters of Bouzy for elite consumers in Belgium and the north of France, as well as the great restaurants of Paris. No gourmet complained that the flavor and *finesse* of this great red was ruined.[7] As the Champenois turned more to white wine, with the market developing for *mousseux* in the eighteenth and becoming dominant in the late nineteenth century, the practice of mixing reds decreased and eventually disappeared.

The general context of the earlier changes is, as Michel Dovaz points out, the evolution of taste and wine between the fourteenth and sixteenth centuries. The growing prestige of the red wines of Burgundy meant a lower social status for white wine. The market trend led to greatly increased planting of the pinot noir in the Champagne, beginning at the end of the fourteenth century. But the pinot noir is happiest in Burgundy; in the Champagne, even macerated, it did not produce a competitor for the wines of Beaune. An interesting switch in strategy based on a different criteria of taste eventually turned the Champagne into a formidable rival of Burgundy. Instead of trying to imitate the fruity and round Burgundian

---

[7] "Recherches sur les origines des vignobles et du vin mousseux de Champagne," *BLEVOM* (*Bulletin du Laboratoire Moët et Chandon*), vol. 1 (1898), etc. On the reds of the Champagne, see L. Mathieu, "Les vins rouges de Champagne," *RV* (Sept. 7, 1895), reprinted in the *Revue de Champagne et de Brie*, 2d ser., 7 (1895): 792–6.

flavors, the Champenois went for delicate floral bouquets.[8] Just pressing
the pinot noir grapes without crushing or macerating them gave birth to
the *blancs de noirs* or *vin gris,* made from the grapes of the Rivière rather
than the Montagne. The Montagne did not play much of a role in producing *vin mousseux* until the nineteenth-century expansion of the market.
What we call champagne (*vin mousseux* – sparkling wine is the slightly
ridiculous English term) is a modern market arrival in the history of alcoholic consumption. Use of the word "champagne" for *vin mousseux* dates
to 1695, according to the *Petit Robert.*[9] This new white wine cannot be
identified with any specific alchemist, although the term *vin gris* appeared
in the second half of the seventeenth century. Because the term *vin
mousseux* appeared about the same time, some writers have argued that it
is reasonable to assume that the two refer to the same wine. Champagne
was born.[10]

The great advantage of *mousseux* over the traditional and fragile *blanc
de blancs,* which yellowed and lost its delicacy in less than a year, was that
the *mousseux* stayed white and kept its impressive qualities for three or
four years. Nature compensated for the crudity of the techniques of vinification, which were then incapable of preventing the oxidation of the wine.
The yield of pinot noir per hectare being less than that of white grapes,
vinification was probably done more carefully with a more precious commodity.

The Craze for Fizz

Saint-Evremond dated the fashion for sparkling wine to about 1661: it was
an appropriate wine for the restoration of Charles II in England. Dovaz
emphasizes that it was acceptable to drink "champagne mousseux" (a
phrase not yet tautological) in high society even before 1674. What is the
explanation for this expensive craze, if drinking can be explained in non-
Dionysian terms? Historical commentary always links the rise of champagne to "decadence" and the old regime's *douceur de vivre,* the sort of
thing that Talleyrand regretted seeing disappear in the French Revolution.
During the regency of Philippe d'Orléans (1715–23) the aristocracy and

---

[8] Dovaz, *L'encyclopédie des vins de Champagne,* p. 13.

[9] Littré's *Dictionnaire de la langue française* classifies champagne as an artificial wine: "Le
champagne est un vin factice qui s'est fait d'abord avec le vin de la Champagne, plus
propre en raison de sa légèreté à être travaillé de la sorte, mais qui a été imité en Bourgogne
et ailleurs. 'Je vide gentiment mes deux bouteilles – peste! – oui vraiment, du champagne
encor, sans qu'il en reste' " (Regnard, *Fol. Amour,* 3:4).

[10] Dovaz, *L'encyclopédie des vins de Champagne,* p. 13, gives 1665 as the approximate date of
its birth. His source is the Chanoine Godinot, *Manière de cultiver la vigne et de faire le vin
en Champagne et ce qu'on peut imiter dans les autres provinces pour perfectionner les vins*
(1718): "Il n'y a guère que cinquante ans qu'ils [les vignerons champenois (N. de l'E.)] se
sont étudiés à faire des vins gris presque blancs."

haute bourgeoisie became avid consumers of the *vin mousseux* of the Champagne. The rapid extension of vineyards to cater to a general increase in the consumption of wine, including this new consumer craze for *mousseux,* caused sober politicians and bureaucrats to worry about the future of fluctuating cereal production – rapidly reproducing Frenchmen lived mostly by bread alone – and in 1731 they prohibited the planting of new vines and ordered untended ones torn up. The order remained as much a dead letter in the Champagne as in Bordeaux.[11] The notorious *soupers* of the regent supposedly made the consumption of champagne more popular; if so, it was the only thing about these affairs that anyone wanted to associate with. The regent got drunk only on champagne.[12] To safeguard his supply, he gave the Abbaye d'Hautvillers to his legitimate son, the chevalier d'Orléans.

Women have been noted consumers of champagne since its invention. The beautiful drinkers of the eighteenth century were Mesdames de Sabran, de Phalaris, and de Parabère, who had a reputation for drinking like a *Landknecht.* Under Louis XV the cult of champagne grew impressively, with the court and nobles leading the way: the comte de Richelieu and mesdames de Pompadour, de Mailly, du Châtelet. Phallic *flûtes,* of ancient pedigree, were popular for champagne drinking before the reappearance of the *coupe,* which, myth says, was remodeled on the breast of the queen, thus perhaps symbolizing a higher status for *mousseux.*[13]

The popularity of champagne with the upper classes spread over Europe with French culture. The English, pioneers in the consumption of many alcoholic liquids, were not slow in putting champagne to excessive use in their *petits soupers.* The *gotter Wein* also penetrated into German courts. Even the frugal Frederick William I of Prussia (1713–40) was an enthusiastic consumer but without enough interest in science to give his academy of sciences any wine to analyze in replying to his request to explain why there are bubbles in champagne. Whatever reasons are given for the enthusiastic reception and continued popularity of this product, its quality and the enormous capacity of its small market of rich consumers must be seen as prime factors. In many ways the appearance of champagne is comparable to the contemporary invention of the new clarets for the eighteenth-century London market, with a similar constellation of forces making for its success.

---

[11] Peiffert-Henriot, *Histoire des champenois,* p. 235.

[12] Philippe, duc d'Orléans, regent from 1715 to 1723, was the son of "Monsieur," brother of Louis XIV; Monsieur was the chief expert at court on women's underwear – for the wrong reason: he liked to put it on.

[13] See Serena Sutcliffe, *A Celebration of Champagne* (London, 1988), "Serving Champagne," pp. 192ff.

The Myth of Dom Pérignon's Mousseux

The reader is probably wondering by now when Dom Pérignon is going to appear in this narrative.[14] He cannot be ignored in the histories of viticulture, vinification, and capitalistic agriculture. And the invention of champagne?

> Mais dix siècles plus tard, le moine Pérignon
> Inventait le champagne et lui donnait son nom.[15]

Alas, not true, or at least not in the historian's history. First there is the problem of definition.[16] "A century or two before champagne was born, effervescent wines were sold in Italy as *Refosco* and *Moscato spumante*. They were cloudy and were not especially prized or even popular."[17] If the word *champagne* means *vin mousseux,* Gaillac (Sud-ouest, the valley of the Tarn) and Limoux (Languedoc) have a much older pedigree: medieval. This southern sparkling wine was identified as *mousseux méthode gaillacoise* or *méthode rurale.* The effervescent *appellations* Blanquette de Limoux, made from the basic mauzac grape and varying proportions of chardonnay and chenin, and Clairette de Die are also made by this method. If the word *champagne* refers to a wine made according to the so-called *méthode champenoise,* with the secondary fermentation taking place in the bottle as a result of the addition of the sugar syrup (*liqueur de tirage*), Dom Pérignon cannot take credit for this invention, for, as far as we know, he was innocent of sinning with sugar. His reputation is based on his wine's natural greatness. If the word *champagne* means "mousseux dégorgé," that is, with the sediment of the refermentation removed, obviously Dom Pérignon had no need to indulge in the practice. Nor did he need to rack his wine. Besides, by the time Dom Pérignon arrived at Hautvillers, people had already acquired the taste for *mousseux.*[18]

Dom Pierre Pérignon, a contemporary of Louis XIV, was for nearly fifty years (1668–1715) the procurator and cellar master of the Benedictine Abbaye d'Hautvillers, a property now owned by Moët et Chandon. This position put him in charge of the monastery's domain of vines (fromenteau or pinot gris, morillon or pinot noir, morillon or pinot meunier, and the lowly goy or gouais that was made into wine separately). The proper-

---

[14] For the French nation, wine is "une boisson-totem." On the mythology of wine, see Roland Barthes, *Mythologies* (1957), "Le vin et le lait," pp. 74–7. "*Le mythe est une parole* . . . c'est un système de communication, c'est un message," p. 193.

[15] Gonzalle, cited in C. Moreau-Bérillon, *Au pays du Champagne* (Reims, 1922), p. 77.

[16] A point raised by Dovaz, *L'encyclopédie des vins de Champagne,* p. 15, who gives the definitions.

[17] L. W. Marrison, *Wines and Spirits,* 3d ed. (Harmondsworth, 1973), p. 95.

[18] Dovaz, *L'encyclopédie des vins de Champagne,* p. 15.

ties in Cumières, Dizy, Champillon, Hautvillers, and Mardeuil (across the Marne), all added up to 25 hectares when he died. Apart from not inventing champagne or *sucrage* (in this case, adding the *liqueur de tirage,* a special sort of chaptalization), it was never clear what Dom Pérignon's oenological achievement actually was until René Gandilhon chronicled *La naissance du champagne* (1968). Max Sutaine summed it up rather well in 1845 by saying that he made vinification into a true art. And if champagne was not born of him, it was born with him.[19] A European demand for "le vin de Pérignon" was testimony to his success. So was its high price. Louis-Perrier, writing during the economic expansion of France in the Second Empire, saw Dom Pérignon as the agent of "a sort of revolution in the commerce" of the wine-producing part of the Champagne, a revolution completed when the "secret" of producing *mousse* in bottles was discovered.[20] (The history of winemaking is full of secrets that generally give disappointing results, as in the case of Dom Pérignon's supposed addition of ripe peaches to his wine.)

Dom Pérignon's reputation as a leading and passionate viticulturalist and vintner emerged only near the end of the seventeenth century. A Newton of viticulture? Some of his principles seem the essence of scientific viticulture and vinification even today; others seem merely practices of his day, now abandoned for a new oenological decalogue. On the basis of the aesthetic principle that the white wine should be as clear as crystal, he used only the pinot noir grape. Wine from white grapes was accused of turning yellow. No wine propagandists existed to praise its golden color. It seems strange to us, victims of an age in which Roederer's Cristal (invented for Tsar Alexander II) is made from 40–50 percent Chardonnay grapes. Moët's 1982 Dom Pérignon – this luxury brand was put on the market in 1936 – was 60 percent Chardonnay.[21] "Blanc comme le cristal" meant 100 percent pinot noir for the seventeenth century; Blanc des Blancs means 100 percent Chardonnay for us. Like Dom Pérignon, later winemakers catered to popular trends among the elite consumers and in so doing created, or at least participated in, the fads of the day.

Max Sutaine placed Dom Pérignon's activities at the center of an oenological revolution – a much abused word, but no synonym seems to measure up to the syntactical demands of the concept. He was a revolutionary oenologist, then, worthy of joining that happy band enjoying mythical

[19] René Gandilhon, *Naissance du champagne: Dom Pierre Pérignon* (1968); Lachiver, *Vins, vignes et vignerons,* "naissance du vin mousseux de Champagne," pp. 275–80 ("le champagne . . . nait avec lui," p. 276); and Max Sutaine, *Essai sur l'Histoire des vins de la Champagne* (Reims, 1845), pp. 44–5, 77. The historian's task was simplified by the destruction inflicted on the Abbaye d'Hautvillers (founded in 660) in 1562 during the wars of religion: only a few documents survived.

[20] Jean-Pierre Louis-Perrier, *Mémoire sur le vin de Champagne* (1865), p. 78.

[21] Sutcliffe, *Champagne,* "The Great Champagne Houses," pp. 80–171.

status such as Pasteur, Ribéreau-Gayon, and Peynaud, because of his ruthless emphasis on quality both in vines planted and those kept in production and on the importance of the quality and maturity of the grapes pressed. Dom Pérignon successfully experimented with assembling batches of grapes he judged compatible for making wine of high quality. Contrary to what is sometimes argued, grapes were *not* picked unripe in order to get a wine that was "mousseux et piquans," as is sometimes said, but in three stages, each a few days apart to ensure optimum maturity – an ideal and labor-intensive procedure that is rarely done. Pérignon has a good claim to having invented this procedure.[22]

We are struck by the curiosity of assembling batches of grapes rather than blending batches of wine, as became the more convenient custom later. The ancient practice of blending wines was a controversial eighteenth-century innovation in Hautvillers in the post-Pérignon period. The winegrowers of Pierry brought a lawsuit against the Abbaye d'Hautvillers in 1779 over the practice, which they regarded as adulteration. Claude Moët pointed the experimental way for his firm when he married five wines in a vat to see if they produced a better *mousse* or broke more bottles.[23] The blending of wines with different qualities in order to produce a certain taste model, today a standard practice in the production of champagne, was long regarded as fraudulent. Béguillet's *Oenologie* (1770) denounced the practice because in Burgundy and Bordeaux the purpose of mixing wines from different regions was often to fool people into buying mediocre wine at the price of good wine.[24] Mixing was also a matter of producing a certain model of wine desired by customers, particularly in the case of color and body derived from the addition of Rhône or southwest wine to good bordeaux. The practice developed in the Champagne in order to integrate desirable characteristics in a specific brand, to create a house style, a distinctive champagne of a particular firm or *maison*.[25] (In Bordeaux, blending wine from cabernet, merlot, and petit verdot grapes has the same aim and has nothing to do with fraud.) As the Californian cult of the 100 percent Cabernet subsides, the blender's art is looking less shamanistic.

Dom Pérignon also succeeded in stabilizing the fickle *mousseux,* which essentially meant keeping its *mousse*-potential without adding sugar. The sugar not changed into alcohol and carbon dioxide during the first fermentation was enough for natural refermentation to occur in the bottle.

---

[22] Dovaz, *L'encyclopédie des vins de Champagne,* p. 16.

[23] Gandilhon, *Naissance du champagne,* p. 129.

[24] Beguillet, *Oenologie,* quoted in Dovaz, *L'encyclopédie des vins de Champagne,* p. 16: "L'art de frelater les vins c'est de les couper à froid."

[25] Blending wines is no longer an inviolable Champenois dogma, for nowadays some prestigious champagnes boast of their single *cru* pedigree, and even Krug departed from the family gospel of blending to make its Clos du Menil.

Dom Pérignon had the talent to master all the details of this complex vertically integrated enterprise and to estimate how much residual sugar would make a clear *vin mousseux* without breaking the bottle most of the time. Scholars are probably right in guessing that Dom Pérignon's "secret" lay in assembling the right mix of different varieties of grapes to give the correct amount of sugar in the must.

Language as well as procedure was different at this time. In order to maintain a decent *mousse* for the life of the wine, the grapes were chosen to ensure having the right amount of "liqueur," a word then referring to the natural sugar content of a wine and still used for *vins liquoreux*. The procedure was not easy, for too much "liqueur" made the *mousse* difficult to control and left deposits in the bottle, whereas too little "liqueur" resulted in a tart wine with no *mousse*. Winemakers knew that the *mousse* depended on the nature of the wine for the year and on bottling in the March after the harvest. In the cold climate of the Champagne there was always sugar left in the must because of slow fermentation. When warm weather returned in the spring, yeast and sugar mated in the bottle to produce undesired bubbles of gas. As Dovaz points out, this is not the "méthode champenoise" but the "méthode rurale," and this was the method that Dom Pérignon perfected and systematized once he went into the bubbles market.[26] Empiricism has its little triumphs. The development of a taste for sparkling wine made the bubbles desirable and led to a systematic production of secondary fermentation rather than just leaving the process to the random action of nature.

Dovaz gives a eulogistic summing-up of Dom Pérignon, this "legendary figure. Above all, he lived in a key period. In the world of Champenois wine, we can distinguish periods of before and after Pérignon. B.P.: few bottles, no or hardly any bubbles, no corks, none of the excellent *blanc de noirs,* no wine that kept, etc. A.P.: thanks to him and also to others, especially brother Oudart, the sparkling wine of the Champagne came into existence, sold well and dear, and was put on the road to a real commercial career. Everything was not simple, [however]."[27] Tradition, technical difficulties, commercial problems, drinking fads, and so on delayed the big splash until late in the eighteenth century and *mousseux's* domination of winemaking in the Champagne until the second half of the nineteenth century.

Perhaps the oenological progress for which Dom Pérignon is given credit was part of a general tradition of scientific learning that existed in

---

[26] Lachiver, *Vins, vignes et vignerons,* pp. 277–8; Dovaz, *L'encyclopédie des vins de Champagne,* pp. 16–17. According to the *Guide GaultMillau: Le Vin,* Gaillac now uses both methods and the Cave de la Clairette de Die uses the Champenoise method for its *Cuvée prestige.* Nothing is sacred, unless consecrated, in the wine business.

[27] Dovaz, *L'encyclopédie des vins de Champagne,* p. 18.

the Abbey of Hautvillers, notorious for its Cartesianism and Jansenism, but also popular for its preservation of the relics of "Madame Sainte Hélène." The monastery was one of the pillars of progress in Western civilization: part winery, it was interested in natural philosophy (science) and technology; part learning factory, it had a rich library. It is certainly tempting to chuckle at the paradox of monks following the rigorous Jansenist way of life also being in the business of producing one of the world's most famous wines, a wine frequently associated with sensual pleasures – unless one wants to see champagne as the most austere of wines and thus a logical Jansenist invention.[28] Of course, most people never heard of it, unless they were lucky enough to work for producers, who in the Champagne traditionally gave domestics and day workers the rough liquid of the last pressing (*la rebèche*); perhaps on festive occasions they got some *vin de pressoir,* something more potable from an earlier pressing. The class basis of wine consumption is as evident at the bottom of the scale of consumption as at the top.

It may be that Dom Pérignon's fame should be shared with Brother Jean Oudart, a contemporary of the Abbey of Saint-Pierre-aux-Monts at Châlons and administrator of the seigneurial domain of Pierry (near Epernay), which also had vines in Chouilly and Cramant. Oudart's *vin mousseux* was appreciated by the wine merchant Bertin du Rocheret, the Parisian bourgeoisie, and foreign clients. One of his innovations was to import cork from Spain. He and Dom Pérignon knew one another for over thirty years. But no one is quite certain what to make of customers ordering "bon vin mousseux de Pierry travaillé selon la méthode du frère Oudart." The documents relating to his activity were probably destroyed in the French Revolution.[29]

Science, Sugar, and Taste in the Commercialization of Mousseux

One of the reasons that production of champagne remained low until the second half of the eighteenth century and did not usurp the place of red wine until even later was the difficulty of storing and transporting this fickle liquid. Bottles were not much used for transport until after 1750, although in 1728 the Conseil du roi had authorized the transport of *mousseux* in bottles. Stronger champagne bottles came into use in the early eighteenth century. This was one of the first French uses of glass made according to the English method, which employed coal rather than wood and had a long cooling period. Champagne would begin its journey in a murky mood and alter to an unstable state before it could be sold. The

---

[28] "La boisson la plus délicieuse, la plus délectable . . . le chef d'oeuvre, d'un sévère milieu janséniste." Gandilhon, *Naissance du champagne,* pp. 205, 208, 219.

[29] Dovaz, *L'encyclopédie des vins de Champagne,* p. 237.

bottle made possible an improvement in the quality of wine; it made champagne a commercial possibility. Although the *mousse* power of the wine was less than it is today, the enormous amount of breakage limited commerce until the end of the eighteenth century. ("Champagne is classified according to the pressure within the bottle. *Grand mousseux* has a pressure of sixty-seven to seventy-five pounds per square inch, *Mousseux* between sixty and sixty-seven, and *Crémant* below sixty.")[30] The profits encouraged risk taking.

In the 1780s Moët produced the astounding number of 50,000 bottles; so did a *négociant* in Pierry. We don't know how much self-destructed; 1776 was a very explosive year. Top vintage years, with great *mousse* power, like 1834 and 1842, might have thirteen to fourteen million bottles reduced to eleven to twelve million by breakage; 1842 was a particularly disastrous year. Different authors give different figures. Moët claimed that he lost 35 percent in 1833 and 25 percent in 1834.[31] In order not to lose the wine from the breakage of anywhere from 15 to from 30–40 percent of production, the storage cellars had a system (a glacis) for recovering the liquid, which would be treated and sold as wine. With average production about seven million bottles (six million of which came from the department of the Marne), this salvage system recovered a lot of marketable wine, even if its fizzle had fled.

Producers of champagne noted that the wine produced at Avize (La Côte des blancs – Chardonnay country) was so effervescent in some years that the "impetuous fluid" broke most of the bottles. The wine of Ay, from the pinot noir, was far less explosive or, as Cadet de Vaux put the principle, "the vehicle of fermentation was more abundant in the white grape than in the black." Clever people thought of blending wines from the chardonnay of Avize, Oger, Mesnil-sur-Oger, and Marceuil with the wine from the pinot noir of Ay in order to get a good *mousse* with a pressure low enough not to break the bottle. The wish for a *vin mousseux* had been reconciled with the aesthetic preference for a white wine that would not turn yellow. Equally important, the great principle of Dom Pérignon, emphasizing the need to give wine the greatest possible stability, had been maintained, in spite of the change of method from mixing grapes to mixing wines.[32] Cadet de Vaux thought that this success would lead to a study of wine according to the "laws of a rigorous theory" to identify the wines that would produce lots, little, or no *mousse*. If science can produce more profit, the profit will produce more science.[33]

[30] Marrison, *Wines and Spirits,* p. 101.
[31] Moreau-Bérillon, *Au pays de Champagne,* p. 123.
[32] Gandilhon, *Naissance du champagne,* p. 132.
[33] Louis Perrier, *Mémoire sur le vin de Champagne,* pp. 80–1.

The exploding bottle was not the only problem that had to be overcome in the production of champagne. Handling bottles was dangerous, but sources, which are heavily firm-inspired, are silent on injuries to employees, who, like nineteenth-century chemists, lost a few eyes to unpredictable liquids. The medical virtue of champagne was associated with its acidity, but this virtue was a great commercial disadvantage because people were not used to such acidulous natural products. (There still is not much of a market for *brut zéro* or *non dosé* champagne.) Using sugar to counteract the piquancy of the wines of the Champagne, both still and *mousseux*, seems to have been an ancient practice, but historical information is difficult to come by. Sugar suppliers are necessarily a silent lot, and rare is the babbler like the old "négociant châlonnais" who told of his thousand and one nights of clandestine transport of sugar to the wineries.[34] The success of champagne in some nineteenth-century markets depended on adding a considerable dose of sugar in a *liqueur d'expédition* to the final product.[35] The theory and justification of adding sugar to champagne are, as in the case of the chaptalization of modern French wines, much easier to document because scientists have produced a vast literature in order to convince colleagues to accept the new theory-practice and to convince producers of the need of scientific guidelines for the process.

The issue of the state of science is part of the debate over why champagne did not develop into a solid, profitable industry before the 1780s. In 1777, Jean-Claude Navier (*fils,* 1751–1828) defended his thesis and joined the faculty of medicine in Reims. He thought that the advanced state of chemical knowledge had at least been useful in dissipating the prejudice that *vins mousseux* are dangerous to health. The "torch of analysis" had dissipated ignorance by shining brightly on numerous products of nature, including champagne. Nearly a century later, Louis-Perrier expressed his skepticism concerning the rosy view of Navier on the benefits of chemistry for the production of champagne. The progress of chemistry had done

---

[34] Anecdote in Moreau-Bérillon, *Au pays du Champagne,* p. 108.
[35] Tsarist Russia, Scandinavia, and South America have been identified as markets that demanded sweet champagne. Present-day *dosages* for the different taste models range from zero for Brut zéro or Brut 100% to over 8% for sweet. The *Guide GaultMillau: Le Vin* (p. 471) gives the following percentages:

Qualificatif	Dosage
Brut 100%	Pas de dosage
Brut	Jusqu'à 1% de liqueur
Extra-sec	1 à 3%
Sec	3 à 5%
Demi-sec	5 à 8%
Doux	8 à 15%

nothing for the consumer, who in buying champagne had to choose between a tart liquid or a sweetened wine, not the delicious product it later became. The issue was no longer health but taste.[36]

Camille Moreau-Bérillon, graduate of the Institut national agronomique, professor of agriculture (1900–14), and a municipal councillor of Reims, echoed this view in declaring that one of the chief causes of the lack of progress up to 1780 was the weak development of science, especially chemistry. Natural philosophers – scientists were not yet invented – had only vague ideas of the composition of wine and the process of fermentation. The influence of various factors on the nature and development of wine could be observed but not explained. The intriguing anomalies of champagne, running the gamut from *terroir* to consumption or explosion of the bottle, were even more puzzling. Champagne makers devised their own empirical techniques for making a controllable product. After the violent first fermentation, if the wine had not stopped fermenting when it ordinarily would, it could be stirred vigorously, drawn off, clarified, and bottled, thus avoiding breakage. In ordinary fermentations, one could add still wine to calm the must or wait until the violent fermenting stopped.[37] Even Dom Pérignon was a groping empiricist. The first scientist to change things was Chaptal, who in the early nineteenth century brought science and heated debate into wine production.

If the new wine was tart, as reports from the 1770s indicate, is it reasonable to conclude that sugar was being routinely used? Louis-Perrier strongly implied so. The probability of losing a substantial percentage of the wine was a good reason for using wine of secondary or even inferior quality to make champagne. There was no problem in selling high-quality still stuff. It is highly unlikely that champagne would have become so popular if sugar had not been added. Louis-Perrier believed that the change, not necessarily an improvement, was the result of successive attempts to get rid of the wine's tartness. Vintners must have sugared in secret, each keeping a secret the other knew. Most of the firms of the Champagne had a factory for preparing sugar syrup (cane sugar in wine) for their champagnes. The science of scientists had little to do with this practice, which was carried out on an empirical basis until at least the 1830s. Moreau-Bérillon dates the use of sugar to about 1700, when the complaints about the tartness and disagreeable flavor of champagne begin to be heard less frequently. There is no evidence that the two Doms (Oudart and Pérignon) sugared their creations. The last procurator of the Abbaye

---

[36] Perrier, *Mémoire sur le vin de Champagne,* pp. 73–4.

[37] Moreau-Bérillon, *Au pays du Champagne,* pp. 107–8. The technique for avoiding *la casse* is taken from E.-J. Maumené, *Traité théorique et pratique du travail des vins;* Maumené quoted the notes of the founder of "one of the best firms of Reims."

d'Hautvillers spoke from retirement in 1821 to declare that the monks had never put sugar in their wine.[38] The virtue of the old regime was soon replaced by revolutionary vice.

Chaptal in the Champagne

The growing use of sugar in wine production meant the inauguration of a new phase in the history of wine production, and nowhere more than in the Champagne, where sugar was much more essential to the market than in other areas.[39] In 1801 Nicolas Perrier of Epernay (1764–1806) wrote to Cadet de Vaux that "one of the signal services that the new oenological doctrine had rendered to the cantons is to have explained the nature of sugar and to have prescribed the way of using it."[40] Sugar seemed to take on the magical properties it had formerly been held to have in medicine. Cane sugar, whose essence was sweeter than that of grape sugar, could not only make wine a more pleasant drink, it could reinforce the action of grape sugar in fermentation.

As it was held that the fermenting wine ate the sugar ("la mousse ronge le sucre"), it was the custom to add sugar after the first fermentation. In 1800 the demand for wine was high enough to cause producers to speed things up, leading to the adding of the sugar during the first fermentation. The results were good, especially at Ay, where second-class wines ("vins de vigneron") sold for the highest prices they had ever reached, even when there was a good harvest. Cadet de Vaux supported Perrier in recommending that sugar be added before fermentation because adding it after simply produced sweetened wine. Adding the sugar before both corrected the tartness of the wine and resulted in a needed increase of alcohol.[41] The modern model of a wine to please all palates had been created as a thought experiment. How to get from unpredictable if frequent empirical success to scientific certainty became a useful obsession inspired by curiosity and the lure of profits.

The Réduction François and the Taming of Mousse

In 1829 the Société d'agriculture, commerce, sciences et arts of the department of the Marne asked one of its members, François, a pharmacist in Châlons-sur-Marne, to be the secretary of a commission to examine a work by Herpin on la graisse, the only disease then particularly affecting champagne (mousseux). After several years of research, François came up

---

[38] Moreau-Bérillon, *Au pays du Champagne,* p. 110.
[39] Perrier, *Mémoire sur le vin de Champagne,* p. 80.
[40] Ibid., p. 81.
[41] Ibid., p. 83.

with the idea of adding tannin to the wine. The notion produced a publication but not a cure. (Tannin does not cure a wine suffering from the bacterial disease *la graisse,* but according to Kayser and Manceau it is of some use in forming a precipitate that clarifies the wine of the filaments giving it an oily appearance. The cure is a careful use of sulfur dioxide.)[42]

In 1836 François published a work of twenty pages in which he established a method for determining the amount of sugar to be added to champagne.[43] His observations on fermentation were soon embarrassingly out of date, but his reduction method of measuring sugar survived. Poor science, good application. The innovation depended upon using technology just made available, the glucometer (hydrometer) devised by Cadet de Vaux. François drew up an approximate "table indicating the amount of sugar to add in order to obtain a desired pressure," that is, a *belle mousse.*[44] It was a step away from empiricism toward precision based on scientific knowledge in harness with technology. Moët in Epernay started using the method, but the Reims *négociants* were perversely content with tasting in order to determine how much *liqueur de tirage* to add to the wine for a seductive *mousse.*[45] It is unclear whether the Rémois had more sensitive palates than the Epernois, or simply that the technology was not aggressively pushed. Probably the latter. And the method used by François was certainly in need of improvement, even if it did continue to be used well into the twentieth century.

François undertook his research to try to reduce losses from breakages in the spring of the year, when secondary fermentation takes place. A basic axiom stated, in the standard scientific language of the day, that fermentation cannot occur unless a liquid contains a sweet or sugar-containing element and another element called a ferment. François became convinced that the best way to avoid breakage was to have a rational procedure for using sugar, which, he believed, produced the carbon dioxide. The correct amount of sugar would produce the correct amount of gas for a pressure that the bottle could tolerate. In spite of his half-

---

[42] "*La graisse* is . . . caused by the growth of mucilaginous lactic bacteria." Without affecting the taste of the wine, the disease causes it "to flow abnormally, like oil." Larousse's *Wines and Vineyards of France* (New York, 1990), p. 618.

[43] *Nouvelles observations sur la fermentation du vin en bouteilles, suivies d'un procédé pour reconnaître la quantité de sucre contenue dans le vin immédiatement avant le tirage* (Châlons, 1836). In 1837 François grew verbose with a work of forty pages: *Traité sur le travail des vins blancs mousseux.*

[44] Loubère, *The Red and the White,* p. 117.

[45] François's *Traité* (1837) is seen by Loubère (p. 117) as incorporating "a technological breakthrough of such importance that his method was still used in the early twentieth century." The technical improvements in making and bottling champagne are covered by Loubère, pp. 108–18: machines for corking, for helping to add the sweetening liqueur, for fastening corks, and for washing bottles. Progress, or at least change, took place in clarification, fermentation, assemblage, remuage or riddling, disgorging, etc. See Moreau-Bérillon, *Au pays du champagne,* chap. 6, "L'industrie des vins mousseux au XIXe siècle."

understanding of fermentation, François concluded that the breakage from too high a pressure in the bottle could be kept to a minimum by controlling the quantity of sugar added (as a nonclarified and unfiltered sweet wine) so as never to go beyond 10 to 12 degrees on the glucometer of Cadet de Vaux. François developed an industrial practice that was used by cellar masters for generations, although some scientists would soon sneer at it as an outdated procedure.

After coming to the conclusion that fermentation required both sugar and an element called ferment, François carried out experiments in which he added increasing doses of sugar to the wine, leading to his conclusion that the *mousse*/pressure increases with the amount of sugar in the wine at the time of *tirage*. Experimenting tirelessly on, he established a correlation between the weight of the sugar and the *mousse*. He did not distinguish between the *mousse* power of different types of wine:[46] the procedure worked anyway. Enthusiastic appreciators of François compared his rigorous experimentation to that of Lavoisier, a comparison less impressive to present-day historians of science than to French commentators in 1900.[47]

Instruments: Approximate Knowledge of Reality in the Vat

The oenological literature on the "réduction François" apologetically notes the approximate nature of his measurements, adequate though they were for production. Secure in his pharmaceutical faith, François was productively ignorant of the problems of relating instrument readings to reality. However, epistemological ignorance does not preclude successful experimentation. As late-nineteenth-century scientists were obsessed by accuracy of measurement, it is understandable that the approximate nature of calculations in fermentation would have been brought to the attention of serious readers. Of course, outside the laboratory measurement was not a problem for the producer, who had no idea that he needed laboratory accuracy – at least until a new generation of oenologists told him he did. In any case, the issue was bogus, for instruments were rarely graduated accurately.

By the end of the nineteenth century an impressive number of analyses could be done on wine. The must could be analyzed for sugar content and acidity. Wine could be tested for its degree of alcohol, tannin, residual sugar, acidity, and volatile acids. The volatile acids were increasingly regarded as being of high practical importance. The method for getting a rough approximation of sugar content was simply to "weigh" the must

---

[46] E. Manceau, "Recherches sur la seconde fermentation ou prise de mousse des vins de Champagne," *BLEVOM* 1 (1898): 287.

[47] See Frederic Lawrence Holmes, *Lavoisier and the Chemistry of Life: An Exploration of Scientific Creativity* (Madison, Wis., 1985), pp. 388–409.

with some sort of hydrometer. A rich variety of gadgets, or at least of words, reflects the fertility of the oenological mind: *aréomètre, mustimètre, pèse-mout,* mostly variations on the theme of Cadet de Vaux's glucometer, and the *densimètre (aréomètre Baumé),* graduated in the Baumé system to indicate the weight of a liter of must. The weight of the liquid and its density are synonymous in ordinary language, and the former specific gravity has been rebaptized as relative density. Happily for the Champagne, determining richness of the must in sugar by measuring the density with the glucometer works well enough for white wines; for red wine musts, the *aréomètre Baumé* must be used, although serious people now use a refractometer.[48]

The utilization of instruments, which could have both weight and density graduations, was simple and quick, especially in the Champagne where the sugar of the must measured in degrees of the *aréomètre Baumé* could be translated into practical figures through the use of a table based on a formula given by Auguste-Pierre Dubrunfaut.[49] In his *Traité complet de l'art de la distillation* (1824) he was the nineteenth-century apostle of the beet-sugar gospel, a distinction often erroneously conferred on Chaptal. Oenologists and instrument makers wanted every winemaker to have an instrument and a table for converting and correlating readings. This dream of instrument makers was not fulfilled until rather recently. A metal glucometer cost 18 francs, rising to 27 francs for a silver model. As the wine journals and newspapers carried more and more advertisements for a greater number of increasingly sophisticated instruments, it may be as unwise to underassess the market as to overstate it. Oenologists still suspect winemakers of conveniently forgetting to read instruments, even simple ones like thermometers, while depending on traditional ignorance.[50]

---

[48] See Emile Peynaud, *Connaissance et travail du vin* (Dunod, 1981), p. 151, and M. A. Amerine and M. A. Joslyn, *Table Wines: The Technology of Their Production* (Berkeley and Los Angeles, 1970), pp. 293, 739. Because of its cost and complexity, the chemical method (using *la liqueur cupropotassique* and *sous-acetate de plomb*) was not used by the average or even nonaverage winegrower.

[49] A sample of this art, based on fifty samples:

1900: Moût de cuvée (Pinot noir)

Densité	Degré Baumé	Glucose déterminé par analyse	Glucose d'après les tables	Différence
1066,5	8,95	147,00	147,00	−0
1069,5	9,35	164,29	155,00	−9,29
				− etc., etc.

See "Conseils pratiques sur l'analyse du vin du Champagne" and "Nouvelles recherches sur la valeur des indications du pèse-moût," *BLEVOM* 1 (1898): 18–19 and 47–50.

[50] Emile Peynaud, *Le vin et les jours* (1988), p. 261, tells about a *maître de chai* who refused to be bothered with reading a thermometer to make sure that the fermenting temperature

Raoul Chandon: The Cult of Champenois Viticulture and Oenology

At the end of the nineteenth century Raoul Chandon de Briailles carried out a set of experiments to verify François's figures and compare them with figures obtained by the chemical method and by Robinet's simple method. Robinet just boiled off the alcohol, replaced it with water, did the weighing with a hydrometer, and made a few simple calculations.[51] Although the three reductions yielded different figures, the approximate "réduction François" was shown to be exact enough, perfectly reliable for making champagne. It did not contain enough of an error to result in breakage because of too much sugar or to produce an insufficient amount of carbon dioxide for a good champagne because of too little sugar. The "réduction François" incorporated a breakage of four to five percent in its system. This level of breakage indicated a successful operation in the process of production, signaling that the time had come to store the wine in a cellar at a temperature of seven to nine degrees Réaumur. At this temperature, fermentation would slow down, forestalling excessive breakage; even in the worst cases no more than 12 to 15 percent would be lost. François deserves a memorial prestige cuvée like Dom Pérignon or Roederer Cristal to remind us of his revolutionary impact on the champagne industry, an impact far greater than that of the elusive Dom Pérignon. But in the real world of business, myths sell better than the reality concocted by historians.

François had solved a major problem in the production of champagne. Technical refinements and improvements in instruments and materials would eliminate breakage as a problem. And scientists could be content with explaining the phenomena until they discovered new problems to solve. Although the production of champagne early reached a level of perfection that could be envied by other wine producers, no serious producer could ignore science, and Moët et Chandon gave it an admirably fetishistic and fruitful role. Raoul Chandon carried out a great number of analyses of wines taken from the vat at the moment of their birth, when fermentation was finished. He wanted to understand the life of wine, its development from birth to death. Most of the attention was given to sugar, which in its union with yeast is the creative spirit of wine. Even after the wine becomes champagne (wine impregnated with carbon dioxide), some sugar remains, which varies from vat to vat and even from bottle to bottle, but is always between a few decigrams and a few grams; it is sometimes as high as eight to ten grams, but usually a few grams per liter. The sugar

did not go above 30°C. The winemaker's guiding principle: "Plus ça chauffe et mieux ça marche."

[51] "Considérations nouvelles sur le tirage des vins de Champagne," *BLEVOM* 1 (1899): 105–17.

rarely ferments, decreasing instead in amount over a two-year period. Having neither the theory nor the technique to find out where it went, Raoul Chandon accounted for the decrease by "a slow etherization of the sugar substances and slow chemical changes, probably helped by variations in temperature."

This ostensibly unimportant residual sugar was of great interest to producers and researchers because it indicated a problem in production as great as breakage – that is, insufficient gas for a good champagne, low *mousse* power. After the "réduction François" and other methods for measuring the sugar remaining in the wine after fermentation became widely known and used, it was easy enough for many producers to calculate the weight of the sugar needed to obtain the correct pressure for a good *mousse*. But general success did not exclude failure of the procedure to produce any *mousse* at all or, more often, a wine sparkling enough to sell as champagne. The failure to produce any fizz still left a wine sweet enough to be held in storage for another try the next year, and a low-fizz wine could be turned into dry white wine.

No one seems to have been interested in predicting the weight of the residual sugar (the unfermented sugar remaining in the wine) until it was taken up as a research problem in Moët's laboratory in 1897. This sugar could throw off the cellar master's calculations and produce wine with no or insufficient gas. It turned out to be a much more complex issue than breakage, although obviously part of the same scientific problem. It involved three varying factors. The composition of the wine in relation to its "dissolving power" in the production of carbon dioxide was simply not known precisely enough. This was evident from the unpredictable existence of a considerable amount of sugar after *tirage*. A theoretical difficulty complicated the problem arising from ignorance of the composition of wine. The volume of gas needed to obtain a determined pressure assumed the applicability of Dalton's law to the solubility of gas in liquids. The law, true for perfect or ideal gases, did not seem to apply to carbon dioxide in wine. And making the champagne maker's scientific nightmare even worse was the fact that the composition of wine changes during its transformation into champagne: an increase of a degree in alcohol with a drop in acidity, and an increase in potassium bitartrate, tannin, and nitrogenous matter. So there was not much point in making calculations on the basis of the *vin à tirer;* the *vin tiré* should be the object of scrutiny.

Two other important factors had to be considered in the low-fizz problem: the activity of the ferment or yeast and temperature conditions (not only the temperature of the wine). The basic problem tended to become obscured by problems growing on problems: it was not enough to calculate the amount of sugar required to produce a so-called determined pressure for a *vin de tirage*. Because all the sugar did not ferment, precise calculations only resulted in an after-fermentation pressure much lower than predicted. The key is to know the proportion of the sugar that will ferment and to add the *liqueur de tirage* on the basis of predicting the amount of residual sugar after the wine has turned into champagne. This problem provided a research subject for a few scientists for a number of years. Meanwhile the "réduction françois" worked well enough – maximum errors did not exceed two grams of sugar per liter of wine – to produce a profit and enough champagne to keep the Belle Epoque afloat.

The importance of some scientific research – the surface tension of soap bubbles, for example – is not immediately evident to outsiders. At the end of the nineteenth-century, oenology dealt with equally important matters – champagne bubbles, for example. In a tasting, the physicist-oenologist defined the best champagne as the one retaining the most carbonic acid (H_2CO_3 = carbon dioxide, CO_2, dissolved in water, H_2O). This aesthetic criterion had the advantage of being a measurable quantity rather than a debatable quality. Basic laws on the solubility of gases in liquids could be used to establish precisely the volume of CO_2 that can be dissolved in a wine of a given degree of alcohol, at a certain temperature, and at a certain pressure. It was recognized that the gas provides the basic structure of champagne in addition to being the vehicle for the bouquet of this delicate wine.

Oenologists also had interesting questions about the nature and behavior of this pleasure-giving gas. Research on *la mousse* was not oenological froth but dealt with questions relevant to consumption and therefore to sales and profit. When a wine is opened – that is, exposed to oxygen in the air – what is its aptitude to produce bubbles and to keep on producing them, both in an open bottle and in a drinking glass? (By varying the shapes of the glasses, the researcher can make a quantum leap to complexity.) What is the length of time and the rate of speed of the escape of the gas? These were questions of experimental physics, which Houdaille and Desmoulins tried to answer in the bulletin published by Moët's laboratory. To measure the speed of release of a gas dissolved in a liquid, carbon dioxide in this case, it was possible to use an instrument (*le calcimètre enregistreur*) that had been devised to study the chlorosis-producing power of soils by calculating the lengths of time it took calcareous soils to kill different varieties of vines. Comparative figures for the release of gas from a Crémant d'Ay rosé, from Moët's White Star, and from Badoit mineral water titillated the measuring mind and satisfied the scientific imagination.[52]

When Henry Junger edited his *Dictionnaire biographique des grands négo-*

---

[52] F. Houdaille and A. Desmoulins, "La mousse des vins de Champagne," *BLEVOM* 1 (1899): 90–100. In 1899 they also published "Dosage de l'acidité par la méthode volumétrique gazeuse" in the *Revue de viticulture*. Desmoulins was known as one of the authors of the *Traité pratique des vins, cidres, spiritueux et vinaigres* (1890).

ciants et industriels (1895) for the Exposition universelle of 1900, he included three Chandon de Briailles: Raoul, Gaston, and Jean-Rémy. Founded in 1743, Moët et Chandon was a great firm by the end of the nineteenth century.[53] "Without doubt," said Junger, "the most famous of the names of our *grand vin national.*" Raoul Chandon (1850–1908) probably had the greatest impact on the company since its foundation, although he is ignored in the well-known books on champagne. The circumstances of his influence were unusual; in fact, the environment, if it did not make the man, provided the opportunity for him to become a major force in the history of the firm and in the history of viticulture and oenology. The context of his success was disease: the spread of cryptogamic attacks on the vine and the steady advance of phylloxera toward the north of France.

Raoul Chandon's education at the Lycée Louis-le-Grand and commercial schools in Paris and Lyon was followed by long stays in England and Germany. In 1875 he entered the firm at a high level (*associé-gérant*) and a few years later became the head of the vineyard division. The company owned about 600 hectares. Given the threats to the vine in this period, the job could have been an opportunity for disaster instead of his striking success in holding the enemies of the vine at bay. Mildew invaded the Champagne in 1884 and had a disastrous effect in 1886, except for the areas protected by Raoul Chandon's ambitious spraying program. He also quickly used sulfur disulfide to treat the soil, a standard weapon against the phylloxera louse, which had been sucking away at vine roots in the Champagne since 1890. But Raoul Chandon had no illusions of the possibility of saving the vine by this palliative. It was typical of his approach to a problem that he carried out a set of experimental treatments on vines of different ages over a number of years. Concluding that the cure, like the disease, would come from America, he turned his attention to American varieties and visited the vineyards already replanted with grafted vines. The great resources of Moët et Chandon permitted replanting the company's vineyards and also helping others, especially small proprietors, to replant. The replanting program included the establishment of the firm's Ecole pratique de viticulture.

After the survival of the vine the most important matter was the quality of the wine, which was threatened by mildew and black rot, and, some alarmists argued, by the planting of grafted vines. Two major research tasks faced the company: first, the study of vineyards of grafted vines to

---

[53] Jean-Rémy Moët's daughter married Pierre-Gabriel Chandon de Briailles in 1815; after 1833 the name Chandon was added to Moët (Patrick Forbes, *Champagne: The Wine, the Land and the People* [London, 1967], pp. 419–20). Along with Nicholas Faith's *The Story of Champagne* (New York, 1989) and André Simon's *History of Champagne* (London, 1962), Forbes's book is one of the standard sources in English on the social, political, and commercial history of champagne.

see if the wines produced were as good as those of the old ungrafted French vines; second, the discovery of the simplest and best means of fighting fungi or preserving the vine from the effects of the cryptogamic diseases. To carry out these complex and costly research programs, Raoul Chandon and the company created an Ecole pratique de viticulture with two research laboratories, one for oenology and one for viticulture, with a special grafting setup for the industrial production of vines.

The oenological laboratory began in 1895, when Raoul Chandon offered Emile Manceau, a teacher in the Collège d'Epernay, a small laboratory that had shrine potential because Moët had collaborated there with François of *réduction* fame. Manceau was not vegetating in Epernay but desperately trying to do research for a doctorate. His work on the role of tannins in vinification admirably fulfilled the firm's need for someone working on a relevant subject and who could also work on a group of problems in Champenois oenology. By 1900 the two small rooms of the laboratory were insufficient to accommodate its activities, and an entire pavilion of the Ecole de viticulture was turned over to it. There was a laboratory of bacteriology and a library receiving the main scientific journals. The laboratory soon found its agricultural analyses to be quite popular with growers: 4000 analyses of soil, wine, and supplies were done in 1899. Some were free, like those for the Société d'horticulture d'Epernay, whose president, Gaston Chandon de Briailles, had made it into perhaps the best horticultural society in France, with about 3000 members. (Gaston was notorious for his right-wing politics, his charity, and his collection of thirteenth-century French enamels.) The company ensured that the oenological laboratory lacked nothing needed for research on champagne.

The resources of the company permitted the laboratory to acquire an expensive legitimation of its research by publishing a journal, the *Bulletin du laboratoire expérimentale de viticulture et d'oenologie de la Maison Moët et Chandon*. The first number carried an explanation by Raoul Chandon, the editor, of why he had decided to inflict yet another such journal upon what looked like a limited and already saturated readership. Given his dozens of scientific, cultural, charitable, and business activities, he did not need another plaything. His reasons for establishing the bulletin were to publish the results of research done in the experimental laboratories of the company, to record tests carried out in the experimental vineyard, and to fight against new pests. All these activities were to be carried out in conjunction with the development of a grape culture adapted to the Champenois region.

One of the bulletin's basic aims was to furnish Champenois winegrowers with information that they could use directly. Existing publications did not do this. The hallmark of the new bulletin was a specific focus on the Champagne. Popular local papers were notorious for recommending methods and products without the guarantees offered by scientific control. Science had to be local because general viticultural procedures often did not work. Thus experimentation had to be local. Nowhere was this more evident than in the search for vine roots that would adapt to local soils. And this local science could be done only by a big property owner or company with the resources to carry out long, expensive tests on plants and procedures before they could be recommended for general use.

What was sorely needed was a means of protection for the small wine-grower, who had no means of scientific control to use against sellers of products to stop phylloxera, merchants who sold plants, and innovators in the areas of viticulture and winemaking. The experimental vineyards of Moët were open to whoever wanted to see them. The small grower would survive by following the practices of the big property growers. And the resources of the company were used paternally to help many small fry survive and replant. Paternalism, social and political conservatism, devotion to the historical heritage, and the best of modern science and technology all jelled into an ideology of viticultural salvation. Raoul Chandon's editorial policy was a great stimulus to success: "Je m'attacherais aussi, de temps à d'autre, à provoquer la collaboration de savants ou de viticulteurs éminents."[54] Science and aristocracy, seemingly natural allies before the Revolution had cooled their ardor, became cooperative bedfellows once again.

Moët et Chandon, Manceau, and the Professionalization of Champenois Oenology

Two oenological projects of the laboratory headed by Emile Manceau had considerable impact on the theory and practice of champagne making: first, the study of the *collage* or clarification of wine; and, second, the study of a subject of ancient vintage, the second fermentation in the bottle. One widely accepted theory of clarification was based on the practice that five grams of fish glue and four grams of gallotannin together make up an insoluble compound. But Manceau's experiments showed that they did not really form this compound at all. The composition of the precipitate of tannin and glue varied with temperatures and the makeup of the wine. So much for the disintegration of the theory and the reason for failures of its application. These matters were of more interest to scientists than to winemakers, who wanted to know what amounts of the products to use in order to produce a saleable wine. A "rational method of *tannisage*" first required knowledge of how much tannin the wine contained naturally. As there was no method by which to measure the amount of tannin in the

---

[54] *BLEVOM* I (1898), introduction by Raoul Chandon de Briailles; no. 9, (1900): 165–85, on the Ecole pratique de viticulture; 3 (1909): 25–53, on the life and work of Raoul Chandon.

wines of the Champagne, one had to be invented. After four years of work on tannin, Manceau devised a rule that worked as well outside as in the laboratory. The search for an answer to a simple question produced answers to complex questions, not even imagined at the beginning, relating to the preparation of pure tannin, the study of metallic composites of tannin and of acids-phenols and phenols, substances chemically close to tannin. Best of all for Manceau, he did his doctoral thesis on the basis of research carried out in Moët's laboratory, thus marrying the utility of industrial research to the academic prestige of the *thèse d'état*.

The second area of research in the early days of the laboratory was the *prise de mousse* or secondary fermentation in the bottle. This subject is important enough to treat in considerable detail. A key research problem at the turn of the century was the intensity of the *mousse* or force of the froth. This problem was separate from the issue of the quality of the *mousse*, which was believed to depend "on the mode of release of the carbon dioxide."

In theory, it was simple to make champagne. The main thing was the *tirage*, consisting essentially of bottling a mixture of old and new wines with a small amount of sugar. The yeasts of the new wine produced the fermentation, giving off the essential CO_2 and the complex deposit characteristic of champagne. Too much gas broke the bottle, and too little gas, the more frequent problem after mid-nineteenth century, produced an insufficient *mousse* for good champagne. Raoul Chandon had written about the problem of low *mousse*.

Manceau carried out a striking piece of research to seize the elusive relationship between the volume of sugar in the bottle at the time of *tirage* and the amount of carbon dioxide and pressure eventually produced in a hermetically sealed bottle. The problem of partial fermentation, in which a weak *mousse* was produced, was at the core of practical concerns in this research. What made the problem intriguing was the fact that the weak *mousse* was not necessarily the result of too little sugar in the wine at the time of *tirage*. And what Manceau wanted to understand was the mechanism of the *prise de mousse*. The complexity of such a task required "a methodical plan of research on the modifying causes of the *mousse*, the deposit, and the bouquet." The nature of the deposit – a bad deposit could spoil the wine – and the bouquet were considered to be factors in the mechanism of the *prise de mousse*. So far so good; but not very far.

Research, often of a groping empirical, practical sort, had been going on in studies on *mousse* and related subjects for over half a century. Manceau took stock of the state of knowledge about *mousse* at the end of the century. No one had been interested in the effect of the *prise de mousse* on the bouquet. Nor had anyone, except Manceau himself, conceived a passion for studying the physical properties of the deposit. Up to 1896 he had studied "only one factor in the 'texture' of the deposit," tannin. Manceau was one of the world's greatest experts on tannin.[55] Work had been done on the production of *mousse,* but even that research had inspired only a few pieces of work. The most important had been done by François, whose empirical gropings in the 1830s made champagne a commercially successful product. At the same time the glucometer came into use to determine the amount of sugar needed for a good *mousse.* The estimation of the weight of the sugar that should register on the instrument was based on the trials of previous *tirages.* Manceau deplored both methods as empirical, the tools of a sort of blind search, even if they had been used for a half-century.

Manceau versus Maumené: Oenology in Reims

The work of Edme-Jules Maumené was of a much higher scientific level than the work of the empirical gropers. Occupant of the municipal chair of chemistry in Reims, he was a major nineteenth-century oenological scribbler. One of his banal distinctions was that he did not jump on the bandwagon of those who accepted Pasteur's idea that fermentation was caused by a living yeast. He also came up with the idea of preparing sparkling wine in large containers rather than using the traditional method of bottle fermentation, although this process of tank fermentation is unjustly named solely after Charmat, the inventor of the tank.

Maumené was probably the first to evaluate *mousse*-power by measuring the pressure of the CO_2 inside the bottle. He constructed an instrument (*aphromètre*) that was really Bourdon's manometer with a tube that could be inserted through the cork of the bottle. Struck by the difference in pressure in bottles of champagne that had received the same amount of sugar, he put forward the "ingenious hypothesis" that different wines have different dissolving powers for CO_2 and therefore required different amounts of CO_2 to acquire the same pressure. To avoid too low a pressure, namely, a weak *mousse,* the dissolving power of the wine had to be measured; the weight of the gas needed for a chosen pressure could be calculated along with the weight of the sugar needed to produce the gas; and results could be guaranteed by a logical, rational explanation. It seemed reasonable enough in light of the different rates of solubility established by physicists and chemists, including Bunsen, for carbonic acid in water, alcohol, and different saline solutions. Wines of the Champagne, not having the same composition, might be assumed to have different dissolving

---

[55] *Sur le tannin de la galle d'Alep et de la galle de Chine* (thèse de doctorat, Faculté des sciences, Paris, 1896; published in Epernay).

potentials. Maumené put forward this explanation in the first edition of his treatise on wine in 1858.[56]

It was ingenious, logical, and wrong. Analogies often lead to nasty surprises. Manceau knew Maumené, but in 1895 this did not prevent him from declaring that attempts to test the "theory of absorbent power" had given pitiful results. Maumené subscribed in principle to the idea of the absorbing power of wine, which seems to have rested on the general dogma that during the *prise de mousse* all of the sugar is fermented.[57] So if different pressures were produced in two different wines having received the same dose of sugar, the difference in pressure had to be attributed to the difference in the absorbing powers of the wine. Maumené reinforced his experiments by showing that his results conformed to Mariotte's (Boyle's) law on the compression and expansion of gas. In a later edition of his work, Maumené enlisted the support of Dalton's law on the total pressure of a mixture of gases.

Manceau was not convinced, for it had been shown that the solubility of CO_2 in wine does not conform to Dalton's law, an approximation only valid for real gases at low pressures and high temperatures.[58] Not that the producer worried about Dalton's law – unless he was Raoul Chandon. But making champagne would always be a bit of a gamble "if the weight of the residual sugar could not be predicted," and especially if the "*prise de mousse* could not be controlled in order to get a negligable residue." Maumené's determination of an average absorbing power was therefore necessarily erroneous; worse, he gave the power of the wine before *tirage,* very different in composition from the *vin tiré* – an extra degree of alcohol would change the absorbing power. Assuming that all the sugar fermented, Maumené was able to calculate exactly the volume of CO_2 in the wine. But in fact the second fermentation is never complete and therefore leaves the volume of CO_2 lower than what is calculated. Maumené admitted to Manceau that he had never advocated the theory of the absorbent power of wine as a model of precision because he had always failed in attempts to apply it. Like the ether in nineteenth-century physics, it was the best rational fantasy available to explain recalcitrant phenomena. It was not long before Maumené's theory became "rational and infallible," discour-

[56] See Edme-Jules Maumené: *Indications théoriques et pratiques sur le travail des vins, et en particulier sur celui des vins mousseux* (1858); *Traité théorique et pratique des vins* (1874; 3d ed., 1890). See also *Comment s'obtient le bon vin. Manuel de vinificateur* (1894).

[57] This applied only to the formation of the mousse: "le sucre se change complètement en alcool et acide carbonique gazeux." For other wines, Maumené said that if the fermentation is complete, all the sugar disappears, but many factors could prevent all the sugar from fermenting. *Indications,* pp. 77, 397.

[58] "Les poids de gaz dissous sont, à température constante, proportionnels aux pressions. Il n'en est rien. . . ." E. Manceau, *Théorie des vins mousseux* (Epernay, 1905), pp. 11, 15–16.

aging further scientific work, until Manceau announced that he had come to replace these "superficial ideas on the *prise de mousse*" by "a really scientific study."

Manceau's Research Program in Champenois Oenology

The producer of champagne had three different methods of evaluating the sugar content of wine ("réduction François," use of a type of hydrometer [*pèse-vin*], and chemical analysis, the most recent, precise, and expensive). Yet he could not regulate the sugar content in order to obtain a specified *mousse* except by trial and error ("essais de tirage, par des tâtonnements"). It was to escape this state of ignorance and impotence that Manceau launched his research program in the laboratory of Moët et Chandon. That program included research on the deposit and the bouquet in "a methodical, analytical study of the chemical transformations that wine undergoes during the *prise de mousse*" and for a few years after. Manceau worked out a plan for three groups of experiments on the *prise de mousse,* that is, specifically on the production of CO_2 and generally on the fermentation of sugar. His first task was to understand the influence of the milieu, which meant not only the constitutive substances of the wine but also the gas dissolved in the wine before the second fermentation, including the influence of CO_2 under pressure. Choosing four samples of wine, ranging from newly fermented wine (after the first fermentation) to two-year-old champagne, Manceau did a series of quantitative analyses on nine elements, including alcohol, sugar, total and volatile acidity, tartar, and glycerine, and followed the development of the wine. His scientific surprises came on sugar and on CO_2 (including pressure).

Manceau found that because the sugar in the *liqueur de tirage* did not invert so quickly as was often supposed, there was always sugar present after the *prise de mousse.* (The inversion of sugar is the process by which sucrose – cane or beet sugar – is split by the enzyme sucrase – invertase – into glucose and fructose.) The weight of the residual sugar in the *grands vins de Champagne* varied greatly, going as high as fifteen grams per liter, but was more usually six to eight grams. So the second fermentation of the *vins de Champagne* was not generally a total fermentation. Manceau's experiments on pressure from the CO_2 showed, by measuring the weight of the sugar really used – the difference in the weight of the sugar before and after the *prise de mousse* – that the dissolving or absorbing power of different wines for CO_2 did not vary very much. He was so surprised to find that "the final pressure was visibly the same when the yeast used the same amount of sugar" that he undertook a careful verification of his results by measuring the gas in bottles of wine (from different vats) with the same pressure. Manceau's interesting and even significant practical research results certainly fulfilled Raoul Chandon's request that his work be useful for the Champagne.[59]

Not many scientists are content with presenting a few research results: they want to invent general theories to which their names will be attached. The ambitious researcher aims at an explanation of important phenomena – fame, glory, a name that will live forever. Manceau's shot at glory was "une théorie de la prise de mousse." Eduard Buchner won the Nobel prize in 1907 for his work on cell-free fermentation, showing that fermentation is "a chemical, enzymatically catalyzed process." The fermentation of carbohydrates results from the action of the enzyme zymase in the yeast cell rather than the activity of the living cell itself.[60] Manceau's theory included a good many well-known elements, although he usually added a titbit of novelty derived from his own research. He played a bit with the fact that the fermenting activity of yeast stops, even though all the sugar has not been fermented, when the degree of alcohol reaches a certain level, generally 13 to 14 degrees, but varying with each fermentation. At the same temperature, different yeasts stopped working at different levels of alcohol, varying from twelve to fifteen degrees. The complex influence of the alcohol varied with the weight of the initial sugar. The yeasts of Ay, of Bouzy, of Verzenay, and of Cramant produced different results. Even the yeasts of the same *cru* from neighboring vines gave different results.

This luxuriant chaos produced by the yeasts of nature's choice would be brought under control by the use of pure or so-called selected yeasts. Different environments influence the activity of the yeast – that is, the amount of sugar used during the *prise de mousse:* acidity, proportions of potassium, phosphoric acid, nitrogenous matter, and so on. It was often erroneously said that a must high in acidity fermented better than one low in acidity. Within their usual limits in the *vins de Champagne,* glycerine (6 to 8 grams per liter) and citric acid (a few grams) had little influence on fermentation. And 1 gr 25 of gallotannin made a wine bitter but did not stop fermentation. Yet three grams of tartaric acid per liter could hinder fermentation. In 1903 the fermentation in the bottle of certain wines of 1902 was slowed up or incomplete because of their acidity. Manceau also did a comparative study of the considerable influence of temperature on fermentation. Most of the research was relevant to general oenology, but Manceau never forgot his major responsibility to solve problems in the production of champagne.

The second fermentation of the wines of the Champagne had a special problem resulting from the fact that the yeast had to grow in an unusually

---

[59] Ibid., p. 14.
[60] Eduard Buchner, *Alkoholische Garüng ohne Hefezellen* (1897); Herbert Schriefers, "Buchner," *DSB,* 2:561.

hostile environment. Not only were there substances resulting from the elaboration of the wine, but also a level of alcohol close to the maximum at which yeast would produce fermentation. The second fermentation, raising the level of alcohol in producing the *mousse,* can stop before all the sugar is used. The wines used by Manceau were *grands vins mousseux,* containing an average dose of 25 grams of sugar per liter at the time of *tirage.* In lesser wines, of lower alcoholic content, fermentation stops because of the lack of sugar. Manceau also concluded that yeast activity slows down, even stops, in the *prise de mousse* because the first fermentation uses up the nutritive elements required by the yeast. Oxygen, for example, is in limited supply during the *prise de mousse.* He found that the availability of air made an important difference in the process.[61]

Nor was it known what influence was exerted by CO_2 under pressure. Experimenting on the *prise de mousse* in very different wines, Manceau concluded from direct measurements of the solubility of CO_2 in wine that at a constant temperature of 0°C "the differences were not so important as had been believed for the previous forty years." Dogma after dogma crumbled, and Manceau increased in oenological stature as his word became law for many champagne makers. Other experimenters had probably come up with different figures because they had not kept the wines at the same temperatures. So the anomalies in the *prise de mousse* of the *grands vins de Champagne* had to be explained in the context of partial fermentations.

The theory advanced by Manceau was a general explanation with lots of "minor" problems for other researchers and even himself to work on. But a major mystery had been analyzed and cleared up enough for a practical advance in making champagne. The weight of the residual sugar after the *prise de mousse* was determined by three groups of factors: the composition of the fermentation environment (the wine), the yeasts, and the temperature. The complexity of life in the bottle. After Manceau's work, solving problems of the *prise de mousse* was more a matter of scientific *Sitzfleisch* than research solving big problems.

Like most scientific conquistadors, Manceau put his own interpretation on how he had changed the area of science in which he worked. Cast in one of the classic forms of historical myth, his gloss has a certain convincing charm: there were not many "old theories" and they were wrong.

The development of champagne making was the work of men who made the wine and were interested in perfecting methods of work, not in explanations of the phenomena. This view was not a matter of hostility to

---

[61] In addition to quoting Pasteur's *Etudes sur le vin* (1866), on the influence of oxygen on the *mousse,* Manceau referred to "l'ingénieuse expérience de M. Denys Cochin" (who worked in Pasteur's laboratory) on the nongrowth of yeast if oxygen is kept out of the fermenting milieux. Manceau, *Théorie des vins mousseux,* pp. 17–179.

theory but rather of ignorance of the world of science. Scientists took a keen interest in the essential phenomenon of fermentation but did not have the mechanisms to transmit either interest or knowledge to the level of production.

Nearly a century later we can see that the transmission of knowledge took place through two major channels. First, the state: through state power and influence, as in the case of Chaptal's oenology, and through state research and educational institutions, whether oenological stations in Narbonne or Beaune or the University of Montpellier. Second, and rarer, through private laboratories in industry, as in the case of Raoul Chandon's laboratory at Moët et Chandon or Ribéreau-Gayon's at the wine firm of Calvet in Bordeaux. With François de Châlons there were glimmers of a theoretical interest in the mysterious *prise de mousse:* an affair of the fertile marriage of sugar and ferment, François said, but the ferment remained vaguely "a special nitrogenous matter." In the 1850s the work of Maumené was an entirely different matter. His *Traité du travail des vins* (1858) takes us into, or at least to, the threshold of the experiment-obsessed Pasteurian age. (It is irrelevant to this argument that he was a nonbeliever in the biological theory of fermentation; his was a quite respectable nonbelief at the time.) From François to Maumené, oenology went from an empirical, pharmaceutical-chemical approach that improved production to a theory-laden chemical oenology. The second approach, like its predecessor, was intellectually defective, and although it had little influence on production, it did try to explain the phenomenon of the *prise de mousse.* Producers could control the *prise de mousse* better than ever, but still badly because no one understood the phenomenon as a piece of experimental science as well as a natural process.

Manceau's new theory, conceived in the post-Pasteurian period of Buchner's new science of fermentation, built upon a long tradition of quantitative analysis of wine from Lavoisier to Pasteur. In 1895 Manceau took up the basic question again, without any preconceived ideas according to his self-proclaimed Baconianism. His painstaking experiments on the secondary fermentation demonstrated the new dispensation. And the new theory was directly related to production, making possible the scientific control of operations from crushing to shipping. (Indeed, scientific concern with vine and wine was partly prenatal, because the grafted vine, the test-tube baby of scientific viticulture, not only required care different from the old vine but gave a different product.)

The new approach to winemaking emphasized systematic analysis of the wine, in order to expose its most intimate life, and intervention in the process of fermentation if any danger appeared. Manceau pointed out that rapid analyses could be useful during fermentation in the bottle. On several occasions he found a sharp lowering of the total acidity (1 gram of H$_2$SO$_4$ per liter), corresponding to a modification of the deposit, as a result of an unexpected malolactic fermentation. The analyses for density, alcohol, acidity, and sugar could be carried out by experienced employees. Scientifically trained personnel would have to do other analyses, including those for volatile acidity (method of Duclaux with operating procedure of Gayon), dry extract (official method), potassium bitartrate (method of Berthelot and Fleurieu), total nitrogen (Kjeldahl's method), and total sulfuric acid (method of Hass). The uncertainty that producers had suffered for over a century in the production of *vins mousseux* had been banished by the practical certainties of oenological science.

And what were the commercial results of all this science? As a scientist at Moët et Chandon, Manceau was necessarily keenly interested in the bottom line of Champenois oenology. In a thirty-year period for the earlier part of the twentieth century, the normal bottle breakage of 5 to 10 percent was reduced to 1 percent.[62] The losses from the failure of the wine to develop a *mousse marchande* – to become a commercially marketable champagne – were also reduced to near zero. Because there was no immediate application of the research and teachings of Pasteur, a general ignorance reigned on causes of abnormal deposits in bottles during and after fermentation, as well as of the all-too-frequent cloudiness. The causes of loss of limpidity, as in the case of the blue *casse* (*bleu*), fatal for modern champagne, were eventually identified along with measures to prevent such occurrences.[63] The early *vins mousseux* were turgid wines – "le boire et le manger" – good for eating as well as drinking, totally unacceptable to the clientele of the twentieth century, faddists of clarity, who would reject the smallest deposit formed during the *prise de mousse*. What science can achieve soon becomes an aesthetic choice, regardless of the dubious means by which the clear object of desire is attained.

The general use of the *réduction François*, the acceptance of the theory of differing sugar-absorbing powers of wines, and the spread of belief in the benefits of tannin all ended up as obstacles to a more rigorous experimental approach to winemaking. Of course, all new science eventually falls victim to the fossilization of novelty, becoming an obstacle to research that threatens it. In the case of champagne, the practical consequences were costly. Even in the preparation of new wine there was a great deal of uncertainty. The producer put his hope in a series of ritual practices – *tannisage, collage,* and so on – without any certainty concern-

---

[62] E. Manceau, *Oenologie champenoise. Contrôle scientifique de la préparation du vin de Champagne en bouteilles* (Extrait des *Annales des Falsifications* [Epernay, 1926]), noted that the glass manufacturers of the special bottles for champagne admitted to a "normal breakage" of 8–10% at the time of *tirage*.

[63] "Blue *casse* is caused by the formation of a tannin-iron complex in the wine. This results in a bluish precipitate, and affected white wines will often turn a brownish colour." Larousse's *Wines and Vineyards of France*, p. 268.

ing the nature of the final product. Manceau claimed that his research program achieved certainty through scientific control of the key process in making champagne.

Manceau found that it was extremely difficult to carry out precise "industrial experiments," especially when working with an unstable must. So he limited himself to weighing the sugar that remained after fermentation. The rest of his scientific work was carried out in the laboratory, where he could construct the artificial milieux essential to experiment rather than be frustrated in working on a rapidly fermenting vat. Yet this distancing from production had major effects upon it: the laboratory made oenology possible through the certainty and stability of experiments. The vat could only be known through the test tube. Much is written about the uncertainty and contingency of experimental science – some of it is probably true – but compared to the "chaos" of life in the vat, life in the laboratory is a model of order and predictability. The great thing about the laboratory is that it is the way to knowledge and control of life in the vat, which would otherwise remain in the realm of accident and freedom, the parents of commercial disasters. The bourgeois world of the late nineteenth century, coming to a clear realization of this paradox, put it to good commercial use.

Much of the research done in oenology was communicated to the members of various viticultural societies, numerous in wine-producing areas by the end of the nineteenth century. Danger and disease promote association, and after the 1840s winegrowers had excessive stimulus to associate. Manceau exposed his theory to a national group in 1902, about three years before it appeared as a pamphlet of thirty pages.[64] Practice may have preceded publication. Perhaps as a result of this openness, most of the champagne companies did better than other wine companies, but it would be difficult to separate the respective roles of scientific and commercial prowess. Perhaps it is better not to apply this academic distinction. Manceau never had any hesitation in identifying Raoul Chandon de Briailles as his patron because of the company laboratory Moët et Chandon had put at his disposal in 1895. The results of the research were published in the laboratory's bulletin. No industrial secrets. Moët et Chandon, the leader of the pack in volume of business and in Champenois oenology, operated in a profitable osmosis of both areas. It still does.

[64] "Théorie des vins mousseux. Communication faite le 4 mars 1902 à la Société des viticulteurs de France et d'ampélographie." The theory had been revealed earlier to a scientific elite: "Sur la fermentation ou prise de moussse des vins de Champagne," *Compte rendu de l'Académie des sciences,* April 22, 1901. Closure was recently brought closer to achievement in *mousse* studies when the question of Frederick William – what makes champagne sparkle – was finally answered by Florence Brissonet in her doctoral thesis done at the University of Reims. See *Le nouvel observateur,* no. 1420 (January 23–9, 1992), p. 63, for a one-page report, which ends by lamenting that the research will be published in the *American Journal of Enology and Viticulture.*

8

Burgundy: The Limits of Empirical Science

Eighteenth-Century Research Schemes for Improving Wine

The old pre-Chaptalian practical viticultural and oenological science had not been without its texts. The third section of the *Théâtre d'agriculture et mesnage des champs* (1600) by Olivier de Serres contained over a hundred pages of "science" and practice on vine and wine. *Le nouveau théâtre d'a-griculture et ménage des champs* . . . (1713), by the Sieur Louis Liger contained five short chapters on the vine and winemaking. The substantial *Mémoire sur la meilleure manière de faire et de gouverner les vins de Provence* (1772) by the abbé Rozier summed up existing wisdom while crit-icizing traditional practices and calling for improvements, with an empha-sis on cleanliness. In *L'art de faire le vin* (1772), Maupin also criticized the weight of tradition and ignorance in winemaking. Criticizing tradition was one of the literary devices of reformist works. Of course, authors were no less wrong in adopting this conventional strategy. Some traditions disap-pear, some are invented, and some undergo so many small modifications that they become a different species of tradition. Burgundian winemaking changed in its details as people's taste in wine altered over the centuries: different epochs, different versions of the same wine.

In the Enlightenment, science was part of the intellectual baggage of the professional classes and cultured men and women. Voltaire wrote on Newton and lesser minds studied the grape. Science is, among other things, an efficient system for ensuring the survival of ideas and practices. An individual of considerable achievement who is outside a scientific group or, worse, is opposed to it in some basic way is assured of oblivion. Maupin, one of the *valets de chambre* to the queen of Louis XV, Marie Leszczynska (1703–68), is even less known than the queen; even his Chris-tian names and the dates of his birth and death are unknown. Yet during the 1770s and 1780s Maupin was one of the most prolific writers on vine and wine in eighteenth-century France, an experimenter of ingenuity, and an indefatigable self-promoter.

Maupin seems to have had a great deal of success in winegrowing and in advising important winegrowers, though his ideas on viticulture were not among those accepted as the basis of scientific viticulture in the nineteenth century, and were even odd in some respects in comparison with eighteenth-century traditional wisdom on quality-wine production. But he probably had as many good viticultural and oenological ideas as anyone else in the eighteenth century and certainly had no monopoly on ideas that later came to be regarded as wrong. His plan for fame and fortune was not helped by his attacks on academies, *savants,* especially chemists – Macquer in particular – and even less by his organization of public experiments to show that all his imagined enemies were wrong. In spite of numerous recommendations of his winemaking gadgets and techniques – heating part of the must to start fermentation quickly and covering the fermenting vat were two of his "secrets" – by the rich and powerful, including Henri Bertin, and even by the Academy of Medicine of Dijon and the Faculty of Medicine in Paris, his name did not long survive his death. If he had collaborated with some scientist or if some scientist had attached one of Maupin's books to his own classic work in viticulture or oenology – as Chaptal did for the abbé Rozier, for example – then it is likely that Maupin would not have so effectively disappeared from history.[1]

In 1770 the lawyer Béguillet published a work on oenology or the best method of making wine and cultivating the vine.[2] He believed that winemaking had to be done on the basis of a theory supported by chemical knowledge and scientific experiments. So all wine provinces should have societies of agriculture whose specific aim would be to study the best methods of cultivating the vine and of making wine in each region. Within this pedagogical organization of oenology, Béguillet advocated assigning specific projects of study and research to scientists (a loose translation of *savants*) in his proposed research operation. He regretted that nothing of this nature existed in Burgundy.

If one is to find Béguillet free of contradiction on the quality of French wines, especially burgundy, it seems best to conclude that the great advantage French wine enjoyed was that so many other wines were inferior. The Dijonnais lawyer also believed that French wine was the only one to go well with food. The foundation of the reputation of a great wine was the opinion that the wines of other countries were inferior, either too heavy or too sweet or too sour. Erasmus could be quoted as being against the wines of the Rhineland and for the wines of Burgundy, which not only agreed with his palate and stomach but cured him of the stone. Even well-

---

[1] On Maupin, meaning his numerous works, see Michaud, *Biographie universelle,* 27:332; *Nouvelle biographie générale,* 33–4:393–5; and the catalog of the Bibliothèque nationale.
[2] *Oenologie ou discours sur la meilleure méthode de faire le vin et de cultiver la vigne* (1770). Béguillet advertised himself as an "avocat au Parlement (Dijon), de la Société d'Agriculture de Lyon, de la Société royale des sciences et des arts de la ville de Metz."

heeled Italians preferred burgundy, or at least Goldoni sang its praises in *La locandiera,* and a literary reference is good for a lot of mileage in publicity in France.[3]

Béguillet was careful to state that French wine was good only if well made, and his belief that it was not well made was the reason for his new theory of winemaking. Oenology took on a civic responsibility, in Béguillet's view, because of the role of good wine in producing happiness. Good wine always being a scarce commodity, it was a matter of the highest importance to develop better methods in viticulture and vinification, especially in Burgundy, prodigally endowed in viticulture by God but not well used by man. Béguillet's treatise would show man how to make a wine worthy of the endowment bestowed by the Creator.

A Lesson from the Champagne

The commercial success attained by the wines of the Champagne in the late seventeenth and early eighteenth centuries provoked a showdown with Burgundy. Although there was little possibility of winning the war on the red front, the Champagne was handsomely compensated by an unchallengeable and original advance in *vins mousseux.* With the growth of the market, the reputation of *mousseux* for quality became increasingly known. Given the greater difficulty of making good *mousseux* than making good still wines, and the even greater difficulty of delivering the *mousseux* to demanding consumers, it is clear that winemakers in the Champagne got an early start on the superb viticulture, vinification, and propaganda for which they are now famous. In his critique of Burgundy (1770), Béguillet identified the Champagne as a model for Burgundy to imitate.

Béguillet seems to be giving an excessively polite response to the rude remarks of the Champenois Nicolas Bidet, whose *Traité sur la culture de la vigne* disparaged the wines of Burgundy while praising the wines of the Champagne, "les premiers vins de l'univers."[4] Experts did not allow objectivity to distort their opinions. Béguillet easily struck a low blow at the proud Champenois by pointing out that history makes no mention of the wines of Champagne. The Champenois used to go to Burgundy to buy wine for trading abroad and for the coronation of kings in Reims. The

---

[3] This ancient gospel of the primacy of France in wine is still preached from distinguished pulpits and recorded in sumptuous volumes, not many of which are as good as Peynaud's on the taste of wine. Emile Peynaud, *Le goût du vin* (1983), "Primauté des vins de France," pp. 214–15, including warnings against profit-obsessed practices.

[4] Nicolas Bidet, *Traité sur la culture . . . de la vigne,* revised, corrected, and expanded by Duhamel du Monceau (1759) as *Traité sur la nature et la culture de la vigne . . .;* cited in a different form by Béguillet, *Oenologie ou discours sur la meilleure méthode de faire le vin,* p. xxv.

Champenois had not concocted so glorious or so ancient a chronicle as that of Burgundy.

In spite of this denigration of the viticultural history of the Champagne, Béguillet recognized that in 1770 the area presented a viticultural and oenological model that Burgundy could well imitate in the care given to the selection of plants, the choice and cultivation of the soil, and the exposure of the vineyards. "Let us study cultivation according to its true principles; let us not leave the making and care of wines to ignorant, uneducated growers." In preserving its wines, Burgundy could also emulate the Champenois techniques, more perfected than those of any other winegrowers. The care with which the Champenois produced wine gave them a better reputation than one would expect in such a rigorous climate. Science and technique can offset the avarice of nature, whose bounty is often wasted in more favorable climes like the Midi. Burgundy seemed to be frittering away the natural advantages it had over the Champagne.

Regardless of the few powerful voices praising Champenois wines, the "vins de Beaune," a generic term for burgundy, had little to fear from their competitors in the red wine market, or at least they did not suffer from competition at the end of the eighteenth century. The "vins de Champagne" never enjoyed a reputation for quality like the red wines of Burgundy.

Peasant Winemaking

There was general agreement among the elite that the existence of an ignorant peasantry was a major problem that had to be faced before the production of quality wine could become a Burgundian ideal. The problem was similar in Provence, where, according to Rozier, a filthy peasantry used empty wine vats as garbage barrels. Over fifty years after Béguillet's lamentations, J.-A. Cavoleau, in one of the standard works on early-nineteenth-century French oenology, was still bewailing the crude way of making wine, even the top wines, in the entire Côte d'Or. In the case of Bordeaux, he made a clear distinction between the way peasants made wine and vinification procedures of the owners of fine wine-producing properties. Essential differences lay in the degree of cleanliness, ullage, choice and use of fertilizers, waiting for the most favorable time to harvest, and keeping wines of different quality separate.[5] Nor, according to Béguillet, did the Burgundian have much knowledge of viticulture, or at least of vines. It was easy to show the ignorance of the vigneron by recalling the amazing viticultural knowledge of agricultural writers of antiquity or to point out that of the 300 species of grapes a large number were cultivated

---

[5] J.-A. Cavoleau, *Oenologie française* (Paris, 1827), pp. 67, 127.

in the gardens of the grand duke of Florence. A typical peasant vigneron knew only a few species, generally producing a flat, tasteless, short-lived table wine for seasonal consumption. The principles of his mixture being always at war, he could not even produce a good harmonious concoction (*vin d'assemblage*) for immediate consumption, let alone a wine capable of some aging.

Before deciding on a wine model, a winemaker had to decide what kinds of grapes could be mixed and in what proportions. Such decisions could not be left up to vignerons or workers. In fact, Béguillet thought that tests to solve the problem should be carried out by a society of agriculture. The Académie de Metz, for example, could lead the way for other provincial societies. Information about good vines should be popularized. Cultivators should switch to better plants. Béguillet was an optimist, firm in his belief that he could change the criteria of a wine's popularity: color enough to stain a cabaret tablecloth and acid enough to tickle the palate. It seems instead that the bourgeoisie perversely moved in a popular direction.

In the late eighteenth century the visually aesthetic part of the models of the wines of Pommard and of Volnay required that they have the light color defined as "oeil de perdrix." The idea of Burgundian blushlike wines disappeared in the more assertive nineteenth century, which dreamed of seizing reality through a more alcoholic wine of intense color. Neither wine nor wit improved as a result. Perhaps one of the reasons for the change was the need to imitate the Bordelais model if Burgundy were to compete with Bordeaux on the English market. French consumers seemed willing to follow the English in demanding the new model wine.

Quality in an Era of Expansion and Increased Production

Part of the discrediting of French wines could be attributed to the practice, well known in Burgundy since the Middle Ages, of replacing good but low-yielding vines with inferior, higher-yielding ones. This penchant for "making the vine piss," was not a problem limited to Burgundy. In 1731 the authorities of the town of Toul (Lorraine) ordered that several "grosses races" of vines (Govaux, Gametes, or Verdunois, Forcans) be pulled up. The Parlement of Metz confirmed the decision.[6] André Jullien, in his *Topographie* (1816), one of the key books of the nineteenth century on French vineyards, pointed out that since the Revolution the cultivation of vines had spread considerably in Burgundy, including to the plains and drained marshland; on the more desirable hills proprietors had introduced fertilizers and had also created new areas of cultivation. (Jullien was a

---

[6] Béguillet, *Oenologie,* pp. 77, 95.

wine merchant and the inventor of powders for treating wine diseases and of several bits of wine technology.) Their object, all sublimely entrepreneurial, was to produce more wine. Other proprietors replaced old vines with young ones and fine varieties with common ones: poorer wine but more of it. The Côte d'Or vineyards expanded from 17,658 hectares (ha) in 1788 to 30,118 ha in 1862, peaking at 34,187 ha in 1880–9. A century later, in 1980–7, they had shrunk to 8354 ha.

The history of vineyard shrinkage is not uniquely Burgundian. Bordeaux has followed a somewhat similar pattern. Apart from a striking decline during the Revolutionary and Napoleonic wars, the Bordeaux vineyards (Gironde) covered much the same area for nearly two centuries. I shall repeat a few basic figures. As a result of post–World War II prosperity, shopping centers and highways, housing and other amenities, the Bordeaux area was strikingly reduced: 135,000 ha in 1788; 126,220 ha in 1862; 141,159 ha in 1880–9; and a low of 98,671 ha in 1980–7. The Burgundian expansion as well as decline was part of a national phenomenon. In 1788 there were 1,576,000 ha of vines in France; in 1822 there were 1,906,000 ha; in 1829 there were over two million ha; and after 1845 the area did not fall below this figure until 1885. In the 1980s the vineyards covered 1,038,744 ha, only about 50 percent of the area of a century before.

Whether production rose or fell, productivity rose. Total wine production in 1870–9 averaged over 51 million hl (hectoliters); in 1980–7, it was up to over 68 million hl. This translates into an increase from 21.7 to 65.8 hl per ha. (The "scandal" of excessive yields seems to be a sort of myth of eternal return. One of the selling strategies for Patrick Dussert-Gerber's *Guide des vins de France 1993* was a "revelation" on the high yields in Bordeaux.) With few exceptions, higher grape yields result in lower wine quality. This is a useful general dogma in need of a large dose of empirical specificity.[7]

Jullien recognized that the customers who bought the products of these postrevolutionary and often degenerated vineyards, or those who were given the wine of a bad year when they had ordered that of a good one, were bound to think that the wines of Burgundy were not what they used to be. In this case, the opinion was founded on real changes rather than misguided nostalgia. But the spirit of commercial objectivity prevailed in Jullien's mind, and he limited the damage he had done to the reputation of Burgundy by pointing out that these few abuses were also found in all other winegrowing areas. And even if Burgundy was producing much more ordinary wine in 1815 than it had thirty years before, the number of

[7] Lachiver, *Vins, vignes et vignerons,* pp. 582ff. See Patrick Dussert-Gerber, *Guide des vins de France 1993,* for criticism of high yields of grapes per hectare in Bordeaux. In his guide to the wines of Bordeaux, Robert Parker is also critical of high yields, but notes some exceptions to the correlation of high yields and lower quality in wine.

bons crus had actually increased on several *coteaux* or hills, whose wines equaled if they did not surpass in both quality and quantity those which greed had destroyed or dishonored in the Revolution. This analysis was retained in succeeding editions of the book, including the fifth edition published by his son in 1866. The bottom line for the consumer was that it was easier to get both good and bad Burgundy – especially the bad.[8]

Quality and volume of wine are inextricably bound up with the profit motive. When Maupin advertised his new method of making wine, the people who followed his experiments and imitated him were abbés, military officers, and nobles, including high-level bureaucrats like Geoffroy de Vandières, Secrétaire du Roi. Maupin's advice for both bourgeois and "distinguished vineyards" was to avoid mixing the juices from the first and second pressings of the grapes, because the second pressing gave a harsher, less alcoholic wine. Maupin assured this entrepreneurial upper crust that following his advice would help them sell their wines better. Improved viticulture and vinification made more sense because they made more profit. Differences of opinion on wine quality often reflect the class divisions of society on the important issue of the type of vine used in production. This was still true in the nineteenth-century post-phylloxera quarrel over hybrids between peasant-vignerons and well-off producers of quality wines.

Capitalism and Coupage (Blending)

Many writers placed strong emphasis on a moral problem that had to be faced before the wines of Burgundy could be improved. This problem was the corruption of the wine merchants, whose profit-inspired alchemy Béguillet denounced as theft. Merchants followed the hallowed custom of *coupage,* blending fine wines with wines from vines producing more but far inferior grapes. This practice could cast eternal discredit on French wines. But Béguillet believed that foreigners were giving up buying bad French wine because they could buy bad wine cheaper in their own countries. This is a reference chiefly to the Germans, who, according to Béguillet, were finding the wines of Swabia as good as those of France. Swift's dictum that good taste is the same everywhere may have some striking exceptions in wine drinking. Other countries were making concoctions from grain or fruit that their inhabitants preferred to French wine. By the mid-

---

[8] Writers are still complaining about the large amount of bad burgundy on the market. See Clive Coates, "The Best of Beaune," *Decanter* 17 (November 1991): 71–4: "Burgundy is the most fascinating, the most complex, and the most intractable wine region in the world. There is fine wine – just occasionally great wine – but there is also a great deal of poor wine."

nineteenth century, French worries would have shifted to the English market.

The issue of blending wines was calmly discussed by André Jullien in the canonical *Manuel du sommelier, ou instruction pratique sur la manière de soigner des vins* (1813). Jullien dedicated early editions of his book to Chaptal, "ce savant oenologue." This seems like a reasonable bit of deferential behavior for a wine merchant: Chaptal, comte de Chanteloup, senator, member of the Institut de France. Jullien begins with the basic principle that wines must be kept as natural as possible, especially those of top quality, whose chief virtue is their bouquet. So much for the ideal. A different reality was more often the case in the wine business. Jullien believed that if a wine has altered or is just naturally of bad quality, there is often no way of making it drinkable other than mixing it with a better wine. In good years nearly all wines are drinkable without blending. But when the grapes have not ripened perfectly, even good *crus* have an unpleasant bitterness to them for a long time: such was the case in 1805, 1809, and especially in 1816, 1817, 1823, and 1824.[9] Common wines must be mixed to obtain an acceptable taste – no issue of bouquet here. A weak wine without any distinguishing quality must be mixed with a full-bodied wine. Mixing is clearly one of the basic principles of the manipulation of wine. This was the gospel of *coupage* according to Jullien.

It is difficult to argue against blending on a priori grounds. A great bordeaux is a blend, a *vin d'assemblage,* a mixture of wines from several varieties of grapes, which in the best of all possible worlds grow on the same estate. In theory, burgundy is the pure blood of the pinot noir. Without venturing into the murky historical issue of bordeaux's being a foreign wine (the sustenance of the English, the Scots, and the Dutch, among others) and burgundy's being the national wine, one can admit that there are good reasons based on taste for the blending of bordeaux. The cabernet sauvignon grape gives a powerful tannic juice that is tamed by the softer but no less potent merlot grape, the predominant variety in Saint-Emilion and Pomerol. The pinot noir grape has no need of such taming, at least in principle, and it would be difficult to think of any grape with which it could enter into a morganatic union without ruining its great qualities. Or so it is often said.[10] A wine made from the pinot noir is not drinker-friendly for several years. Jullien noted that the consumer often

---

[9] A. Jullien, *Nouveau manuel complet du sommelier,* rev. ed. published by C. E. Jullien fils, 1860.

[10] The exception to this pious dogma is the gamay grape (*gamay noir à jus blanc*), which makes up two-thirds of the wine called Passetoutgrains. Generally a mediocre concoction, with at least one-third pinot noir, this wine can succeed brilliantly, as Maréchal-Jacquet showed in 1990.

prefers blended wines to "pure wines" because an unmixed wine, even in the best *crus,* keeps for a certain time the taste of its *terroir* – a word that later acquired a pejorative meaning – and a tartness that is hard on the palate. The pure is not pleasing to the corrupted palate.

Jullien accepted the judgment of Grimold de la Reynière, famous if bizarre gastronome, that in Paris a good wine cellar was as rare as a good poem. Jullien emphasized repeatedly that the young *vin de cru* is less pleasing than a mixed and cheaper wine. People bought wine for immediate consumption, which meant buying mixed wines, because an unmixed wine would keep a bit of its tartness – "une légère âpreté, nommée *grain*" – until it reached full maturity, defined as the point at which the wine started to lose its bouquet and taste. Few denizens of Paris had the leisure to follow this evolution. Parisians also wanted cheap wine, and that meant a mixture. (Only in recent years has this situation partially changed, with the development of a new style of vinification and the growth of a large class of people willing to spend money on good wine). In the early nineteenth century, according to Jullien, the wine merchants of Paris did a great business in selling their good wines to Normandy, Picardy, and Flanders, where a great number of good cellars were maintained by lovers of "natural wines of top quality."

The ego of the author prevailed over the wine merchant's desire for secrecy when Jullien revealed some tricks of the trade: a new wine of good color, even a good *cru,* is not a pleasant drink; by adding to it some white wine of an inferior *cru* but clean in taste and very soft in the mouth, the merchant creates a wine that one can drink with pleasure. Winegrowers were just as bad or as good as the merchants in mixing wines, especially bad years with the good ones. If a vigneron has a white wine whose color is off, turning yellow, he can mix it with his most colored red, thus concocting an agreeable, apparently older red wine. Jullien proudly remarked that it would take a connoisseur to detect the mixture. This blending of wines considerably alleviated the suffering of the wine drinker.

There are other circumstances in which mixing becomes necessary and "contributes to the improvement of the wines": delicate, pleasant wines, which do not keep or cannot travel great distances without deteriorating, can get the qualities needed, without being denatured, from a wisely proportioned mixture. An example: the excellent wines of the hamlet of Torins, in the Mâconnais, lasted longer and became better if mixed with first-class wines of Chénas or Romanèche. Nor was this a "spéculation mercantile," Jullien quickly added, for the wines were about the same price. Of course, secret blending was not an exclusively Burgundian practice. The light, pleasant wines of the canton of la Chalosse, in the Landes, were keenly appreciated in northern Europe, but they could only survive the trip if mixed with the hearty and tannic Madiran wine, made from the tannat grape. Blending was a widespread practice, part of traditional winemaking.

To clinch his case, Jullien cited the sparkling wine of the Champagne, which was rarely the wine of a single variety of grape. Making champagne from one variety would result in a wine that either foamed too much – and broke bottles – or not enough, or was too tart or too sweet, too alcoholic or not enough, had too much body or was too light, and so on. What emerges from Jullien's comments is a picture of the nineteenth-century wine merchant as a skillful technician creating a product suited to the tastes of the populations to which he sold. Champagne sent to Russia was different from that sent to England; that sent to Paris different from that sent to the provinces. Champagne does not or did not taste the same in France as abroad. This was no accident but part of a careful commercial policy based on the consumer taste model assumed to apply to each targeted group. A suspicious mind might conclude that Jullien felt slightly guilty about diluting the pure blood of the *cru* with liquids of lesser pedigree, for he insisted that the process was in the best interests of the consumer. A wine was blended to conform to a certain consumer's taste model. In the best of all possible worlds that model corresponded with commercial convenience.

A National Call for Research in Oenology

In 1842 the Société royale et centrale d'agriculture came to the conclusion that the poor condition of French viticulture required missionary action: the society set forth the guidelines for "la mission des oenologues." The excessive expansion of vineyards was encroaching on grazing land and even wheat fields, obviously an unfortunate development in light of the growing food demands of an increasing population, which already had too much wine to go with too little bread and meat. Even industrial crops were feeling competition for land and manure from the aggressive vine. The vines planted were more productive than distinguished, for it was assumed that bigger crops and more wine per hectare would be an effective way to counteract low prices. In the hungry 1840s the working class had started to consume less wine; what was supposedly good for the workers turned out to be bad for big producers of ordinary wines. The worse commercial effect abroad was that France's fine wines sometimes found themselves discredited through guilt by association with the massive surplus of plonk. This national society of agriculture concluded that in the export market the only way to increase business was to improve quality.

The Royal Agricultural Society did not entertain the illusion that the problems of the wine industry could be cured by legislative action, the easy fix often desired by desperate producers. The whole vine culture had to be reexamined in the light of contemporary consumer mentality. Scientists and winemakers would have to cooperate to change the situation, presumably including a change of the consumer's taste model. Agronomists could not avoid learning about the entire rural economy or supervising the coordination of its varied branches. The nature of oenology and the quality of its research may be inferred from the people subsequently mentioned as sources of hope for the future. One of the distinctive and basic strengths of oenology from its early days was its close contact with production. Indeed, it often seemed to be defined in terms of the interest of the producer. Only later would the word be monopolized by the newly arrived academic clan.

In the 1840s the Royal Society saw oenology represented in Maine-et-Loire by M. Guillory, president of the Société industrielle d'Angers. Everyone was impatiently waiting for the work on ampelography by the count Odart. Pastor Castor of Rochefort deserved inclusion in the elect because his experiments had produced a new and very good grape. All efforts were elements of progress in agricultural science, but only husbandry could give them to France as a whole. The Society liked reports on what had actually been achieved: so Vilmorin and Sageret wrote a report on obtaining varieties of plants by planting seeds; Vibert did the same thing for varieties of vines. A sort of collective wisdom, a pool of research knowledge, could be built up by the elite producers who brought their knowledge to congresses for integration with the knowledge of others; in this way the materials of science were formed. It was not quite the house envisaged in Henri Poincaré's architectural plan for science, but there was here a foundation of intellectual empiricism. It was a situation nearly ready for the annunciation of a scientific oenology by a professor in a faculty of science. But it would come only after yet another complaint about viticultural and oenological ignorance, and this time the alarm was seductively attired in contemporary conventional science.[11]

A Call to Arms by a Physician-Proprietor

In 1846 Apollinaire Bouchardat, hygienist and oenophile, professor at the faculty of medicine in Paris, and proud possessor of a vineyard in Burgundy, published his *Etudes sur les produits des cépages de la Bourgogne.* He began with the lament that there were few scientific works on vines and their products, an important part of France's natural wealth. Bouchardat did not expect winegrowers to exert any grass-roots pressure for scientific research, which they regarded as of little use. Because no one had showed them that it could be useful, their skepticism was not surprising, if

---

[11] *Bulletin des séances de la Société royale et centrale d'agriculture* 3 (1842–3): 335–7.

they thought about the issue at all. In the golden age, before oidium, mildew, and phylloxera, and the related chemical fix for vines, there was also less need for enlightenment of the peasantry by science.

Bouchardat was hardly a typical vigneron. His grandfather Baudoin had planted a vineyard in the late eighteenth century with the aim of collecting all the vine types of Burgundy, a task that it was possible to contemplate and even accomplish before the fissioning of the vine took place under the aegis of scientific viticulture. Yet it is hard, and often undesirable, to change basics: as Bouchardat said, the pinaux (pinots) are the glory of Burgundy. The greatest triumph of science, saving the classic vines from destruction by American diseases, was far in the future. Nor did it appear that the scientific research done on the grape could be of much use except for the satisfaction of the scientist's ego.

Even Bouchardat saw little use for the fascinating discovery that two substances in the grape have a rotary molecular power on polarized light. Biot had shown that natural grape sugar causes polarized light to deviate to the left; tartaric acid in a free state or in combination with potassium bitartrate was found to have the same effect. The alcoholometer (hydrometer) of Gay-Lussac, much more useful, was employed by Bouchardat himself in his experiments. In spite of the ostensible unity proclaimed and practiced by Pasteur between research in physical science and research in biological science, it would not be until well into the twentieth century that such work would become an essential part of oenology.

The Faculty of Sciences (Dijon) and Other Institutions

In its mid-nineteenth-century adolescence oenology was quite different from it what it is today. It was much closer to the actual working concerns of the winemakers and was therefore endowed with far less potential for changing both winemaking and wine. Ten years after Bouchardat lamented the paucity of scientific works on wine, C. Ladrey, professor of chemistry in the faculty of sciences in Dijon, established a course in the application of chemistry to viticulture and oenology. The lectures, published in 1858, provided a sort of pioneering textbook spreading the good word. In 1858 vineyard owners, not so unenlightened as some people thought, asked Ladrey to give his lectures in Beaune: "Au coeur du vignoble bourguignon . . . prestigieuse cité du vin," in the immortal clichés of the *Guide vert*. In 1860 the Société impériale (formerly royale) et centrale d'agriculture de France decided to award Ladrey its *grande médaille d'or* for his work.

One of the reasons for the imperial society's honoring Ladrey was that in 1859 he had started publishing *La Bourgogne,* a monthly oenological and viticultural review. The enthusiastic recommendation from the *Section des cultures spéciales* to the general society came from Bouchardat, who noted the need for a permanent organ of discussion and information for viticulture and oenology. The pragmatic aim of the journal was to enlighten and instruct the vigneron, the property owner, and the *négociant*. To this viticultural trinity was added a fourth target, the consumer. This was an original idea, one that would be integrated into the strikingly successful course for the popularization of oenology devised at Bordeaux a century later. Practical viticultural and vinicultural methods were mixed freely with theoretical considerations on fermentation in the presentation of various operations and their results. Science and education would lead to an improvement of procedures traditionally protected by ignorance and routine, as well as a suppression of fraud.

The journal was not exclusively devoted to Burgundy. The first issue carried a translation of work by Swallow on the cultivation of the vine in Missouri, research originally published in the transactions of the Academy of Sciences of Saint Louis. The wines of the United States got considerable attention, though only the wines of California were considered able to compete in the far distant future with those of Europe. The French noted, with a certain satisfaction, the high cost of labor in California. Australian wine was not ignored by *La Bourgogne*, but it was not viewed as even a distant threat to French preeminence in the wine world. (This view would be reasonable for both potentially competing wines during the next hundred years or so; it was not yet fashionable to damn these wines because of their bland perfection based on the latest sci-tech.) Ladrey saw viticulture in the context of general agriculture, a perspective indicating to him the most important function for his review. By 1860 the general trend in agriculture could be summed up by optimistic phrases such as "on the road of progress and improvement," and Ladrey thought that viticulture and the wine industry were trying to follow the same route. Neither the results of such changes nor their happy consequences could be foreseen, but the continued extension of the domains of science and technology provoked only optimism and euphoria.

Ladrey viewed agricultural science as a bit of a laggard in the striking progress made by the other sciences in the nineteenth century, much as medicine had been in the eighteenth century. This was not a surprising claim from someone who was establishing a new journal, though the fact that the comment might have been self-serving does not mean that it was necessarily wrong. The sciences were traditionally judged according to the criteria of certainty and the degree to which they contributed to the improvement of the human condition. Ladrey held a straightforward, no-nonsense view of the relations between theory and facts. The problem with agricultural science had been that it was behind the harder sciences in having its applications indicated by theories, themselves based on experience. Ladrey had no basic problem in connecting theory with fact, because theories are "nothing else but the expression and linking together of all the known, verified, and incontestable facts." The aim of the journal could thus be justified in terms of presenting only the facts concerning viticulture and oenology. Ladrey believed in giving useful advice to the farmer on the use of chemical fertilizer: it should be combined with manure and compost. The experimental work of agronomists such as Georges Ville also had its place in works on agricultural science. The half-brother of Napoleon III, Ville was one of the pioneer advocates of the use of chemical fertilizers: "L'azote, c'est moi."[12] Ladrey's science reserved a large space for the practical and empirical.

In 1861 Ladrey confessed to having had a feeling of indecision back in 1856, when he started the journal. By the third year of success he was confident enough to venture into a little socioeconomic history, which also served as good propaganda for French wine. He published Cobden's brutal judgment on the bad effects of high English tariffs on the consumption of French wine. In 1703 the English had concluded a treaty with the Portuguese admitting their wine to England at a greatly reduced tariff, while the Portuguese reciprocated for English textiles. The English soon dominated Portuguese trade, but at a great cost to their own good taste, for the wines of Portugal were not very good and, as a result of their privileged position granted by the Methuen treaty, soon got worse. Cobden denounced an absurd treaty that perverted English taste while depriving Englishmen of the best wine in the world. When the English drank French wine they were gay and sang, but when they drank sherry or port, they became stupid and sleepy. It was not until 1860 that the Chevalier-Cobden treaty reduced tariffs on French wines to one-twentieth of the tariffs of 1815, thus making "Gladstone claret" available to people poorer than the chancellor of the exchequer. In little over a decade the consumption of French wine in the U.K. increased eightfold.[13] It was a market Burgundy did not want to ignore. And interested scientists were willing to ensure that burgundies had their share of this market.

In the history of wine, the years 1835 to 1846 were classified as disastrous by a dour Bordeaux chamber of commerce. As the period from 1786 to 1828 was one of declining or stationary prices for ordinary wine, one wonders why vineyard acreage continued to expand, or why there was not more quality wine, whose price usually rose, even if not fast enough to keep up with increasing costs.[14] In this situation it is not surprising to

---

[12] C. Ladrey, *Traité de viticulture et d'oenologie,* 2d ed. (1872), includes a revised edition of *Chimie appliquée à la viticulture,* published in 1857. On Ville, see McCosh, *Boussingault,* pp. 110–14, 124–9.

[13] Johnson, *Vintage,* pp. 379–80.

[14] See Loubère, *The Red and the White,* pp. 133–4.

find producers flirting with science and to discover scientists advocating scientific, or at least rational, solutions to economic problems. As late as 1872 Ladrey followed Henri Mares in taking an optimistic view of ostensible disasters. In 1852 oidium caused great alarm in viticulture, but in the Midi the introduction of the treatment of vines with sulfur began a period of considerable progress as the conquest of this disease led to prosperity and improvement. Ladrey urged this consolatory view on people worried by phylloxera. In the end science is always right, and the rise in the cost of production that results from salvation through science becomes unimportant in the context of saving the means of production itself.[15]

On reading the oenological literature of the professors at Bordeaux – from Gayon and Millardet (1880s to the early twentieth century) to Jean Ribéreau-Gayon, Peynaud, and Pascal Ribéreau-Gayon, among others (1930s to today) – one gets the strong feeling that they have had and continue to have an input, often decisive, into viticulture and even into the creation of taste models in Bordeaux wine, especially the *crus,* particularly in times of crisis. Not that this has prevented scientists like Peynaud from complaining about their lack of influence on winemakers. A very different situation prevailed in Burgundy, where there has been no Peynaud phenomenon. Indeed, in some commentaries there is a distinct despair that science could have any input at all into traditional practices of vinification, which have been successful for so long. It is easy to exaggerate the resistance of Burgundy to scientific input. The famous house of Latour advertises the fact that the fourth Louis Latour was a winemaker with a strong belief in science and technology, a combination that led him to adopt pasteurization when it was the latest scientific fad in vinification. Gayon would have been happy to find a Latour in Bordeaux. Bordeaux winemakers were as suspicious of scientists as were the Burgundians.

The extent of wine-oriented education and research in Burgundy in the nineteenth century gives no basis for predicting a difference in the relations between scientists and winemakers in the two areas. And it may be that the key divergence came about well into the twentieth century, chiefly as a result of the different commercial structure in Bordeaux – Ribéreau-Gayon and Peynaud worked for the big wine firm of Calvet – combined with the scientific superiority of the University of Bordeaux over its Burgundian counterpart. It is clear that after the 1880s – perhaps even earlier – the Bordelais faculty of science had a keen interest in oenology, even if the producers were not eager to encourage intimate relations except in times of disease and disaster. Anthony Hanson has commented on this division between science and production, still evident to a keen observer in the early 1980s. "It is perhaps regrettable that Burgundy's *Station oeno-*

---

[15] C. Ladrey, *Traité de viticulture et d'oenologie,* 2d ed. (1872).

logique should be located on the boulevards in Beaune, so much a part of the commercial life of the town. Under pressure from local merchants in need of Côte d'Or wines which can be turned over quickly, it has often, since the 1930s, advocated methods which are at odds with the region's destiny to produce great red *vins de garde*. A little academic isolation as part of the University of Dijon might not be a bad thing." Not that Hanson had much hope that the "rough-hewn, hidebound band of Burgundian individualists" would listen to scientists. Obviously the truth is stated with Shavian exaggeration here.[16]

Burgundy did not lack personnel and institutions. Ladrey was professor of chemistry at the faculty of sciences in Dijon from 1850 to 1880. He was succeeded by Margottet, 1880–93, who moved up the academic ladder in becoming dean of the faculty of sciences in Lille.[17] Dijon had a very small faculty, with three professors of physics from 1845 to 1900. When Curtel resigned as director of the Institut oenologique de Bourgogne, the departmental general council of the Côte d'Or reorganized the institute, putting it under the permanent supervision of the dean of the faculty of sciences of Dijon. The rector of the academy was given a considerable influence through his power to nominate the heads of three public services: the Station oenologique, the Station agronomique, and the Station Pasteur. The prefect of the Côte d'Or appointed the head of the Station agronomique, which was responsible for the analyses required by the Service de la repression des fraudes alimentaires et agricoles. The Conseil général (departmental council) showed its confidence in the Institut oenologique by giving it a subsidy of 6250 francs – a modest sum compared to the subsidies for practical science provided by local governments in many parts of France since the late nineteenth century.

In addition to the faculty, other scientific institutions also served agricultural and viticultural education. Under the rubric of "établissements scientifiques," according to Vermorel and Danguy, we can include the Ecole pratique d'agriculture d'Ecully (near Lyon) and the Station viticole de Villefranche-sur-Saône.[18] The school at Ecully was run by the well-known ampelographer, Victor Pulliat, a property owner in Chiroubles, where he now has his statue for his pioneering in grafting the gamay to

[16] Anthony Hanson, *Burgundy* (London, 1982), pp. 104–5. Documentation in the Archives départementales of the Côte d'Or: M 13, J – Station oenologique de Bourgogne. 1. Laboratoires de Beaune, correspondance, 1886–1901; 2. Laboratoires de Dijon, rapport 1918. M 13, 1, on the Institut regional oenologique et agronomique de Bourgogne. The Station d'agronomie was created in 1874. It now has as a neighbor the Centre de recherches agronomiques de Dijon, on the campus of the university and one of the 23 centers of the Institut national de la recherche agronomique (INRA) in France.
[17] Margottet's successors were Recoura and Pigeon. Emery taught botany, 1870–1900.
[18] V. Vermorel and R. Danguy, *Les vins du Beaujolais, du Maconnais et Chalonnais* (Dijon, 1893).

American rootstock. The school claimed to turn out educated farmers, especially the grape growers among them, who would know how to profit from their school learning and would also spread the gospel of profit through "the new methods related to scientific progress." The methods were unspecified, but in viticulture they were certainly related to grafting and its problems, and in vinification they were probably matters of relevant basic chemistry, of conservation, and of sanitation. Neither anticlerical republican bourgeois nor Catholic capitalist could object to this ruthless practicability.

The Station viticole de Villefranche was founded in 1890 by Victor Vermorel, who became its director, as a point of scientific penetration into the Beaujolais. Vermorel, a property owner in several communes, proudly and legitimately advertised his operation as a "station de recherche." Its assets were impressive: several buildings, including a laboratory of micrography with a hothouse to provide plants for study, a laboratory for chemistry and oenology, a room for photography, and rooms for the library and botanical collections. Next door was a meteorological observatory. Its big collection of varieties of vines was complemented by an experimental vineyard of 50 hectares. Proof of the station's serious scientific status was enshrined in periodical publications that diffused the results of the work done. Among the studies both practical and scientific, as Vermorel liked to emphasize, were "important works" on ferments. Yeast was a subject of passionate interest to winemakers and oenologists, most of whom were interested in controlling the fickleness of fermentation. And some may have entertained the illusion that with the magic of the right yeasts an ordinary wine might be metamorphosized into a *cru*. Most other research was more realistic.

Oenology and Tradition in Burgundy

In studying the history of oenology in Bordeaux it is difficult to avoid flirting with the great-man theory of history: Gayon, Millardet, the Ribéreau-Gayon dynasty, Peynaud. Hookians will find no oenological heroes to knock in Burgundy.[19] True, even without the support of an institution of national reputation, Ladrey did respectable work in oenology early in the Second Empire. But neither of two of the leading twentieth-century scientific experts on the wine of Burgundy, Louis Ferré and Max Léglise, have made much of a case for the role of scientific research in wine production.

In 1937 Louis Ferré, the head of the oenological station in Beaune,

---

[19] See Sidney Hook, *The Hero in History* (Boston, 1943).

made a classic Burgundian statement about the weight of tradition in winemaking.[20] He seems simply to have been recognizing, however reluctantly, the existing stagnant situation rather than making an indirect criticism of Burgundian indifference to oenological research. Because the old methods of vinification were solidly based on centuries of observation, Ferré did not believe in changing them without good reason, especially if they later appeared to have a reasonable scientific foundation. This was a very different approach to winemaking from the constant reexamination of vinification that had become characteristic of oenology in Bordeaux.

It is tempting to suspect that it may be the scientist, not science, who has not had a major role in winemaking at the level of production. How did scientific knowledge pass from scientist to winemaker? Organizations and their congresses played a key role in times of crisis, as in the preservation and eventual reconstitution of the phylloxeric vineyards. Sometimes the scientifically educated winemaker himself was the agent of change, as when the Maison Latour adopted pasteurization. Eventually, university-trained oenologists would be hired in the wine industry, for it became accepted that scientific knowledge and understanding can help when things go right and become nearly indispensable when they go wrong.

Léglise seems a bit ambivalent, for he recognizes that a radical change in our knowledge and understanding of vinification has occurred in the twentieth century, with a feedback into production. Léglise, who has a certain image as the venerable oenologist of Beaune, has identified some questions that are frequently asked on the vinification of the wines of Burgundy.[21] Has technological progress changed their character or nature? Is the wine we drink the same as that of our ancestors? Do modern wines keep for a shorter period than wines of former times? These are questions that can only be answered by examining the history of the making of wine as it is legally defined. The law does not permit much change in practices or allow much latitude for innovations. Léglise points out that white wine has been made in the same classic way since the ancient Egyptians: pressing the grapes, fermenting the juice, racking the wine. A certain flexibility exists in the sulfuring process, in the size of the containers used, in the selection of yeasts, and in time allowed for fermentation. Now for Léglise

---

[20] Louis Ferré, "La vinification et la conservation des vins à appellation d'origine en Bourgogne," Congrès international de viticulture (Paris, 1937), p. 197.

[21] Max Léglise has recently published Les méthodes biologiques appliquées à la vinification et à l'oenologie, vol. 1: Vinifications et fermentations; vol. 2: Conservation, traitements, embouteillage (forthcoming). In Le Monde, March 26, 1994, Jean-Yves Nau approvingly cites Léglise's anxiety concerning the disappearance of artisanal know-how and traditional methods as industrial methods take over French, and even Burgundian, oenology. The excessive use of sulfur dioxide is denounced as incompatible with the biological option.

these parameters, as he calls them, can only produce minor consequences in the process of vinification; unable to change the basic type of the wine, they can only improve certain characteristics.

This in itself is perhaps a typical Burgundian outlook, based on the idea that great white wines have always been made in Burgundy. Skeptics might argue that the improvements made in the details have produced a totally different wine from the viewpoint of the consumer, even if it is still made from the same variety of grape in the same basic way. And certainly, in some other cases, like the radical change in making Italian and Bordeaux whites in recent years, a change of detail, like the amount of time the wine is kept in wood, can produce a quite different wine. If a sufficient number of the characteristics of a wine are changed, it is not unreasonable to say that the wine has basically changed.

Léglise also believes that, because of the limitations inherent in the process, not much basic change has occurred in the making of red wines. During fermentation the grape juice must enter into fertile union with the "characteristic substances of the skin," the tannins and coloring matter (polyphenols and anthocyanin pigments). In the twentieth century, which is so eager to speed up the creative act, several attempts have been made to modify the extraction of the substances that give great wines their seductive cachet. Nature has stubbornly refused to relinquish her time-consuming embrace of the process, whether for improvement, intensification, acceleration, or isolation of the substances. If the winemaker wants to make the typical traditional and basic *vins de cru,* and especially the *vins de grand cru,* he has no choice but to stay within the parameters of the standard process of fermentation and extraction.

Léglise does not tire of repeating that the most important development in vinification has been the tremendous growth in knowledge and understanding of the phenomena rather than any basic change at the practical level of winemaking. Pasteur, like Mendel, was "rediscovered" at the beginning of the twentieth century; his pioneering microbiological studies had been sidelined during the fight against phylloxera. Later understanding of the dozen stages of the process of alcoholic fermentation along with the responsible enzymes led to a knowledge of the physicochemical requirements of a correct fermentation as well as of what could go wrong. The progress in microbiology shattered the illusion that adding the yeast from Chambertin to the juice of Aramon grapes would produce a *grand vin,* but the precise amounts of oxygen the yeasts require, their resistance to the alcohol they produce, and their adaptation to the milieu in which they develop all became known.

By the 1960s the grape itself was being analyzed so precisely that it was no longer necessary to guess at such important matters as when the sugar content is high enough for picking and crushing. Some people's guesses may be as accurate as measures by instruments, at least for sugar content, but the comfort given by instrumentation counts for something, and the new precision avoids bad guesses by ungifted winemakers. Only the researcher or passionate oenophile will pursue the details of the analysis of the thousands of elements in wine that can be detected by chromatographic analysis. Léglise doggedly emphasizes that this big body of knowledge has not changed anything substantial in basic winemaking techniques, except to establish a mastery over vinification in order to avoid developments that could ruin the wine. This message of oenological modesty, in which science confirms empiricism, is not acceptable to ambitious oenologists.

Once fermentation begins, the Bernardian idea of milieu becomes more important than the Pasteurian concept of ferments. For Léglise, the parameters of these "environmental conditions" are the leading physicochemical factors, including temperature, aeration, pH, and oxidation potential. Because winemakers traditionally feared high temperatures during fermentation, various means were used to keep the temperature below 35°C. A more recent discovery is that fermentation below 20°C remains incomplete. This scientific knowledge leads to the adaptation of existing technologies to solve problems. Knowing the temperature alone turns out to be of limited use: it is also necessary to know both the stage to which the fermentation has advanced and its rate of advancement. Plotting both density and temperature curves on a graph makes it possible to predict the state of advancement and the changing temperature of the fermentation. This can be done automatically, thus making possible a machine-made wine. The machines do as well as many winemakers, when it comes to making wine by numbers.

The great advantage of measurement by precision instruments is that any interruption of the fermentation can be noticed immediately and a series of analyses and a microscopic examination carried out to identify the cause of the interruption. Measures can then be taken to prevent a deviation arising from the domination of bad microbes over good ones, and normal fermentation can restart. As Léglise points out, winemakers used to be practically helpless in this sort of situation. Along with temperature, oxygen is a key factor in the multiplication of yeast cells. The oxygen is also responsible for oxidation, a frightening word for the wine drinker; but the process is easily controlled by the use of big closed containers for fermentation and the subsequent protection of the wine from the possible ravages of too much air.

Léglise may admit that scientific knowledge and technology are the foundation of the winemaker's art, but he resists the confident whiggism characteristic of Bordelais oenology. His argument seems to be that, although oenological knowledge has greatly increased, the basis of winemaking practices has not and cannot be changed. This seems a paradox, even a trifle perverse. True, the basics of the process have necessarily retained a core of identity since Noah, or at least the Romans, but as any Bordelais oenologist would be quick to note, it is the differences in winemaking over time that are important, not the similarities.

Burgundy seems to have a certain genius for losing its way in winemaking, or at least every few years wine writers discover "the new burgundies." In the final analysis, the problem seems to be an oenological inability in Burgundy to distinguish the essential from the peripheral, let alone to develop a general program for quality wine production. The controversies surrounding issues like destemming, pasteurization, and, more important, chaptalization take on a life of their own instead of being relegated to the historical graveyard of scientific triviality. In these squabbles, the main issue of wine quality can easily become forgotten. This rarely happens in Bordeaux, where the issues are discussed, problems are given scientific solutions – and these solutions can change over time – while wine quality remains the main priority. It seems only fair to give a good deal of credit for this sense of direction and emphasis on priorities to Bordeaux's school of oenology, a growing viticultural and oenological force since the late nineteenth century. This was precisely what was lacking in Burgundy.

Some of the classic issues in winemaking have been chaptalization and pasteurization – words that signify two empirical revolutions in oenological science – and, a more minor matter, destemming of the grapes before pressing. These issues can be used to illustrate the different approaches of Burgundy and Bordeaux to the same general problems. This discussion of oenology and tradition in Burgundy will end with an analysis of the debate over destemming. Bordeaux's conceptual mastery of issues is built on a scientific basis, itself dependent on the resources of university science. In the end, Bordeaux's powerful oenology became the source of the theoria and praxis of a quality winemaking program. Nothing comparable existed in Dijon. This absence of a decision-making mechanism opens the door to long quarrels of dubious usefulness to winemaking and the perpetuation of confusion on the nature of quality wine. Destemming provides a classic minor case of this indecisiveness.

Confusion: The Debate Over Stems

Unlike in Bordeaux and California, there is still some debate in Burgundy over *égrappage,* destemming the grapes so that the stalks are not part of the must. Old processes of vinification did not involve removal of the stalks or stems, which make up to between two and six percent of the total weight of grapes at maturity.[22] As vinification lasted for two to three weeks,

---

[22] Figures from Amerine and Joslyn, *Table Wines,* p. 234.

the tannins and other astringent substances made it necessary to soften the wines in wood and then in bottles for several years. To destem or not to destem is an ancient debate in Burgundy. In his *Oenologie* of 1770, Béguillet warned that letting the must stay too long in contact with the stems would give the wine a bitter taste from acquired harsh and crude elements; worse, the wine would not keep. Béguillet also noted that letting the must ferment on the stems for a month was one of the ways used to obtain the dark-red color favored by the Lorrainers and the Germans, whose pleasure and trade seemed worth the risk of ruining the wine through this "overfermentation," which made it more likely to fall victim to the cloudiness of *la pousse* or *la tourne.*[23]

Maupin's *Art de faire le vin rouge* (1775) agreed that stems made a wine rougher but also noted that the stems could have good effects, especially in weak wines of rainy years, and that it is therefore good to leave at least part of the stems in the must. Maupin criticized the abbé Rozier for teaching that the stems are always harmful, never necessary, never useful, whereas they are vital in making a solid wine capable of lasting for some time. He believed that stems prevent wine disease and aid conservation; so in years when grapes were fully ripe he recommended that one-quarter to one-third of the harvest should not be destemmed. In his *Cours complet de chymie économique* (1779), Maupin confessed that he realized the dangers of excessive *égrappage* when he discovered that his well-made wine of 1766, made from mostly destemmed grapes, did not last as long as his own vigneron's rougher wine. Where stems were favored, generally not more than one-half of the grapes were crushed with their stems still on, although more could be used if the must needed them to produce a solid and lasting wine. A new-model eighteenth-century wine was perhaps being conceived here.

One of the best-known treatises on vinification in the nineteenth century was Henri Machard's *Traité complet de vinification* (1845), which devoted several pages to the debate over destemming.[24] There was still no agreement on the value or danger of the procedure, for in some vineyards all the grapes were destemmed before crushing, in some one-quarter, and in others one-third or one-half. In a few rare cases, all of the stems were left on when defective grapes needed as much tannin as they could get naturally; this view assumed that getting elements from stems is natural and the direct addition of tannin is not.

Machard was confident that his opinions on the value of using stems in the must were solid because they were based on numerous comparative experiments done at different times, all showing the great advantages of

---

[23] *La pousse* or *tourne* (tartaric fermentation) is caused by the organism *Bacterium tartarophthorum.* See ibid. p. 784.
[24] Machard's treatise went into its fifth edition in 1874.

using the stems – the latest science at mid-nineteenth century. Stems could affect even the aroma of wine because of the boost they gave to fermentation by a liberating decomposition of its elements. At the same time, the stems forced the must to produce all of the alcohol its nature allowed. Some of these mythical functions would later be assigned to sulfur, once its use came under attack.

Machard had to answer the question of why the valuable qualities of stems were generally not appreciated enough, at least in the opinion of their defenders, who made up a numerous group wanting unanimity on the virtue of their oenology. Machard's explanation was that the opponents were confusing astringence with tartness, two different principles. The tartness or harsh taste of unripeness did not come from the stems. This opinion was given in the second edition of 1849; by the fourth edition of 1865, he was using a contrast of tastes based on astringence (good) and acid (bad). Stems were even recommended as the best item for correcting the bitter taste of a defective wine, by counteracting the excess acid and for giving body to a thin wine. Certain regions should take note: the lower Jura, the Haute-Saône, the Doubs, the Haute-Marne, Alsace, the Orléanais, Burgundy, and so on. In fact, most of France produced wine lacking in astringence and needing the tannin of the stems to escape diseases like *la pousse*. By helping to decompose the ferment abundant in these wines that was responsible for the disease, tannin assumed the role of preventive medicine. Any bad effects from the use of the stems resulted not from the stems but from the action of inexperienced winemakers who did not know when to separate the stems from the liquid. To leave the must and stems in contact too long would put too much tannin in the wine and give it a bad aroma. The use of stems in making the best wines of Burgundy and the Beaujolais was evidence enough for Machard that it was a beneficial procedure.

For those still unconvinced by his arguments, Machard introduced the example of Bordeaux. Stems made wine more healthful by giving it a certain resemblance to the wine of Bordeaux, "one of the most wholesome wines in the world." In the 1860s, because of the notorious ease with which burgundies spoiled, Burgundy was perhaps more willing to listen to such disagreeable analogies than it had ever been before or has been since. This bad-export reputation led to quite different judgments on its vinification process. The storm over stems would soon be replaced by the drama of Pasteur's microbes. Machard admitted to one disadvantage of destemming: the *eaux-de-vie* distilled from the wine of unstemmed grapes was not as good as that made from the wine of destemmed grapes. But the better quality and body of the wine made with stems, along with its careful distillation, could compensate for such a minor disadvantage in this one area of potential vice from the use of stems. Thus the general dogma was saved.

We must assume that Machard was advocating a moderate level of tannin; for nowadays we are more likely to follow modern oenologists in recognizing that excessively high tannin will produce an astringent wine. In describing one type of unpleasant wine, with too much tannin coming from vinification with unstemmed grapes, Emile Peynaud uses the same vocabulary that Machard used to describe a disagreeable wine resulting from low tannin as a result of vinification without the stems. "With high levels of tannin, wine is harsh. It has the taste given by vinifying with stems, or the taste of *vin de presse;* it is bitter; it promotes this sensation of a contraction of the mucous members called astringence."[25]

Machard believed in the local model of vinification. Each locality had to proceed according to laws applying only to it. General principles were not sufficient. There were three regions for which rules could be established: the north, the center, and the Midi. There were certain oenological and even some regional practices that could be followed in all areas. These were items found in most of the books on wine by the 1860s: great care in keeping the casks clean, frequent sulfuring, racking in March, and frequent topping up of casks. Whether invented by God or Koch and Pasteur, microbes were the great enemy of wine as well as a necessary friend in some processes. As in medicine, in oenological theory and practice the actions of unknown factors and how to control some of them were known before the germ theory got established. Hence the emphasis that Machard placed on the need for cleanliness and on the importance of sulfuring the wine. The year 1858, when much of a fine vintage was spoiled while still in the casks, stood as a warning to careless winemakers. To counteract such disasters, Machard also recommended vinification with stems and, not unrelated, a short fermentation. He believed that this style of winemaking provided an excellent model for all to imitate in Burgundy.

In recent times Léglise has reiterated the traditional view that the reason for including the stems in the must is to prolong the life of the wine. He emphasizes that the wines are not better: taste is not improved, and the period during which the wine is best to drink is not lengthened. It was in the 1930s that the custom of total *égrappage* was established, with the object of producing wines that could be drunk soon after being bottled, thus eliminating potentially heavy storage expenses. Between 1945 and 1949, total *égrappage* became widespread to relaunch the wine economy through the sale of fruity young wines. A classic example of this wine model can be found in Georges Duboeuf's impressive list of beaujolais

[25] Peynaud, *Connaissance et travail du vin,* p. 17.

wines, unlikely to alarm even the most wine-shy person. It is a good thing in its category, but as Léglise laments, any viticultural region can produce that kind of wine, whereas only a few areas can produce long-lived wines, the *vins de garde*.

In their little classic on wine, Amerine and Singleton justify the nearly universal practice of destemming wine grapes on the grounds that the tannin and other substances in the stems make wines astringent, bitter, resinous, or even add peppery flavors.[26] (Presumably the wine tasters who wax ecstatic over the peppery prickle of some reds would have a problem with this justification of common Californian practice.) Of course, if tannin is needed in the wine, then the stems can be left in or added to the must, although it is much easier to pump and handle the must without the stems.

The issue is not so simple in Burgundy. Léglise's approach to the subject neatly inverts the Californian oenological gospel. He admits that excellent *vins de garde* can be made from destemmed grapes if the maximum amounts of substances necessary for such wines are extracted from the skins. The skins of red grapes, like the seeds, have a high tannin content. But he points out that certain estates of the Côte d'Or that produce *grands crus* have never practiced destemming even when it was the fashion, while continuing to produce wines of distinction. In any case, whether to destem or not is a false problem for Léglise. Because it is the young unripe stem that is very acid (i.e., tannic) and bitter, what counts is the degree of maturity of the stems, not simply the use of stems. In this empirical oenology, science defers to tradition, often to the detriment of both wine and science.

The oenological gospel of Bordeaux is condescendingly hostile to stems in the must because they can ruin fine wines. Emile Peynaud recommends total destemming for the production of wines of *souplesse* and *finesse*, more supple and subtle wines with a better and a higher acidity in the must. Pacottet had shown early in the twentieth century that, contrary to a widely believed idea, vinification with stems, poor in acid and rich in potassium, reduced the acidity of the wine. But fermentation with stems is easier, and this is probably why the practice retained a certain popularity before a more precise technological control – for example, a cooler temperature – made it possible to obtain the same results without the disadvantages of stems in the must. The only exception Peynaud makes is for certain types of grapes like the Aramon, which do not have a very woody stem, or for grapes from young vines, where the stems would give a little backbone to the wine, or for grapes suffering from rot, a case in which oxidative *casse* would be less likely in the wine if stems were used.[27]

---

[26] M. A. Amerine and V. L. Singleton, *Wine: An Introduction,* 2d ed. (Berkeley and Los Angeles, 1976), pp. 97–8.

[27] Emile Peynaud, *Connaissance et travail du vin,* new ed. (Paris, 1981), pp. 133–5.

Peynaud coyly revealed, in *Le vin et les jours,* that careful destemming has always been one of his professional secrets for producing great wine.[28] Most of the bordelais châteaux destem, but Pétrus and Château Ausone add stems (10 to 30 percent) to the must.[29] Total exclusion of stems is not a universally followed dogma, even in Bordeaux, but the use of stems seems nowadays to be more ritual than functional.

Anthony Hanson has taken notice of the persistence of the adding of at least some stems to the must by certain Burgundian winemakers. Some believe that the stems work against "a jamlike consistency in the must," whereas others believe that the old proportions of skins, seeds, stems, and juice are no longer the same because grapes are now bigger than they used to be. In this minibelief system, "stalks are necessary to redress the balance" essential for the model of Burgundian wine subscribed to by the believers in at least some of the older practices. Hanson, however, gives more space to elaborating the advantages of destemming: less volume to the must, less *marc* to press after fermentation, more alcohol and color because of the absence of water from the stems, and elimination of the problems that could come from an excess of tannin if that of the stems were added. One still gets the impression that Hanson suspects the stem adders have something in their favor, but he does not complicate the issue in the same way that Léglise does with his scholastic distinction between young and old stems. Hanson ends up partly in the Peynaudian camp in noting that destemming in years of imperfect maturity of the grapes will produce a wine of greater mellowness, finesse, and quality than will be the case if stems are used.

Reinventing a Tradition: The Post-Chaptalian Burgundy

There is an implication in the squabbles over destemming and chaptalization that a heritage may have been betrayed or, more likely, traded for a higher profit margin. Unlike in modern capitalistic farming, in historic wine country a tension exists between the need for maximum profit and the desire to remain true to the traditional faith – living still, in spite of capitalism, chemistry, and technology. Many wine merchants have recently shown that profit and tradition can be happy bedfellows. A large number of consumers appreciate the revival of traditional virtues; science and technology have got rid of traditional vices. And with a little luck and about twenty dollars it is possible to find a good generic burgundy – from Hubert de Montille, for example – and even an unsulfured wine: the *bourgogne grand ordinaire* of Domaine Prieuré Roch, for example. Of course,

---

[28] Emile Peynaud, *Le vin et ses jours* (Paris, 1988), p. 236. See the excellent section "De l'égrappage," pp. 228–36.

[29] Robert M. Parker, *Bordeaux* (New York, 1991), p. 947.

it is still much easier to find a bad or mediocre burgundy than a bordeaux for the same price.

Only in recent years has a new, small consumer demand, presaging perhaps even a new model of taste, led producers to supply unchaptalized wines having the qualities that Jullien valued in the *grand crus*. One can even find nonchaptalized Beaujolais – for example, the *Cuvée traditionnelle* of the Domaine du Vissoux, produced by Pierre-Marie Chermette, who identifies himself as a "viticulteur-oenologue." And for once the line on tradition is not postmodern hype: no sugar added, several months in wood and bottled after Easter, nonfiltered but fined with egg whites – a "natural living product" with a possible light deposit in the bottle. Across the Rhine, some German producers, seeing the folly of killing wine with sweetness, are returning to the *trocken* tradition. But in the mass market we shouldn't expect too many producers not to take advantage of their right to chaptalize and to acidify or deacidify according to the rules governing the five ecological zones into which European Union rules divide France on the basis of climate in order to regulate minimum and maximum alcohol, sugar, and acid contents. A climatic or geographical determinism justifies convenient practices in vinification. An ecological rubric is good public relations.[30] And the market for high-quality wine remains small.

Many of the problems with burgundies have been blamed on the *négociants* and the way the wines are sold. Does this reflect the much more important role of the wine merchant in Burgundy? Does the power of the château in the Bordelais, with a greater role for the proprietor-producer and pride in the product, give Bordeaux an advantage in a quality-oriented market? Whatever the sins and crimes of the *négociants* in the past, it now seems simplistic and just plain wrong to blame the wine merchants for the dissatisfaction with burgundy.[31] In the 1930s the big commercial houses were criticized for selling blended wines under generic names like Bourgogne Tartempion rather than under the old Burgundian names like Pommard and Gevrey-Chambertin, which authorities could control in comparing the amount bought with the amount sold. These clever blends, including foreign as well as domestic wines, sold for 6000

---

[30] See Lothar A. Kreck, "The EEC's ecological zones," *Wines and Vines* 61 (1980): 21.

[31] The wine merchants of Beaune seem to contain as high a proportion of scoundrels as you would expect in any sample of a general population. In "The Best of Beaune," *Decanter* 17 (November 1991): 71–4, Clive Coates does not name the bad, but the good: Bouchard père et fils, Champy, Chanson père et fils, Drouhin, Jadot, Latour, etc. The best of Nuits Saint-George were named in the May 1991 issue of *Decanter*. The complete drinker will consult Hanson, *Burgundy,* and Robert Parker's mind-boggling and not uncritical catalog, *Burgundy* (New York, 1990), as well as his periodical guide, *The Wine Advocate*. For a "massive tasting of 1989 red burgundies," see *The Wine Spectator,* Jan. 31, 1992. We may soon have more good guides than good wines.

francs per *pièce* (about 25 dozen bottles), while the wines of Gevrey-Chambertin remained unsold at 1500 to 1800 francs.[32] In spite of the power of Bordelais proprietors – a fluctuating entity, especially in touchy economic times – "Les négoces" are still powerful in Bordeaux, lording it over a good part of the Gironde, through owning châteaux, working with cooperatives, and possessing vast export power.[33] These ups and downs do not seem to affect the general reputation of bordeaux, though it is easy enough to single out falls and rises in the quality of individual estates. With a few notable exceptions, the *négociant* has a more sinister reputation in Burgundy. So it has been argued that there is a connection between the increased role of a new generation of Burgundian proprietors in the selling of wine and the better quality of wines available.

The news from Burgundy is often optimistic. Patrick Dussert-Gerber's *Guide des vins de France 1987* was upbeat about better wines being made in Burgundy – "Le coup de foudre" – and the guide for 1988 continues to celebrate "La force de l'homme" revealing itself in the world of the vigneron. Rigorous selection is the order of the day, but Dussert-Gerber in 1988 was still friendly to big houses like Drouhin, Jadot, and Latour: "Sélections de premier ordre." Latour was dropped from the guide in 1993, presumably for not sending in his samples to the guide. A curious system, as one can tell from reading the lists of the banished. It seems that Burgundy has serious problems: image; stuffy, sometimes sexist publicity; high prices; an evolution of taste away from red burgundy. Sooner or later Burgundy must emulate the genius of Bordeaux in maintaining its wine as the universal criterion of *the* great red, if it is not too late; as the wine-drinking public, perhaps tired of expensive mediocre reds, increasingly hopes for something better from white wine, whose models seem more in tune with less gourmand times.[34] Some drinkers derive consolation from the reds of Michel Juillot, Bruno Clair, and Georges Roumier, among others, while waiting for the Burgundian millennium.

In his guide for 1988, Patrick Dussert-Gerber waxes ecstatic over the new Burgundian *vignerons* and the implied threat to the trade (*le négoce*). As some of the *grandes maisons* or big firms like Drouhin, Moillard-Grivot, Latour, and Bouchard are big proprietors, this may seem like premature ecstasy. Louis Trocard has categorized the Bordelais *négociant* as a commercializer working in partnership with the winegrower and also a

---

[32] "L'appellation Bourgogne," *BIV,* 4e année, no. 36 (May 1931): 78–83, esp. pp. 8off.

[33] *Guide GaultMillau: Le vin* (1989), pp. 276–7.

[34] "La Bourgogne existe-t-elle?" *GaultMillau Spécial Vins 88,* pp. 112–13. See also the *Magazine GaultMillau,* no. 259 (January 1991): 73, for the results of a selection of wines for the hotels Mercure: many good burgundies were found, but many also had all the Burgundian defects: "beaucoup d'alcool et pas de terroir." A blind tasting of wines considered for purchase by the Swedish State Monopoly was a catastrophe, including even bad bordeaux.

force for quality. As nearly 75 percent of the production of Bordeaux is sold by *négociants,* this is a good thing.[35] While waiting for the millennium, we are warned by our guide to choose very carefully the producer, the *cru,* and the year. He might also have added that one should only buy wine that has been transported at the proper temperature from the cellar in France to the wine store. Burgundy is a wine easily ruined by heat and light in transportation and storage, if it has not already been damaged by overfiltering and overchaptalization, the greatest Burgundian vices, which now show some signs of disappearing.

The *maisons* select, prepare, and sell about two-thirds of the wine produced in Burgundy and sell more than 90 percent of the wine exported. Small growers, unable to provide the volume and range of wines demanded by buyers, sell to the big houses, which assemble and sell the wines that the market will accept. Thus the social structure of property determines the nature of the product furnished to the customer. And few can nowadays follow Jefferson, one of the first customers to cite the *cru,* the vintage, and the grower in stating his preference for "du Meursault goutte d'or 1784" of M. Bachez.[36] Owners of the *premiers crus* of Bordeaux sell through a few *négociants,* who distribute the wines but handle their own publicity and maintain a key role in the commercialization of their wines. The participation of the producer in this process explains why the idea of *cru* has taken on such importance in the viti-vinicultural world for over a century.[37] It is an excellent marketing concept. Bourdieu's brilliant metaphysical dogma seems to tie it all together: "Taste is the mysterious instinct that enables producers and consumers to find one another in the dark night of the free market."[38]

From the 1970s to the economic recession of the early 1990s, Burgundians lived in happy times, when a growing, well-heeled, and often ignorant population drank according to labels, followed ratings of pseudoprecision, and, most important, believed that there is a connection between quality and high prices. As Béguillet lamented at the end of the eighteenth century and P.-M. Doutrelant and Anthony Hanson recently, blending burgundy with southern wines produced the hearty red and stable "burgundy" demanded by wine writers, merchants, and customers.[39] The state of the art

[35] Patrick Dussert-Gerber, *Guide des vins de Bordeaux. Bordeaux et Bordeaux supérieurs* (1990), p. 9, preface by Louis Trocard. In 1946–55, one-third of the AOC wine of the Gironde was AOC bordeaux and bordeaux supérieur; in 1990 the amount of AOC bordeaux and bordeaux supérieur had risen to 52% of production.

[36] Félix Crestin-Billet, *Vins de Bourgogne. Les grandes maisons de Bourgogne* (1990), pp. 13–15.

[37] Dominique Denis, *La vigne et le vin – régime juridique* (Paris, 1989), p. 63.

[38] Malcolm's Bull's exegesis, in a review of two books on taste, one by Stephen Bayley and another by Peter Lloyd Jones, *The Times Literary Supplement,* Dec. 27, 1991, p. 14.

[39] See Hanson, *Burgundy,* chap. 8, "Magician's Hands," and P.-M. Doutrelant, *Les bons vins et les autres* (1976).

in the early nineteenth century was probably not good enough to fool the small and discriminating band of *cru* worshipers, but others seemed satisfied. It is probable that today's technology and a bit of basic science make it much easier than it was even a century ago to produce a striking mediocrity that will make most people happy. In the present cycle of Western civilization the dominant taste model still emphasizes striking color and a degree of alcohol that often requires more sugar than is provided by a fickle nature. The nineteenth century is still with us in many ways.

9

Languedoc-Roussillon: Innovations in Traditional Oenology

Montpellier is the big name in viticultural and oenological research in the Midi, though it is not the only well-known "school" in the Languedoc. Any survey of oenological literature soon reveals the existence of "the Narbonne school of enologists."[1] In 1895, a year notorious for its *vins cassés,* the ministry of agriculture founded two oenological stations, one in Narbonne and the other in Montpellier. It was a difficult time for the wine business, and a decade later it would be worse, culminating in the riots of 1907, more bloody in Narbonne than in many parts of the Midi.[2] The government hoped that an injection of science into viticulture would help an important but sick industry. This belief was part of the powerful Pasteurian ideology prescribing science to cure national ills.

In an analysis of the viticultural revolution in the Aude during the late eighteenth and early nineteenth centuries, Jean Valentin points to a basic change in emphasis, from quality to quantity. High wine prices, better growing practices, and greed pushed yields from twelve hl/ha in the early nineteenth century to eighteen hl/ha in mid-century.[3] By 1850 so much wine was being produced at such low prices that one-third of it was distilled. The competition of cheap Russian wheat was an important factor in this unwise rush to economic overdependence on the vine in the period from the 1850s to the 1880s, when viticulture flourished in the Midi. Cereals were still important nationally, accounting for 73 percent of agricultural production as compared with 58 percent for the four departments making up the viticultural Midi.[4] Prosperity was promoted by the coming

---

[1] The title is used by M. A. Amerine and M. A. Joslyn, *Table Wines,* 2d ed. (Berkeley and Los Angeles, 1970), p. 304.

[2] Paul Carbonnel, *Histoire de Narbonne* (Marseille, 1956). The Midi, defined viticulturally, includes the four departments of the Aude, the Gard, the Hérault, and the Pyrénées-orientales.

[3] Jean Valentin, *La révolution viticole dans l'Aude, 1789–1907* (Carcassonne, 1977), fascicule 1.

[4] Rémy Pech, *Entreprise viticole et capitalisme en Languedoc-Roussillon, du Phylloxera aux crises de mévente* (Toulouse, n.d.), pp. 36–8. Pech exhumed and used the work of Dr. Jules Guyot, *Etude des vignobles de France* (1868), vol. 1.

of the railroad, which allowed wine to be shipped to thirsty and undiscriminating urban consumers in the north. Other people's disasters also helped to sell more wine. Mildew was less important in the dry Midi than elsewhere. And up to 1884 phylloxera spread faster in other wine-growing areas. The crash in wine prices was brutal, beginning in the 1890s and bottoming out in the early twentieth century. Revenue from wine fell from an annual average of 135 million francs in the 1880s to 18.6 million in 1906. The time of troubles was at hand.

It has been argued that in the late nineteenth and early twentieth centuries artificial wine – made from raisins and worse – nearly drove natural grape wine off the market. It is an argument that was often made by wine interests that did not want to face up to the fact of overproduction, but it is doubtful that fraud, including wine from raisins, and *sucrage* alone accounted for the crisis. In the first decade of the twentieth century, production averaged nearly 56 million hl annually, nearly 20 million more than for the previous decade; the 1920s–1930s would see similar bumper crops of grapes and similar surpluses of wine. Algerian wine was also coming heavily into the market, with a production of over 5 million hl by 1900; by 1939 it was nearly 32 million hl, little of which was drunk by the French settlers, and less by the Muslims.[5] Wine producers in the Midi, more angry and riotous than they had been since 1789, blamed frauders, not themselves. Many oenologists were seduced into making technical and economic studies on fraud, *mouillage* (adding water), and chaptalization, presumably in the hope of helping to replace the artificial wine on the market by the vast surplus of natural wine held by producers and sellers. Even if they had been aware of the complexity of the crisis, neither scientist nor politician would have been eager to tell misguided but furious peasants to turn their vineyards into truck gardens. (Governments have gotten round to doing it only recently.)

In 1936 the socialist prime minister, Léon Blum, who since 1929 was also the deputy from Narbonne, inaugurated the Foire du Languedoc, with its Exposition vinicole and the new oenological institute of Narbonne (Station de recherches oenologiques de France). Historians of the Third Republic have perhaps overemphasized the importance in Blum's career of the Ecole normale supérieure, less significant politically than "la fête des vins de l'Aude." Why should Narbonne have been chosen as a center for the application of science to grapes and wine? Apart from its days of

[5] For figures, see Lachiver, *Vins, vignes et vignerons,* "annexes." For an analysis of the crisis, see Valentin, who points out the limitation of the crisis to the Languedoc and the growing unprofitability of *mouillage,* because wine was so cheap and there was a big drop in the tax on wine after 1900; nor was it profitable to buy sugar to make wine after 1904, as it was difficult to sell it at a high enough price to recover costs. Valentin is right: it was a complex crisis.

Roman glory and medieval splendor, a splendid legacy ruined by subjection to the French monarchy and Catholic orthodoxy, and its exciting modern politics, Narbonne's fame does not elicit long entries in encyclopedias, even French ones. Like Montpellier, Narbonne represented all that was wrong with the wine industry: the mass production of table wine in an economy already burdened by a great surplus of mediocre wine. Unlike Montpellier, or Bordeaux and Dijon, Narbonne had no science, or at least no scientific institutions, and no hope of becoming a "grande cité universitaire." The distinction of Narbonne was that it overcame this disadvantage to establish a unique viticultural-oenological reputation for contact between the grape-growing population and practical science.

This success was the work of Lucien Sémichon, who was for nearly half a century the pontifical voice of oenology in the Languedoc outside Montpellier, perhaps even in the Midi and beyond. Most oenologists flatter themselves as being the voice of science. Although not immune to this occupational disease, Sémichon was proud of his role as the voice of the vigneron. Of course, taking science to the grower and the producer must be one of the primary roles of the scientist in an oenological station or, higher in the social hierarchy of science, an institute of oenology, but getting input back from the small producer is rare, and few have ever been so successful at it as Sémichon.

Lucien Sémichon began studying wine scientifically at the Institut national agronomique de Paris, where he had the good luck to be enlightened by Emile Duclaux, Eugène Risler, and Achille Müntz. His first job (1891) as a lowly *répétiteur-préparateur* was at the Ecole nationale d'agriculture of Montpellier, where he was able to do research and publish with scientists as good as Bouffard. After a year at the school of agriculture in Rethel (Ardennes), he successfully competed for the post of head of the Station oenologique de Narbonne, where he stayed for the rest of his long career. His retirement coincided with the establishment of the new palace of oenological science, the Institut oenologique de Narbonne. The new director was Michel Flanzy, a scientist whose training corresponded more to the Bordeaux oenological model than to the agricultural school model, although these models overlap, especially in their practical components.

There have been arguments over which is better for winemaking, but there is no doubt that the Bordeaux model is better for the prestige of the oenologist. It is tempting to see the careers of both Sémichon and Flanzy as good illustrations of Peynaud's sally that the wine makes the oenologist. It is difficult to be a great oenologist in the land of *vin de pays*. Sémichon had a brilliant strategy in this difficult situation: elevate the status of the *vins de pays* and that of the oenologist also rises. It worked to a certain degree, but success with a Fitou or a Hautes Corbières could never bring the same scientific recognition as success with a château in the Haut-

Médoc. Nor, curiously, does the science seem equal. This is no doubt due to the imperialistic presence of the faculty of sciences and the invasion of the oenology by the latest research in the relevant sciences.

Sémichon had a respectable oenological career, crowned by the big prize given by the Office international du vin, and he even strongly influenced the winemaking techniques of the region. At a banquet celebrating his life in science, it could be said that he had put his mark on, if not actually invented, "l'oenologie méridionale." If Narbonne was more creative in winemaking than Montpellier, its input into oenology seems of minor interest except in the production of *vins de pays.* Unable to develop a universal model of a great wine, which is paradoxically based on local conditions, Narbonne sought to fulfill its own paradoxical aim, the production of great "petits vins."

Sémichon's analysis of local wines and their methods of vinification established his expertise and power at his home base. In 1905 he published a book on wine diseases that greatly increased his reputation in the worlds of wine and vine. (It appeared in Turkish in 1948.) Since 1897 he had made a local reputation by advocating the use of potassium bisulfite in vinification as the most practical and rational way for small producers to reap the benefits of sulfur dioxide. His book on the *Maladies des vins* emulated the practical side of Pasteurian *Etudes* while also being respectably scientific, which meant keeping up on the latest research. Ulysse Gayon, on whom the oenological mantle of Pasteur had fallen, honored the author and Narbonne with a preface to the book, calling it "a true treatise of oenology . . . filling an important gap in the literature."[6] Giving Sémichon's book the Pasteurian *nihil obstat,* Gayon declared that it was inspired by the master's thought of joining scientific explanation to techniques. ("Je me plais à rattacher aux explications de la science les usages techniques.") The book gave Sémichon a solid national reputation in viticultural and oenological science.

Sémichon even tackled the problem of what to do with the big surpluses of unsaleable table wine produced in the Midi. While waiting for governmental action, he became a major force in introducing and supporting automated distillation of wine by cooperatives in the Midi. No one knew more about the making and selling of *vins de liqueur* than Sémichon. He also promoted wine cooperatives. During World War I, Sémichon developed a system of analysis for large volumes of wine shipped to the front: good wine for the soldier would inspire him to kill more Germans.

Sémichon was a keen methods man. He developed new methods of analysis and microanalysis for different alcohols and acids. Many sciences,

---

[6] Lucien Sémichon, *Traité des maladies des vins. Description – étude – traitement* (Montpellier, 1905). The Station oenologique is identified here as departmental, "de l'Aude" rather than the usual "de Narbonne."

many methods; even many methods for one science and, worse, several methods of analysis giving different answers to the same research question. Fortunately, in most problems the approximate is accurate enough to create a reality that can be manipulated. The method depends on the accuracy and quickness desired as well as the reality to be defined; nor are these desiderata separable in practice. Sémichon has an important place in the purgatory of oenological methods.

What about originality? Basic discoveries? Sémichon has two major claims to fame, according to himself and his admirers. These discoveries were important enough to be communicated to the academies of science and of agriculture. In the area of the biology of fermentation Sémichon discovered that alcohol selects yeasts. This gospel of liquid Darwinism is based on the principle that the genus *Saccharomyces cerevisiae* (*Sacch. ellipsoideus*), the yeasts that are found on ripe grapes and are best for fermentation, are the only ones capable of growth in a sugared environment of 4 percent alcohol. Sémichon was not slow in leaping to the corollary of this theorem of natural physiological selection: starting fermentation with a 4–5 percent solution of alcohol automatically kills off all the wild (bad) yeasts of other genera like *Sacch. apiculatus,* thus leaving more sugar for the good yeasts to change into alcohol (more or less 0.5 percent) and gets rid of the bad tastes and odors that wild yeasts sometimes produce in wine. Mixing musts and wine, either new or old, had long been done to activate and regulate fermentation. Sémichon insisted that the winemakers did not know why this practice worked, only that it did.

Science came to the cognitive rescue, with precise instructions. Sémichon insisted that his originality was in discovering the role of alcohol in the physiological selection of the yeasts, which gave the possibility of fulfilling the Pasteurian dream of pure fermentation from pure yeasts in a sterile sweet juice. This "remarkable property of alcohol could be used to great advantage by vinifying in a system of continuous fermentation," in which a fermenting must of at least 4–5 percent alcohol is put into the vat of virgin must. It looked as if vinification could have fermentations as pure as brewing and distilling – a minor miracle and not a total illusion. Sémichon presented this fermentation (soon given the impressive name of "fermentation superquatre," or superfour, by his friends) as the magic bullet in vinification, with positive as well as negative virtues: more *finesse,* more *moelleux* in many cases.[7] All over the Midi the rave re-

---

[7] "La sélection physiologique des ferments par l'alcool," *Annales de la Brasserie et de la Distillerie,* 28e année, no. 7 (Nov. 10, 1929), pp. 97–101, reprinted as "Les fermentations pures en vinification par le système de sélection dit de 'fermentation superquatre,' " *Annales des Fermentations et des Fraudes* (1929), pp. 466–71. See also "Résultats pratiques de la fermentation superquatre en vinification," *Bulletin international du vin* (*BIV*), 4e année, no. 36 (May 1931): 101–4.

views rolled in. The process spread to the Magreb, to Spain, to Portugal, to South Africa, to Argentina.

And was this work relevant to the *grands crus?* No French oenologist feels successful unless his discovery or invention is important for vinification in the Champagne, Burgundy, or Bordeaux. Sémichon had a winning discovery. His system was used on a few estates in the Médoc and in less prestigious Saint-Emilion. (It was also used at Château Ausone, which even the Médoc-oriented British liked and which had a good reputation down to 1945, lost it, and has recovered it since 1975.) Was this procedure the major triumph it was thought to be? Sémichon retired confident that his claim to fame was secure, that his great discovery had become a permanent part of vinification. Superfour turned out to be more useful in warm than cool climates, because it keeps down fermenting temperatures. In cooler climes it has little to recommend it. And in spite of the oenological hoopla of the 1930s about Superfour, it is not much heard of today. With the advent of the easy cooling of vats, it has even lost its significance in warm climates. Practices derived from theoretical advances are more permanent in basic processes than those based on a *bricolage* concerned with symptoms and fixes. And this obsession with the biochemical mechanisms of the fermentation and aging of wine was the genius of oenology in Bordeaux after the 1920s. Of course, the Bordelais oenologists often had the advantage of working with a great product.

In his work Sémichon paid proper homage to the patron saint of the life sciences in France. In communicating to national academies he could hardly begin without a salute to some god, preferably French. Down to mid-century, and even well after, it would probably have been Lavoisier; by the end of the century it was inevitably Pasteur. Sémichon's science, his new theorem of vinification, showed that Pasteur's principles had reached the point of diminishing returns. Pasteur's brilliant work had now only a negative significance, incapable of stimulating new, original work in oenology. Dead germs had become less important than live ones. And for a more basic understanding of the living wine an entirely different kind of science was needed, one radically different from traditional bacteriology. This was ionic theory. Sémichon's enthusiasm for Superfour was based partly on the prevalent dissatisfaction with both heating and the use of massive doses of sulfur dioxide to control bacteria. With the *crus* these processes threatened quality; with ordinary wine there was the cost of processes and material. No doubt he was proud of his contribution to winemaking in general, but improving *crus* was not of much importance in the Aude, or even the rest of the Midi. And it would hardly improve his scientific status to become the guru of an oenology of *vin ordinaire.*

Sémichon's second claim to fame, local perhaps, intersected the world of the *cru* and that of the *ordinaire:* a campaign to restore the *vins de pays* to their presumed former status. This might even be seen as a potentially more valuable contribution to oenology than Superfour. He who supports the slogan "Narbonne, capitale du vin pur" must also support the battle against the industry of standardized wines ("vins omnibus"). In the wine business, if you cannot sell *grand cru,* you may as well sell nature and purity, concepts that tap into the myth of purification, which may, however, be less powerful than the appeal of snobbism. Sémichon feared that the spread of wine factories would turn vignerons into slaves of capitalistic viticulture. Among the myths centered on the vine is the idea that French character and spirit are stimulated by wine, the national drink, which must be a natural product. "Wine will only remain this national drink that gives birth to something of the French character and spirit if it remains the natural product of the vine." Unfortunately, the production of *vins de pays* is all too often based on the same standardized model as industrial wine, a liquid banality having little to do with the hopes of Sémichon.

Sémichon hoped that one day history would contain at least a line on "a new school founded at Narbonne." This seems likely, as oenologists recognize Narbonne as the center of a school of carbonic maceration, a widely used practice of ancient origins. In the wine texts the practice is treated under the rubrics of "noncrushing" and "complete anerobic fermentation. Two processes seem to take place in this procedure: an intracellular fermentation of malic acid (and possibly of some tartaric) and an improved malo-lactic fermentation."[8] Michel Flanzy, Sémichon's successor as head of the Station oenologique de Narbonne, has been a great defender of this type of vinification. Sémichon was an agricultural engineer; Flanzy was a man with a doctorate in science, a university one rather than the more academically prestigious state doctorate. But Flanzy's thesis was born under the right star. His *maître,* Paul Sabatier, academic politician, master builder, and great scientist, was ruler of the chemical empire that he himself had established at the University of Toulouse.[9] Sabatier pushed agricultural studies in a scientific direction at Toulouse, but his catalytic talent was more relevant to oil than to wine. Or perhaps he didn't think that the wine of the region was important enough for his talent.

Flanzy became the disciple and successor of Sémichon at Narbonne. As early as 1939 he assumed the existence of "l'oenologie méridionale."[10] Very early Flanzy became an apostle of the Narbonnais doctrine of pure

---

[8] Amerine and Joslyn, *Table Wines,* p. 609; see also p. 304 on "noncrushing."
[9] Greatness in this case means having nearly two pages in the *DSB* (article by Mary Jo Nye). Any doubts can be stifled by reference to Sabatier's Nobel prize. Flanzy's *thèse doctorat de l'Université* was *L'alcool méthylique dans les liquides alcooliques naturelles* (1934).
[10] Michel Flanzy, "Oenologie méridionale," *BIV,* no. 135 (1939): 60–1.

wine, although it is certain that this model would not pass muster today as an organic wine ("vin biologique"). Excessive use of sulfur dioxide and other worse oenological products seemed to be turning wine into an excipient for chemical products. But what was distinctive about the oenology of the Midi according to Flanzy's doctrine? Not much in 1939. True, Flanzy did not believe that in the making of red wine the healthy stems, rich in tannins, should be separated from the grapes before vinification. But this was merely to take a position in a very old debate over destemming, which had supporters and opponents with various subtle variations on the theme. Any position on the issue would depend on many factors, including geographical determinants, and especially the model of wine the producer wants to create. Flanzy also thought that white wines could rest on healthy lees – better known as a technique of the "pays nantais," used for Muscadet and Gros-Plant – which would make them more *moelleux* and help in clarification. Arguments in favor of leaving the muscadet on its lees are that it promotes a better development of bouquet and a longer preservation of its fresh taste. The issue is hardly a major one in the history of oenology, or even of winemaking.[11] Not that to leave on lees or not to leave on lees is unimportant; it is just a technical point significant for the gustatory model of the wine desired rather than a theory-laden oenological debate. Flanzy believed that making wine is an operation of great simplicity.

Flanzy's modest niche in the history of oenology is probably due more than anything else to his stubborn advocacy of carbonic maceration (CM) as the best type of vinification.[12] CM is yet another process whose origin is attributed to a speculation of Pasteur's fertile mind. Suspicious of the destructive influence of oxygen on wine, he wondered about the advantages of fermenting the grapes in a protective shroud of carbon dioxide. Grapes kept in carbon dioxide had a better flavor than those kept in air; so why not use the gas to make a better wine, better being defined as a wine with more flavor? The preservation of flavor was the last aspect to be considered in the range of problems afflicting viticulture and vinification in the second half of the nineteenth century. So the idea of CM sank low on the scale of worthwhile experiments in winemaking. By the 1930s the old problems had been solved or replaced by different, less serious ones.

[11] "La mise en bouteilles sur lie," *Guide GaultMillau: Le Vin*, p. 755. Now officially regulated, the technique specifies that the wine must spend only one winter on the lees of vinification before being bottled before the 30th of June following the harvest.

[12] Flanzy did considerable work on acidity, as is clear from my discussion of the acid craze of the 1930s. On CM, see Claude Flanzy, Michel Flanzy, and Pierre Benard, *La vinification par macération carbonique* (INRA, 1987; 1st ed., 1973). Amerine and Singleton, *Wine*, p. 161, point out that "under carbon-dioxide pressure . . . the grapeskin cells die and then the anthocyanin pigment readily diffuses out" – more color in red wine.

The CM experiment was also easier to do, or at least easier to carry out on a large scale. In the Midi scientists seemed much more concerned with large-scale results than those of the laboratory alone. Instead of constructing nature, the scientist must change it, for the better if he is lucky, which is the only way of convincing suspicious vignerons of the value of science and of scientists.

In CM the grapes are not squeezed, except for those at the bottom of the vat that are crushed by the grapes on top, which means that up to 20 percent of the fruit does not undergo CM. Because the fruit is covered by carbon dioxide, the fermentation is anaerobic, produced by the enzymes within the grape rather than by the traditional action of yeast enzymes on the squeezed juice. When, after about ten days, the free-run juice is pumped off and the unbroken grapes pressed, about four-fifths of the sugar has been converted into alcohol; juices, skins, and stems are put together for the finishing of the vinification, including the traditional yeast action during the remaining third of the process, which includes malolactic fermentation.

Flanzy's experiments of the early 1930s were repeated by Gallay and Vuichoud in the Vaud and reported in the *Revue de viticulture* in 1938.[13] This testing of Flanzy's method of vinification confirmed many of his claims for CM, especially that it gives a better bouquet and a more *moelleux* wine than that produced by traditional vinification. Gallay and Vuichoud vinified varieties of grapes from the Loire, Bordeaux, and Sainte-Foix to show that the theory was good for grapes other than the lowly, widespread aramon. A vat filled with sulfur dioxide gas, instead of carbon dioxide, produced a better colored wine. The wine from the CM vat had 1 percent less alcohol and less acid but more volatile acidity than both the sulfured vat and a control vat fermented in the traditional way. Sulfur hinders natural deacidification; CM makes it easy. The easy fermentation of CM wine had the disadvantage of producing wine of less density, with less dry extract, less fixed acidity, and less residual sugar than traditionally vinified wine. The counterbalancing virtues of CM wine were its aromatic richness, softness, and "harmony in the mouth." According to Flanzy, color and tannins depend on the temperature and on the duration of the CM fermentation. And he carefully distinguished his method from the type of CM used in the Beaujolais, where no carbon dioxide is used and vats are not sealed. Pure CM produced organoleptic differences in aroma and taste.

It is usually assumed that CM wines have to be drunk young. Flanzy

---

[13] R. Gallay and A. Vuichoud, "Premiers essais de vinification en rouge d'après la méthode Flanzy," *RV* 88 (1938): 238–42, reprinted from the *Annales agricoles vaudoises des écoles et stations agricoles du canton de Vaud.* For a review of CM or "noncrushing" or "complete anerobic fermentation," see Amerine and Joslyn, *Table Wines,* pp. 304, 609–11.

vigorously denied this. An oenologist will not derive much scientific prestige from any research that is not relevant to a wine model including considerable aging, a distinguishing feature of the *cru* and especially the *grand cru.* Flanzy argued that CM wines keep better than those vinified traditionally. "Better" in this context means an aromatic complexity and balance in taste close to the original qualities. This judgment is unlikely to impress lovers of the *grands crus,* one of whose characteristics is development from a youth that might be described in CM terms to a seductive maturity having quite different qualities. In an aggressive mood, Flanzy would admit that CM wines won't keep their qualities for a long time if defenders of tradition would admit that traditionally vinified wines don't either. A good point, for most non-CM wines. "It's a relative question," says Flanzy. So is the answer.

Carbonic maceration is going through a period of popularity in Languedoc-Roussillon, where research, commercial, and political institutions promoted it enthusiastically in the 1980s. It has had an enthusiastic reception in Italy for the production of "Beaujolais-style" *vino novello.* In 1988 eight major Italian producers were selling two million bottles of this stuff.[14] CM has also penetrated Burgundy, with the arrival in Nuits-Saint-Georges of Guy Accad, who promotes his oenological variation – CM means cold maceration in his repertoire – on a theme by Flanzy to a number of Burgundian winemakers. The results are praised by Michel Bettane and Robert Parker; Matt Kramer is less impressed, especially when the wine is decently aged: "the wines lacked detail, nuance, and subtlety." Nor can CM be used to vinify the great wines of Bordeaux without eliminating the distinctive characteristics of the grape varieties in the wines, as well as character and structure when the wines are aged.[15] In Flanzy's research dream, CM is associated with the ideology of purity and the natural qualities that Sémichon found in well-made *vins de pays.*

Why the commercial enthusiasm for CM wines? *Wines and Vines,* a voice of oenophilic capitalism, summed it up nicely: "The low tannin means the wine usually lacks bite and length of finish. For the producer and wine trade, such wines have an obvious appeal in that they free capital

---

[14] Philip Dallas, *Italian Wines,* new ed. (London, 1989), pp. 23–4.

[15] Matt Kramer, *Making Sense of Burgundy* (New York, 1990), pp. 72–6. The curious can see what the fuss is about by trying wine of the Domaine Jean Grivot (Vosne-Romanée), although Parker found the 1987 Echézeaux smelling like Syrah, of Jacky Confuron-Cotetidot (Vosne-Romanée), of the Domaine Daniel Senard (Aloxe Corton), and of the Château de la Tour (Clos de Vougeot), etc., "all of whom have gone over to the Dark Side." P. Martinière reexperimented for 15 years on "thermovinification et vinification par macération carbonique en Bordelais" before arriving at his negative assessment of CM. See Pascal Ribéreau-Gayon and Pierre Sudraud, eds., *Actualités oenologiques et viticoles* (1981), pp. 303–10.

much quicker and assist cash-flow. Caterers have to do less binning and there is no decant for deposit. . . . The wine is for early consumption."[16] The fruit and freshness obviously appeal to the consumer seduced by the taste model successfully used for a long time for Beaujolais and some of the Loire wines. Some of the *crus* of the Beaujolais (Morgon, Juliénas, Moulin-à-Vent) can survive a few years; some wines of Roussillon, like the Côtes du Roussillon-Villages and Collioure (an in-wine of knowledgeable, impecunious drinkers) have a possible life span of four to five years and more.[17] But age is usually only a privilege nature grants to the sweet stuff (*vins doux naturels*) of the Roussillon, and for the regular wines there is usually less point in waiting for the possible virtues of age to reveal themselves than in the case of the wine of Bordeaux and Burgundy.

Flanzy and his colleagues agree that there are economic reasons for the spread of the oenological gospel of CM. An increase of 20 percent in sale price is enough to destroy any tradition. Improvement through replanting with better-quality vines or grafting requires a long investment compared to the quick results achieved by using CM. Quality improvement in CM wine is clearly attained for certain varieties of grape like the carignan, whose "habituelle rudesse," as Woutaz kindly puts it, must be tamed by CM. It can be the same for other varieties (grenache, mourvèdre, syrah, cinsaut) "when the technological objective is correctly specified." CM makes available a range of products from *vin primeur* to *vin de garde*. It even taps into the health-conscious market, because a careful CM vinification uses less sulfur dioxide than the traditional one, not a bad argument to make as international legislation on SO_2 becomes "more and more draconian." Like Sémichon with his panacea of Superfour, Flanzy cites the growing use of CM in France for red and rosé wines; one can add Australia and even California. A high point in the history of CM wines was the serving of a CM Napa Valley Gamay to the leading politicians of the industrial powers meeting in Williamsburg (Virginia) in 1983. There is no report yet of CM wines at the Elysée, and even if the Socialists had retained the presidency, the Blumian imperative of presidential *crus du Roussillon* was unlikely to have been fulfilled. (Mitterrand served Château Latour 1978 to the queen of England during her state visit.) CM supporters have a final argument for anyone not convinced by these practical arguments: the intellectual challenge of producing novelty, the pleasure of creating, through reasoning, another product tending to have a definite

[16] Conal R. Gregory, M. W., "Europe's Look at Carbonic Maceration," *Vines and Wines* 63 (1982).
[17] Collioure (same *terroir* as Banyuls) was elevated into the ranks of AOC wines in 1971. In *L'Atlas des vins de France*, p. 192, Fernand Woutaz describes the Collioure as a "vin rouge corsé et capiteux, étoffé, riche en bouquet." The grapes used are Grenache (minimum of 60%), Carignan (minimum of 25%), Cinsaut, and Mourvèdre – fairly long carbonic maceration.

aesthetic appeal.[18] Like Omar Khayyám, the Frenchman clings to a last illusion, belief in wine.[19]

[18] Flanzy et al., *La vinification par macération carbonique,* pp. 108 ff. The vinopolitical powers supporting Flanzy's spreading of the CM gospel are given in his book: ONIVINS, CEVI-LAR, INRA, the Institut des produits de la vigne, etc., and the Conseil régional Languedoc-Roussillon. ONIVINS and the Conseil financed the distribution of brochures to wine producers on MC, control of time in oenology, tartric stabilization, malolactic fermentation, and the production of white wines.

[19] Cioran, *Syllogismes de l'amertume* (Paris, 1952), p. 26: "Omar Khayyam . . . *croyait* encore au vin."

PART IV

Oenology in Bordeaux

10

The Pasteurian Oenology of Ulysse Gayon

Emile Peynaud once remarked that the value of oenologists is a function of the quality of the regional wines in which they make some improvement. The reputation of the Bordeaux school of oenology certainly owes much to the fame of the wines of Bordeaux; it is also true that the quality of the *grands vins* owes much to the work of the school's oenologists since the middle of the nineteenth century, especially since the 1930s. Several years ago Pascal Ribéreau-Gayon, then director of the Institut d'oenologie (made a faculty in 1995) of the University of Bordeaux, observed that the grands *crus* of Bordeaux, far from being a gift from nature, are "the fruit of a discipline imposed by man upon nature." Not that past generations had let nature pursue her wanton ways – to continue this Baconian metaphor – but winemaking was the result of centuries of practice resolutely based on empiricism. Ampelology and oenology brought improvements in the quality of wine that were based on the comforting certainties of experimental science. True, the cognitive basis for these scientific specialties already existed, but it was their institutionalization with an input into production that made the difference.

In all the provincial universities of the nineteenth century there was a strong connection between faculties of science and local economies.[1] Chemists were often at the forefront in forging the connections. The rare chemist who was uninterested would sometimes find himself being pushed or pulled in the direction of agriculture or industry by governments or enterprises in need of expert help. It was clear in Bordeaux from the beginning of the faculty of science that chemists and botanists could be of immense help to the wine industry, if only the producers could be made to see the potential profit of science. A notorious exception was Auguste Laurent, holder of the chair in chemistry at Bordeaux from 1838 to 1847, who preferred to starve in Paris rather than flourish in Bordeaux. Laurent's polymath successor, A.-E. Baudrimont, forged a solid connec-

---

[1] See Mary Jo Nye, *Science in the Provinces* . . . (Berkeley and Los Angeles, 1986), and Harry W. Paul, *From Knowledge to Power: The Rise of the Science Empire in France, 1860–1939* (New York, 1985).

tion between chemistry and agriculture through his publications and his official duties. Baudrimont did research on fertilizers, the soils of the Gironde, the vine and its diseases, and the coloring matter in red wine. The official connection of science with agriculture and commerce was consecrated in the appointment of Baudrimont as chief governmental analyst for fertilizers in the southwest. With the increasing use of chemical fertilizers came clever forms of fraud that could only be detected by simple scientific analyses. Baudrimont was also the first head of the Station agronomique de Bordeaux. Government, chemistry, botany, and agricultural production were intellectually and institutionally linked in a cooperative partnership.

In the explanation of the emergence of oenology at Bordeaux, the nature of the research interests of its scientists is as important as the *crus* to whose fame they contributed. If any one man deserves credit for laying the foundation of oenology, it is Ulysse Gayon, student, collaborator, and friend of Pasteur, who thus emerges – and one hates to give Pasteur credit for anything else in science – as a powerful influence in early oenological research in the faculty of sciences and the Station agronomique of Bordeaux. In one sense this is not surprising, for Pasteur's theory that disease in wine is due to the presence of microscopic parasitic organisms, although only partly true, can be seen as a basic discovery in oenology, with a practical corollary for the preservation of wine by heating it. When he succeeded Baudrimont in the chair of chemistry in 1884, Gayon declared that he was not going to change the direction of his research in order to do pure chemistry. Perhaps he felt a twinge of guilt that he was not going to give a theoretical fillip to a faculty that should have been much more ambitious than it really was.

Gayon described himself as the student of a "maître illustre" who had created a new science – Pasteur also invented the term "microbiology" – in whose path he would follow, not in the hope of keeping up with the great man's giant steps but rather of collecting some booty left on the route and gathering some fruits Paster had not deigned to pick. Most of Pasteur's students manifested signs of the Mertonian "on-the-shoulders-of-giants" syndrome. Gayon's visits to the Pasteurs for Sunday dinner were always welcome because of his cheerfulness and good humor. During a visit by the austere Pasteur to Bordeaux in 1881, Gayon even managed to pry him loose from the library to take a canoe excursion on the river. There is no doubt that Pasteur enjoyed Gayon's company, and Madame Pasteur perhaps enjoyed it even more because it sometimes briefly interrupted her husband's relentless devotion to work. In addition, each man had an interest in the other's work, and Pasteur was especially curious about Gayon's research on wine, some of which he had asked his disciple to undertake.

Gayon came from a modest farming family in Bouëx (Charente), went to the school of the commune, showed himself to be a bright lad, was sent to the lycée in Angoulême, passed his baccalaureate examinations brilliantly, and then got serious by going to the lycée in Poitiers to prepare for one of three *grandes écoles,* the Ecole polytechnique, the Ecole normale, and the military school Saint-Cyr. His father, an ex-soldier, was quite keen to have an officer in the family. Ulysse chose to enter the Ecole normale rather than the Polytechnique, and his father might have thought that he had named his son after the wrong Greek hero. The Ecole normale was an excellent choice. As head of the scientific section, Pasteur had transformed the school into an important research center, in competition with the Polytechnique for France's brightest minds.

In the grim year of 1871 Gayon finished his studies at the Ecole normale and took the examinations for the teaching degree of *agrégation* in the physical sciences. About to be assigned to a respectable position as a lycée physics teacher, he was saved from this not unusual stage in his career by Pasteur's decision to give him a job as assistant in his laboratory of physiological chemistry at the Ecole normale. Gayon did research on supposedly spontaneous changes in eggs and earned his doctoral degree in 1875. His subject was directly related to Pasteur's attempt to destroy the idea of spontaneous generation. Gayon found that changes in eggs came from microbial invasions. This research supported Pasteurian theory and also refuted a counter-Pasteurian theory positing the development of microorganisms from the transformation of organic matter, which was advanced by one of Pasteur's enemies, Antoine Béchamp of the medical faculty in Montpellier. Gayon's career was beautifully launched into the mainstream of French scientific research on the side of the winning paradigm.

French education and research are notorious for the existence of numerous systems administratively separate, but in the cases of research and of laboratory analysis they are often linked because one person heads both operations. In 1875 Gayon went to Bordeaux as the head of one of the regional laboratories established in France by the ministry of finances for the customs administration. He remained as head of this laboratory for 44 years. With E. Dubourg he did scientific research on raw cane sugar, one of the many products Bordeaux received from the French colonies. Their work on alcoholic fermentation established the first steps in the amylo process for the industrial manufacture of alcohol. Gayon took over Baudrimont's faculty course in chemistry in 1878 and became a professor in 1881. When Baudrimont died in 1880, Dumas and Pasteur urged Gayon to become also the director of the Station agronomique de Bordeaux. But Gayon did not want to leave higher education. An agreement by Dumas, Pasteur, and Berthelot gave Gayon the task of studying phylloxera, still a fascinating problem of scientific research, although it was being brought under control in the vineyards. If two jobs seemed too much, a third position appeared to be the perfect solution to the problem, especially when added to another research assignment. But Gayon would not get very far in fulfilling the politician's hope that scientists would find a quick fix for phylloxera. Alexis Millardet's work on grafting turned out to be an important part of the solution to the problem. Several other researchers, especially Pierre Viala, were also working successfully on the large scale needed to get vineyards replanted and wine back on a solid commercial footing. Although his work on sprays and in soil science was important, Gayon's had more of an influence on wine than on the vine.

The arrival of Gayon in Bordeaux meant the establishment of the microbiology of both wine and soil.[2] Historians of science give Pasteur the honor of being the founder of microbiology and, with Koch, a cofounder of medical microbiology. After Lavoisier, understanding and utilizing the chemistry of life were basic aims of many chemists. The discovery of microorganisms opened up a new experimental area in science. Pasteur often asked Gayon for analyses of samples of beer and wine, while Gayon kept Pasteur informed on the nature and results of his research. So Pasteur was a real force in stimulating the development of oenology at Bordeaux. And Gayon soon found out that having his research praised by Pasteur did more than give a valuable boost to his scientific ego: it gave him a scientific empire in Bordeaux, a gold medal from the Académie d'agriculture, and even the offer of a dual appointment in Paris. Gayon's science justified the smile of fortune, a goddess also noted for responding quickly to the prayers of the establishment.

Chemistry and botany were the two main scientific supports of nineteenth-century viticulture and oenology. It was clear as early as the 1870s that botanists had a key role to play in oenology. The first professor of botany at Bordeaux was Alexis Millardet, whose notoriety derives from his viticultural research as well as from his classic studies on mildew and his work with Gayon on *la bouillie bordelaise,* the copper sulfate-based Bordeaux mixture used to protect vines against diseases and parasites, including mildew and black rot. A double doctor in medicine and in science, more botanist-viticulturist than oenologist – it was during his tenure at Bordeaux, from 1876 to 1902, that this specialty became clearly defined – Millardet did some remarkable research on the grafting of vines. His work on rootstocks resistant to phylloxera is still not entirely obsolete. I shall briefly recall his viticultural triumphs. Also a man of action and

---

[2] René Marcard, *Ulysse Gayon (1845–1929),* p. 12, published on the occasion of the international symposium in oenology, Station agronomique et oenologique, Bordeaux, June 10–15, 1963.

property, he planted 30 hectares of the new rootstocks in the limestone soil of the Grande Champagne area, near Cognac, thus showing skeptics that the certainty of science is practical as well as epistemological. Before he died in 1902, Millardet enjoyed seeing his revived vineyard, once famous for the quality of its *eaux-de-vie,* again become the object of wooing by the Charentais distillers and merchants. Baudrimont, Millardet, and Gayon all have streets named after them in Bordeaux. Millardet and Gayon both have monuments erected in the public garden of Bordeaux by "a grateful viticulture."

Gayon was fortunate in having Millardet as his colleague in the faculty of sciences. Along with researchers in other institutions, they showed that science could save French vineyards from destruction by downy mildew, the most powerful enemy of the vine since the appearance of phylloxera. The dreaded fungus produced an infected plant whose fruit usually produced undrinkable wines. A startling decline in the quality of wine took place: eight vintages of Château Latour fell victim to mildew between 1880 and 1891, when the wine turned sour after bottling.[3] Finding a way of eliminating or at least controlling mildew quickly became the highest priority of an adolescent but ambitious oenology. J.-E. Planchon (1823–88), pharmacist, botanist, and much more, at the University of Montpellier, was perhaps the first person to call attention to mildew in France. This was in 1878, and in two years it spread from a few areas to most of grape-growing Europe and Algeria. In 1884 publications gave news of the efficacy of various concoctions in preventing mildew. The common element in the liquids used was copper sulfate. *Progrès agricole et viticole,* the *Journal de Beaune,* and a communication to the Academy of Sciences carried the good news in the fall of 1884.

The work of Millardet and Gayon on Bordeaux mixture shows that, when science enters the picture, purely empirical procedures must take a backseat and soon end up as only a source of anecdote. Mildew was widespread in the southwest of France by the 1880s, but by 1888 spraying the vines with copper sulfate had produced a complete recovery of the vineyards. Who should be given credit for this minor miracle? Proprietors used to spray a mixture containing copper sulfate on the vines next to roads and paths, in the hope that the poison would prevent people from stealing grapes. They were happily surprised to notice an unintended consequence: no mildew on the sprayed vines. Ernest David, the manager at Ducru-Beaucaillou and Dauzac, is supposed to have been one of the observant empiricists who stumbled onto this discovery. But widespread systematic application of the spray at the required strength could hardly be done

---

[3] Loubère, *The Red and the White,* p. 163.

quickly by hit-and-miss methods. This was the work of a few scientists, especially Millardet and Gayon.[4]

Millardet seems to enjoy priority in having noticed the fatal effect that copper sulfate has on mildew, or rather he noticed the connection between spraying a poisonous concoction on grapes and the fact that the sprayed vines kept their leaves whereas the unsprayed vines lost theirs. Millardet also observed that the disease could be prevented by using water from a copper pump in a well on his property. An analysis of the water done by Gayon found five mg of copper per liter of water, nearly twenty times the level required to kill the spores of the *plasmopara viticola,* the guilty mildew. Millardet related his observations to David, who was willing to carry out large-scale, comparative experiments on the properties (Ducru-Beaucaillou and Dauzac) he managed for Johnston. The laboratory experiments were done by Gayon. But the years 1883 and 1884 were not favorable for the development of mildew. In 1885 Millardet's note to the Academy of Sciences admitted that his *bouillie bordelaise* was not the result of a methodical study of different metallic salts but of observing the effects of the traditional spraying to prevent stealing. He was excessively modest.

By 1887 Millardet, with the collaboration of Gayon, had developed the first formula for his Bordeaux mixture: eight kilos of commercial copper sulfate and fifteen kilos of lime per 100 liters of water. This concoction would change over the years, as the two scientists carried out experiments to determine the exact formula that would produce the desired protection for the plants. By 1928 laboratory tests found that the copper sulfate could be reduced to one kilo and the lime to less than half a kilo per 100 liters of water. In practice, far greater amounts were used, because conditions in the vineyards were different from those in the laboratory. Experience often seeks agricultural security by far exceeding laboratory guidelines.[5]

Winegrowers, used to regarding copper sulfate as a poison, were a bit leery of using it on their own grapes, especially if they were going to drink wine made from them. Some alarmists even claimed that their wine was poisoned by the copper. In 1888 peasant producers in the Vivarais put on a violent show against the newest weapon in the scientific arsenal of pesticides. To head off a *jacquerie vivaroise,* Pasteur recommended Gayon as the best-qualified person to reassure growers about the innocuousness of the treatment against mildew; having devised such a clever way of stopping the killer mold, scientists were not about to be cheated out of their prize by idle speculations about public health. Numerous experiments – a comforting phrase often used in science – justified the quickly dissemin-

---

[4] Edmund Penning-Rowsell, *The Wines of Bordeaux,* 4th ed. (1979 and 1981; 1969), p. 31.
[5] A. Millardet, *Traitement du mildiou et du rot* (Paris, 1886); M. Lafforgue, "La bouillie bordelaise," *BIV,* no. 6 (1928): 25–9; E. Carrière, "Actions des bouillies cuivriques . . . ," *BIV,* no. 158 (1943):104–7.

ated conclusion that Bordeaux mixture was absolutely safe. Desperate winegrowers were not hard to convince. The issue remained alive, at least as a topic of debate, because of a growing political concern about public health, and perhaps also because of a nascent healthy skepticism of the general public about guarantees resting on self-serving experiments that accompanied the new role of science in daily life.

Oenology, viticulture, and even soil science were integrated into the Pasteurian research empire. In 1893 Serge Winogradsky succeeded in isolating nitrogen-fixing bacteria in the soil, and by the end of the nineteenth century agronomists had worked out a theory of plant nutrition. Schloesing and Muntz attributed the production of nitrates to a "ferment nitrique."[6] Winegrowers were keenly interested in fertilization, although their knowledge was based mostly on traditional empirical practices. Gayon studied the role of anaerobic microbes in the transformation of the soil's nitrates into nitrites and nitrogen. Like many agronomists, he contemplated the mystery of the manure pile and in a study of its fermentation process showed the role of an anaerobic microorganism in the production of gases. Most important for winegrowers was the systematic study he began in 1882 on fertilizers for the vine. His experiments were designed to show, for certain types of soil and vines, what form of application of the chief fertilizing elements best suited the vine's development. To avoid the distortions of data that would come from the vine's infection with phylloxera, Gayon used the sandy soil of the Landes, avoided by the plant louse. He chose to test the cabernet sauvignon grape because of its importance in the production of fine wines in the Gironde. Gayon found that growers were not paying enough attention to the vine's requirements in potash and magnesium. Agronomists often discover deficiencies and excesses in soils. (Guy Accad, agronomist turned oenologist, apostle of cold maceration vinification, contends that at the present time "many Burgundian vineyards are deficient in magnesium and overendowed with potassium" because of an "excessive use of commercial fertilizers.")[7]

Public courses offered by faculties of science in agricultural chemistry provided a means of transmitting knowledge generated at the faculty level to growers curious enough to come to class. They also encouraged a great increase in the use of fertilizers. It was the clear function of the director of the Station agronomique et oenologique, which was not part of the university, to establish collaboration between science and agriculture, ardently desired by some faculty members and less ardently by others. Most growers were uninterested until they became convinced of the profitability of science and of its importance in avoiding disasters.

[6] See F. W. J. McCosh, *Boussingault, Chemist and Agriculturist* (Dordrecht, 1984), chaps. 10 and 11.
[7] Matt Kramer, *Making Sense of Burgundy* (New York, 1990), p. 74.

The public course in agricultural chemistry at Bordeaux was started by Baudrimont, a believer in the trickle-down approach to science, whose mysteries and practical applications to agriculture had to be put in a palatable form for the relevant public and not just kept in the academic towers of the the faculty. Continued by Gayon, the course included plant physiology, the study of arable land, fertilizers, and especially viticulture and oenology. The Pasteurian Gayon was certainly one of the few top researchers and teachers in this new area for university science. Basics and the latest developments in winemaking were given a prominent place on the professorial menu. One full semester was devoted to the study of alcoholic fermentation, including its theory and practice, the role of yeasts, the use of selected yeasts, and the influence of oxygen from the air. Also included were studies of fermentation in vats, including the accidents that can occur, studies of musts, and, finally, scientific scrutiny of the wine itself.

In the second semester, the course concentrated on disease. Of key interest were the modifications caused in the chemical structure of wine by microorganisms, whose characteristics and role were identified after microscopic examination. Gayon also analyzed the volatile acids (acetic, formic, and others) found in wines suffering from different diseases – "les vins aigris, tournés, gras, amers, mannités." He emphasized the means of preventing disease: filtration, freezing, pasteurization, and other violence done to wines, mainly to cover up human errors. A great expert on the pasteurization of wines, Gayon gave a special course on this process applied to the wines of the Gironde; it included a discussion of the apparatuses available for this hotly debated new technique. Daniel Jouet, an agricultural engineer who was manager of Château Latour, used to visit Gayon's laboratory to have the château's wines analyzed and also to participate in tastings of pasteurized wines, as well as to debate their virtues and vices.[8] Bordeaux had an impressive new program in viticulture, and especially oenology.

The public course in agricultural chemistry met in the evening, thus allowing wine wholesalers, warehouse foremen, constructors of viticultural machinery, and winegrowers to attend. Chemistry students also attended; for the course offered an excellent opportunity for them to escape the oppressive tutelage of the bourgeois Bordelais family, which rarely permitted its pubescent members any nocturnal sorties without the proper moral guarantees. Once liberated, they merely enjoyed ragging the demonstrator of experiments if he made any mistakes. The important thing was that oenology had a firm foundation in the educational establishment of Bordeaux, with a diversity of theoretical and practical teaching, all solidly

---

[8] R. Pijassou, "La notion de qualité des grands vins du Médoc de 1880 à nos jours," in Pascal Ribéreau-Gayon and Pierre Sudraud, eds., *Actualités oenologiques et viticoles* (1981), p. 97.

anchored in the midst of the world's greatest wine country. It was a hard combination to beat, and it was not long before the faculty began attracting serious French and foreign students to oenological studies.

A more structured program became possible in 1891, after the Ecole de chimie appliquée à l'industrie et à l'agriculture de Bordeaux was established by Gayon and Alexandre Joannis, who was especially interested in industrial chemistry. Although the chief official function of the school was to prepare students to take the examination for the *licence ès sciences,* its higher purpose was to orient students toward the theoretical and practical study of fermentation. E. Dubourg, Gayon's collaborator in the customs laboratory, the first director of the school, introduced physiological chemistry into the faculty. He had done his doctoral research on urine before moving on to doing joint research with Gayon on oenological subjects, including mildewed wines, the nature of grape sugars, and mannitic fermentation. Dubourg's work showed that yeasts differ not only in form but also in their power of affinity for glucose and fructose, the principal fermentable sugars in the must. Along with Gayon and Charles Blarez, he clearly defined rules, with the limits of their application, for recognizing certain frauds, especially the watering of wine (*le mouillage*). Public health issues were becoming an important part of studies on wine.

In 1928, the year before Gayon's death, Bordeaux hosted the first international congress on vine and wine, including a session to celebrate the creation of Bordeaux mixture. In the excessive rhetoric generated by the occasion, Gayon was hailed as the founder of world oenology, while a bust of Pasteur, the faculty icon, might be imagined to show smug satisfaction with this recognition of his student and of *la science française.* Emile Peynaud has recognized for Gayon a more precise but not less important scientific role, that of creator of the oenology of fine wines because of his work on (1) the difficulties of fermentation in warm years, (2) the stopping of fermentation by high temperatures, and (3) the need for aeration and *remontage* (pumping wine over the cap of the must). Gayon also placed a wise emphasis on the eternal danger of bacteria, a warning to be expected from a Pasteurian devoted to the feasibility of preserving wines through heating. Back in 1928, the abbé J. Dubaquié, who carried on Gayon's work from 1923 to 1947, had made the same point as Peynaud: Gayon, a worker in pure science, created "our oenology of fine wines."

Faithful to the Pasteurian gospel, Gayon still seemed to have his doubts, or at least to have ignored the theory in favor of common sense and traditional practice in the case of the serious failure of the Pasteurian paradigm to account for one key phenomenon in winemaking. In 1897 Gayon observed a sharp fall at the end of six to seven days in the total acidity of a fermenting *premier cru* of the Médoc, a drop from 4.2 to 3.2 grams per liter. At the same time there appeared a great number of microbes with a slight increase of volatile acidity. Jean Ribéreau-Gayon points out that this is a clear case of malolactic fermentation. But Gayon attributed the fall in acidity to the precipitation of potassium hydrogen tartrate (cream of tartar) and interpreted the appearance of microbes and the increase in volatile acidity as indications of a basic change in the must, the beginning of disease. This Pasteurian conclusion could lead to the dubious winemaking practices required by attempts to achieve pure or germ-free alcoholic fermentations along with short fermentation periods. It was not part of Pasteur's theory that a natural limitation exists on the appearance and disappearance of malic and citric acids, the substances attacked by bacteria. Yet it was known that the *grands vins* have a comparatively high volatile acidity. After testing thirty wines, Gayon concluded that among the well-kept wines it was the *grands crus* of great years that had the highest volatile acidity, about a gram per liter. In concluding that these imprecise observations were clearly opposed to Pasteurian principles, Jean Ribéreau-Gayon, grandson of Ulysse Gayon, gave the origins of his new oenology a respectable origin in the Pasteurian camp itself. It also became a family affair.[9]

After the death of Millardet in 1902, Gayon's doctoral students played an important part in the development of oenology at Bordeaux. Jules Laborde did his work in Gayon's laboratory on the physiology of a new mold (inevitably named *Eurotiopsis Gayoni*) and defended his thesis at the Sorbonne for the added lustre of a defense before a jury of scientists of the University of Paris. As assistant director of the Station agronomique et oenologique, Laborde published a large number of works on oenological subjects: tannins, the biochemistry of noble rot (*Botrytis cinerea*) in Sauternes, esters and the various precipitations in wines and fermentations, including the then enigmatic malolactic fermentation. In 1907 Laborde published his *Cours d'oenologie,* which still retains the admiration of Pascal Ribéreau-Gayon because of its modern analytical spirit. Perhaps Laborde's reputation should be greater than it is, but it was difficult for his contemporaries to appreciate some of his strikingly original research, such as that on the identification of the oxidase and the dextrane of *Botrytis* and on the structure of tannins. His works could only be fully developed long after their publication, when they could be integrated into a body of scientific knowledge created later.[10]

Perhaps oenology had gone as far as it could, at least in its effect on wine production, without further advances in the basic chemistry of wine.

[9] J. Ribéreau-Gayon, *Traité d'oenologie. Transformations et traitements des vins* (Paris and Liège: Librairie polytechnique Ch. Béranger, 1947), pp. 115–16.

[10] Pascal Ribéreau-Gayon, "L'école bordelaise d'oenologie," *Regards sur l'institut d'oenologie, Université de Bordeaux II,* p. 12.

Change did not come from the Bordeaux station itself. L. Mathieu, head of the station between 1920 and 1923, continued to work on methods of developing and preserving wines of high quality. The former director of the Station agronomique et oenologique of Burgundy, Mathieu at least showed that science is not only international but that it can cross the difficult, invisible barrier between Bordeaux and Burgundy. But Bordeaux returned to the faith with Mathieu's successor, the abbé J. Dubaquié, one of Gayon's students. Even in conservative Bordeaux after the anticlerical passions of the prewar years had abated, the appointment of an ecclesiastic to such a position was a bit of a surprise. But religion, like science, has its intimate relations with wine, and Dubaquié remained head of the station until 1947. He had a great reputation as a wine taster, perhaps in part due to his classic formula of approval: "ça se pète."[11] Meanwhile, radical oenological changes were fermenting in the university and in the private sector, outside the traditional academic wine establishment.

In the 1930s "a new oenology" – this phrase is used by Pascal Ribéreau-Gayon – was developing indirectly in the faculty of sciences and directly in the professional laboratory of Calvet, the big wine shippers of Bordeaux, where Jean Ribéreau-Gayon and Emile Peynaud closely collaborated from 1927 to 1949, when they shifted their research to the university itself. Peynaud went to work in the laboratory at the age of fifteen and was lucky enough to be supervised by Ribéreau-Gayon. The novelty in their research was that it was closely linked to work being done in physical chemistry, especially the type going on in the faculty laboratory of biological chemistry headed by Louis Genevois. In 1947 Jean Ribéreau-Gayon and Genevois published their classic treatise, Le vin.[12]

One of the great pieces of research done by Peynaud and Jean Ribéreau-Gayon was on the importance of malolactic fermentation for the great red wines. After the alcoholic fermentation carried out by the yeasts, lactic acid bacteria (LAB) change the malic acid into lactic acid, reducing the acidity of the wine, raising its pH, and introducing the complexities into its flavor that cause the dégustateur to wax eloquent on a wine's "astonishing flavors." Many of the phenomena occurring in wine were understood for the first time because of their clarification through the concepts of physical chemistry. Colloids, oxidation-reduction, and the pH value of aqueous solutions were among the phenomena whose activity was seen as vital to the understanding of the nature and evolution of wine. Oenologists learned a new scientific language. The problem was that the Bordelais reigning paradigm, given hieratic status in the teachings of Pasteur and

---

[11] Se péter means s'enivrer (to get drunk); péter is to fart. One of its metaphorical uses given by Littré: "Ce vin fait péter les bouteilles."

[12] Le vin was in the series Actualités scientifiques et industrielles, no. 1017, Nutrition.

Gayon, was unable to absorb this anomaly. Neither could it ignore the anomaly introduced by the idea of a specific bacteria's action being essential to making great wine.

The concept of pasteurization had been part of Gayon's oenology, but it was the rejection of pasteurization as a panacea that signaled the arrival of a new way of thinking about the structure of the *grand vin*. An epistemological revolution in oenology centered on the microbe. Pasteur's privileged status changed radically once it became clear that it was wrong to believe that the only good microbe is a dead one.

The head of the oenological station during this critical period of new paradigm formation and all that sort of thing (occurring outside the station, as prescribed by Kuhn's law) was the abbé J. Dubaquié. He had done his research under Gayon on intracellular exchanges, with resulting chemical changes, in the lower plants. The thesis (1909) was in the tradition of Pasteurian research, especially in its development of Raulin's work on the microscopic fungus *Aspergillus glaucus* cultivated in a mineral environment. The *Eurotiopsis Gayoni* was given a place of honor in the study.[13] Early in 1944 Dubaquié published in the *Bulletin de l'office international du vin* an article on "Les vins du Sud-Ouest de la France et les microbes," which rejected the main conclusions of the new work on malolactic fermentation and launched a general warning against admiring, on a priori grounds, this sensational contribution of the latest oenology. If there can be a good microbe that improves wine, what then is the value of "all our oenology and this war unto death against microbes?" Not only did this malolactic novelty question traditional practices; it also cast doubt on the doctrine of scientific vinification according to the Pasteurian school of Gayon and his students.

Tradition, authority, science – a trinity that should not be doubted lightly. The general dogma of this successful oenology was "pas de microbes," which required that fermentation ensure only a minimum multiplication of bacteria, thus limiting their action. Dubaquié saw the old oenology as a total package that had proved itself by producing an uninterrupted series of fine results, and to ignore these rules would bring disaster. No matter what contrary theories were being aired, he was confident that the results of the old oenology would stand firm against any assault.

Not that Jean Ribéreau-Gayon found much to quarrel with in this swan song of the director of the Station oenologique, for Dubaquié, speaking ex cathedra, did not produce any "fact" either to contradict the new oenological theory and practice or to support his own conclusions. A worse

---

[13] The *Aspergillus* is a genus of microscopic fungi that takes its name from its shape, which resembles a holy-water sprinkler; the blue mold *Aspergillus glaucus* gives delight to some lovers of infected cheese.

scientific sin was the fact that his references were to well-known old works. The one recent reference was to an anonymous note in the *Bulletin de l'Agriculture* in 1939 that simply identified malolactic fermentation as a disease to be avoided, which was sometimes the case for certain ordinary wines of the Midi whose low acidity should not be lowered further. Dubaquié emphasized that the recommendation carried the weight of ministerial authority. Ribéreau-Gayon fell back on a higher authority: "to prefer the teachings of authority, whether that of Pasteur or of an official organism, to those of experimentation would clearly be the very negation of the scientific method." Even anti-Pasteurian in spirit, he might have added.

It was easy enough for Ribéreau-Gayon to make mincemeat out of poor old Dubaquié's article, but what of the principles of Pasteur and Gayon, which had supposedly proved their worth in common use and now comprised the defense shield from behind which Dubaquié was taking pot-shots at the newfangled theory? Ribéreau-Gayon did not believe that "the old principles of Pasteurian oenology" were validated by the results of common practice, because these principles had never received anything other than a very incomplete application; in particular, no one, fortunately, had tried to obtain pure or aseptic alcoholic fermentations. Even when one searched for these principles, one did not find them. And Ribéreau-Gayon declared bluntly that in reality the methods used to produce and finish the *grands vins* were pretty much the same as they had been for more than a century. The new oenology had yet to penetrate the university, let alone begin the more difficult task of influencing producers.[14]

[14] J. Ribéreau-Gayon, "Sur la fermentation de l'acide malique dans les grands vins rouges," *Bulletin de l'office international du vin,* no. 182 (1946): pp. 26–9.

11

The Ionic Gospel of the New Oenology

Who created modern oenology? In his published discourse for the centenary (1980) of the Station agronomique de Bordeaux, Emile Peynaud included a photograph iconistically labeled "Professeur Jean Ribéreau-Gayon, le fondateur de l'oenologie moderne." A summary of Ribéreau-Gayon's brilliant achievements in oenology during his long career in Bordeaux makes a convincing case.[1] When did oenology come to Bordeaux? One obvious and solid answer is that it came in the 1870s with the arrival (1875) and rooting of Ulysse Gayon in the viticultural and university communities of the Bordelais. The paternity of Pasteur, Gayon's own oenological work, his organization of laboratories, his promotion of physiological chemistry, his forging of connections with viticulture and "agro-industry," along with a modest acquiescence in late veneration of himself by a large number of disciple-admirers – all make the answer a reasonable one. The accompanying temptation to label this series of events as the arrival of modern oenology in Bordeaux is nearly irresistible, even if heuristic confusion is introduced by the concept of modernity.

Another reasonable approach is to postpone the arrival of modern oenology in Bordeaux until after World War I, when Jean Ribéreau-Gayon and his associate, Emile Peynaud, appear as the two major stars in the Bordelais oenological firmament. We may add a contextual judgment to Peynaud's statement on the revolutionary impact of Ribéreau-Gayon's work and *methods:* an institutional transformation took place in the subject when the two oenologists joined the faculty of sciences. The problems with these judgments is that they plunge us into the hopeless general problem of origins. There is also the matter of seeing predecessors doing science right and doing it wrong, even if they were right once upon a time, thus accepting the linear model of scientific progress, a model that has been condemned but never banished. Scientists writing on the history of oenology are particularly susceptible to this epistemological infection and

---

[1] Emile Peynaud, "Discours prononcé à l'occasion du centenaire de la fondation de la Station agronomique de Bordeaux (1880–1980)," *Le vin et les jours* (1988), pp. 152–4.

like it because of the importance it gives their own work. The scientist as historian using the Whig model – Peynaud on Pasteur, for example – establishes an ambiguous and critical relationship with the glorious ancestors of his discipline. Historians often like the model because it puts them on the side of winners, possessors of the truth, and gives them improved social status from associating with greatness.

The new oenology of Ribéreau-Gayon and Peynaud was based on a new approach to reality, that of the methods of the physical chemistry of Gibbs, Duhem, van t'Hoff, and Berthelot, and especially of Otto Meyerhof's physiological chemistry or, to use a more modern term, biochemistry. The aim of this new science was to understand the chemical mechanism of biological processes, to answer the question posed in the title of Schrödinger's biochemically disappointing book, *What Is Life?* So the original question about the introduction of oenology in Bordeaux, whether Gayon's old oenology or Ribéreau-Gayon's and Peynaud's new oenology, only hints at the important questions: what was this science whose methods were used, when was this science introduced, and by whom?

The Biochemical Research Program of Louis Genevois

A key figure in the transformation of life studies in Bordeaux between the two world wars was the biochemist Louis Genevois. It is not surprising that he gets short shrift in the history of oenology as recounted by oenologists, for they have a special interest in either establishing a virgin birth for their discipline *as a science,* or in presenting the new oenology as a redemptive gospel, with Pasteur and Gayon cast in roles redolent of that of John the Baptist. Nor can one overlook the interest of the oenologist in advertising his own creative role in the new science, with its practical consequences for a major sector of the economy. And Genevois's reputation as a *mauvais coucheur* in the faculty has not helped his historical resurrection, especially by people who knew him. It is a bit like the old reputation of Pierre Duhem at Bordeaux. But the time may be ripe to resurrect Genevois. After receiving his *licence* in physical science at the faculty of sciences in Bordeaux (1919), Genevois joined an elite of science students at the Ecole normale in Paris. In 1928 he obtained his doctorate in the natural sciences from the science faculty in Paris. It looked as if his life would follow the banal pattern of hundreds of other scientists: research, publication, teaching, and a few lines in Larousse if he were lucky. (Genevois does not have his lines yet.)[2] Then, in 1926–7, after being awarded a Rockefeller scholarship, he went from Eugène Aubel's laboratory of physiological chemistry in Bordeaux and the Station biologique

2 See Joseph S. Fruton, *A Bio-Bibliography for the History of the Biochemical Sciences since 1800* (Philadelphia, 1982), for one reference, and *Who's Who in Europe* (1972).

d'Arcachon to Otto Meyerhof's laboratory of physiological chemistry at the Kaiser-Wilhelm-Institut in Berlin. After a stint in Aubel's laboratory in 1928, he returned to Bordeaux to teach in the faculty of sciences.

Historically, the biochemical sciences have covered a large area, including everything from animal chemistry to molecular biology.[3] From time to time these sciences have been given a sort of unity through the influence of various theories. Perhaps no theory had a more profound experimental influence than ionic or Arrhenius theory, which eventually changed the definition of acids and bases. In 1887 the Swedish physical chemist Svante Arrhenius had published his theory of electrolytic dissociation in its quantitative formulation, the definitive statement of his notorious thesis (1887) that alarmed the professors of Uppsala.[4] By the turn of the century, ionic theory had migrated through Germany, and in the interwar period had begun to penetrate oenological research in France.

Bordeaux already had a solid basis for biochemistry when Genevois started his research. Henri Devaux (1862–1956) was well known for introducing into biological science the idea of an exchange of ions.[5] An ion is "an atom or group of atoms that has either lost one or more electrons, making it positively charged [a cation], or gained one or more electrons, making it negatively charged [an anion]." The prevalent natural process of ion exchange, which became a fruitful axiom of the new oenology, refers to an "exchange of ions of the same charge between a solution (usually aqueous) and a solid in contact with it."[6] In the 1930s, the use of ionic theory conferred a cachet of distinction on oenological research, linking it to high-level science, to novelty, to new methods, to a new understanding of the basic processes of wine the living liquid. Some older oenologists, often made quarrelsome and defensive by the new approaches, stubbornly soldiered on, armed only with traditional chemical analysis. Genevois distinguished between the old biochemistry and the new way of Devaux and Otto Warburg. Instead of isolating the chemical reactions of the living

---

[3] Fruton, *A Bio-Bibliography,* preface; and Claude Debru, *L'Esprit des protéines. Histoire et philosophie biochimiques* (1983).

[4] In the theory of ionization, liquids (water solutions of acids, bases, and salts) that conduct electricity are called electrolytes. Distilled water and water solutions of organic compounds (alcohol and glycerine, e.g.) are nonelectrolytes, i.e., they do not conduct electricity. In a recent version of this theory, "acids are defined as substances that dissociate in water to yield electrically charged atoms or molecules called ions, one of which is a hydrogen ion (H+) . . . , and bases ionize in water to yield hydroxide ions (OH−). . . . The acidic behaviour of many well-known acids (e.g., sulfuric, hydrochloric, nitric, and acetic acids) and the basic properties of well-known hydroxides (e.g., sodium, potassium, and calcium hydroxides) are explained in terms of their ability to yield hydrogen and hydroxide ions, respectively, in solution." *Encyclopaedia Britannica* (Chicago, 1989), 1:588.

[5] On Devaux, see articles by J. G. Kaplan in *Science* 124 (1956): 1017–18; by Genevois in the *Revue générale de botanique* 63 (1956), 341–6; and the *DSB,* vol. 4.

[6] See Oxford's *Concise Science Dictionary* (1987), pp. 357–60, for a brief introduction to all about ions; hereafter referred to as *CSD.*

cell, separating the different ferments from the cell, and determining its metabolism, the new method intervened in the cell in order to change the working of certain mechanisms without altering its life and to understand the role of these mechanisms in the functioning of the whole rather than the parts.

Genevois translated into French works by the biochemists Otto Meyerhof and Otto Warburg. Their work on human physiology was linked directly to oenology. Meyerhof succeeded "in extracting from muscle the group of enzymes responsible for the conversion of glycogen to lactic acid." The "preparation of a cell-free glycolytic system was a counterpart of the earlier successful extraction from yeast of the enzyme system (zymase) that converts glucose to alcohol and carbon dioxide during fermentation." It became clear that "the chemical pathway in the breakdown of glucose by muscle and by yeast was . . . similar." What had been shown was "the unity of biochemical processes amid the manifold diversity of the forms of life": "cette grande idée de Claude Bernard."[7] Genevois had antivitalist ideas similar to Meyerhof's: all living activity is regulated by physicochemical laws, most clearly seen in simple systems. Common scientific currency since the Enlightenment, this ideology was now experimental dogma rather than a philosophy of science. It was not yet popular to ask if there is much epistemological difference between the two. Hence the interest in studying "the basic unit of biology and biochemistry," the world of the living cell (yeasts, bacteria, and so on).

Genevois spent seven years on three biochemical problems: ionic equilibrium, oxidoreduction (oxidation-reduction reaction), and the relations between fermentation and respiration in plant cells. He proudly claimed to have published, in collaboration with E. Aubel, "the first article to appear in France focussing on the phenomena of oxidoreduction."[8] This process is a key to the oenology of the 1930s. *The Encyclopedia of Chemistry* notes that the term "oxidation," originally meaning a reaction in which oxygen was introduced into another substance, as in the aeration of wine, has long had its usage broadened to include any reaction in which electrons are transferred. Oxidation and reduction always take place simultaneously. The substance that gains electrons is called the oxidizing agent. In thermodynamically reversible systems, which refer to conditions in which electrons are easily transferred, the quantitative measure of the "oxidizing power" of a substance is called the "oxidation-reduction potential." Un-

---

[7] Joseph S. Fruton, "Otto Meyerhof," in *DSB* (1974), 9:359; Genevois, *Notice* (1932), pp. 9–10.

[8] "Les oxydoréductions," *Revue générale de botanique* 49 (1928). Genevois collaborated with Aubel, J. Salabartan, and R. Wurmser on oxidoreduction; see, e.g., "Phénomènes d'autoréduction produits par la levure en milieu anaérobie," *Procès verbal, Société des sciences physiques de Bordeaux,* 1925–6, p. 93 (March 25, 1926).

fortunately for people who have old wine, organic oxidation reactions are in general thermodynamically irreversible.[9]

In addition to all his research, publication, and translation, Genevois was active in changing the teaching of physiological chemistry at Bordeaux. The certificate program was completely recast. As in the other faculties of science in France, the certificate became one in biological chemistry, covering the chemistry of the living cell, fermentations, plant life, and foodstuffs. Biochemical modernity was the touchstone in subjects (for example, vitamins) and in methods. Doctoral students had access to laboratories outside the faculty of science: the Centre anticancéreux, the Ecole de santé navale, the Ecole de boulangerie de Bordeaux, and the laboratory of marine biology at Arcachon. Ribéreau-Gayon had his laboratory at Calvet. The extent and nature of research in Bordeaux cannot be judged solely on the basis of the limited facilities in the faculty.

In the 1930s the laboratory of physiological chemistry was an important center of biochemical research. Genevois has identified three groups totaling twenty-five researchers doing work for diplomas and theses. One group included pharmacists and physicians from the Ecole de santé navale de Bordeaux who joined the students from the Ecole de chimie de Bordeaux – they were destined to serve local industry – to do research on acetic acid bacteria, the respiratory mechanism of yeasts, lactic acid in the metabolism of cancerous cells, and the exchange of ions between yeasts and their enveloping solutions. A second group of researchers was non-French: six or so from the Soviet Union, Bulgaria, Hungary, Egypt, Iran, and the United States worked on lactic acid bacteria, the application of the latest advances in pedology, stimulation of plant growth, analyses of pectic and ascorbic acids in wine, and vitamin C. A third group of researchers was made up of graduates of the Ecole de chimie who came to the laboratory because of their high-level research interests.

The young scientists from the Ecole de chimie, who were able to maintain research contacts with the laboratory after graduating because they worked in the southwest, often faced problems best handled in Genevois's laboratory. Genevois named nine researchers in this category, including Ribéreau-Gayon and Peynaud ("son élève" – Peynaud would have various designations before he rose to the elevated rank of collaborator of Ribéreau-Gayon).[10] All these people did good work, but in Genevois's opinion two towered above all the rest: Ribéreau-Gayon and Peynaud (given

---

[9] Clifford A. Hampel and Gessner G. Hawley, *The Encyclopedia of Chemistry,* 3d ed. (New York, 1973), pp. 775–7.

[10] There were two women (Suzanne Camlong and Marguerite Cozic) in the group from the Ecole de chimie, and one among the foreigners. The biological sciences, especially those areas related to agriculture, were more open to women than other sciences. It is easy to compile a long list of women who published research articles in standard journals, but finding out more about them still remains to be done.

higher status as "laboratory assistant of Ribéreau-Gayon and, later, his private assistant"). Because of the work of many of these researchers the grape became "the best-known fruit in world literature," and wine was not far behind as the best-known liquid, although beer was an equal if plebeian competitor.

War put an end to all laboratory work in 1942. Peynaud lost six years as a prisoner of war. The laboratory, a nest of resistance to German occupation, lost two of its top researchers in the deportations to Germany: Laure Gatet, whose thesis was on the ripening of grapes in Cognac and the Cher, and Pierre Cayrol, known for his work on yeasts and on oxidoreduction.[11] Even Genevois gave up science for guerrilla activities. The death rate of young scientists in the Second World War was not comparable to the wiping out of a generation in the First World War, but it was not insignificant. And the considerable biochemical research operation in Bordeaux came to an end until after the war.

In 1945, when Genevois reviewed his own scientific work for the period 1933–42, he made a clear division between the research done before 1933 and that done after. Before 1933 his work was that of a naturalist and physiologist: the measurement of obvious phenomena like respiration and fermentation, the connections between them, and the action of some reagents on them. Then the study of ions became all the rage. The purely physical theory of permeability based on ionic activity, elaborated between 1925 and 1931, changed the nature of research on living matter. Henri Devaux, Genevois's predecessor at Bordeaux, had established a research tradition in this area. To be at the cutting edge of research – a mythical entity of which nonresearching administrators are excessively fond – meant an end to measuring the consumption of oxygen, amounts of CO_2 released, lactic acid or alcohol produced, all of which can give only a very approximate picture, a caricature of the metabolism of tissues. Genevois believed that the change in scientific literature, including the disappearance of the type of memoirs of the 1920s on cellular metabolism, meant that the subject was exhausted.

The 1930s saw another change in biochemical literature, the appearance of studies on water-soluble vitamins (B2, C, and so on) and the announcement of their synthesis. In the heyday of Otto Warburg and Albert Szent-Györgyi, when the study of vitamins and ferments (for example, niacinamide), along with their nutritional roles, became the domain of chemistry, "the biologist had to become a clever organic chemist if he wanted to make decisive contributions to knowledge." It was this radical change in the direction of biochemical research that led Genevois to look for "new

[11] Laure Gatet (Pharmacien, boursière de la fondation Schutzenberger, 1938–9), *Recherches biochimiques sur la maturation des fruits* (thesis, University of Bordeaux, 1940).

chemical methods of titration and of investigation," including analysis of the oxidation-reduction potential in living systems and analyses using polynitriles and fluorescent compounds.[12] The gospel of Genevois made it clear that advances in biochemistry could only come with new analytical techniques.

The new oenology was part of the new biochemistry. The transformations that take place in wine, especially the fermentations, the elements it contains, and its economic importance all made it a prime target for scientific scrutiny. Genevois declared, in the imperial professorial style of the day, that he set his student Jean Ribéreau-Gayon, a chemist in Bordeaux wines, and his assistant, Emile Peynaud, to reexamine all the analytical techniques used in the study of wine. After 1933 Léo Espil, assistant to Genevois, helped them in producing an enormous amount of high-quality analytical work, including new techniques and variations on accepted techniques. Included in the most important work done on wine was the titration of glycerine, lactic acid, malic acid (perhaps the most important for practical oenology), citric acid, neutral esters, coloring matter, and tannin. A new method of measuring the pH of wine was also developed. The complex composition of wine ensured that chemists would not run out of work (see Table 7).

Genevois's research program tried to explain the mechanism of oxidation-reduction in wine. The beneficial effects and the destructive potential of air (oxygen), which had been known for centuries and impressively explained in limited Pasteurian terms, was now made clear in terms of the behavior of the basic constituents of matter as manipulated according to the latest techniques of biochemistry. Useful new knowledge also came through these techniques. Ribéreau-Gayon and Peynaud showed that the sour taste of *piqûre* was due to ethyl acetate, not acetic acid as previously believed. The new analytical methods were not interesting to Genevois as tricks of the chemist's trade but because they opened a window on the very relationships of living matter and of more or less unknown molecules. And, of course, wine was the living matter in which oenologists were interested. Ribéreau-Gayon took about twenty pages in the influential *Revue de viticulture* in 1935 to explain how the physical chemistry of solutions applied to wine.[13] He emphasized that the introduction of physical chemistry did not mean that analytical or basic chemistry could be thrown out; far from it. The two approaches are complementary, each with its own virtues and vices. Chemists are bilingual, their forked

---

[12] "Titration is the process by which an unknown quantity of a particular substance is determined by adding to it a standard reagent [titrant] which it reacts with in a definite and known proportion" (Hampel and Hawley, *Encyclopedia of Chemistry*, pp. 1103–10).

[13] J. Ribéreau-Gayon, "Application au vin de la chimie physique des solutions," *RV* 83 (1935): 133–9, 155–7, 171–5, 181–90.

Table 7. *Wine Composition*

Component	Proportions per Liter	Comments
Dissolved Gases		
Carbon dioxide	0–50 cc	
Sulfur dioxide		
Total	80–200 mg	More in some sweet wines
Free	10–50 mg	Some in some unstable wines
Volatile Substances		
Water	700–900 mg	
Ethanol (alcohol)	8.5–15% by vol.	More in fortified, less in low-alcohol wines
Higher alcohols	0.15–0.5 g	
Acetaldehyde	0.005–0.5 g	
Esters	0.5–1.5 g	
Acetic acid	0.3–0.5 g	Expressed as sulphuric acid
Fixed Substances		
Residual sugar	0.8–180 g	According to type of wine; more in botrytized wines
Glycerol	5–12 g	
Phenolics	0.4–4 g	
Gums and pectins	1–3 g	Depending on the vintage
Organic Acids		
Tartaric acid	5–10 g	Depending on grape origin
Malic acid	0–? g	According to extent of malolactic fermentation
Lactic acid	0–1 g	
Succinic acid	1–3 g	
Citric acid	0–1 g	
Mineral Salts		
Sulphates	0.1–0.4 g	Expressed as potassium salts
Chlorides	0.25–0.85 g	
Phosphates		
Mineral Elements		
Potassium	0.7–1.5 g	
Calcium	0.06–0.9 g	
Copper	0.001–0.003 g	
Iron	0.002–0.005 g	

Source: Reproduced with permission from Jancis Robinson, ed., *The Oxford Companion to Wine* (New York, 1994), p. 1068. For a more complete listing of the elements in the "general composition of musts and wines," see Larousse's *Wine and Vineyards of France* (New York, 1990), pp. 546–7.

tongues glibly spouting both chemical and ionic equations for the same processes. Basic chemical analysis identifies the components of wine through titration: alcohol, tartaric acid, potassium, and so on. It tells us what the wine is made of; but it can only do so by decomposing it, or at least altering it, destroying its balance. Basic chemical analysis cannot not take into account the states in which the various elements are distributed *in* the wine, for these states are different when they are extracted in their pure state. Moreover, the same substance is often found in two or more states. Nor do the diverse fractions in equilibrium with one another all have the same role or the same reactions. Only with the help of physical chemistry can a "real representation" of the wine be obtained and a total correlation established between this representation and the qualities or characteristics of the wine. Instead of telling us of what the wine is made, physical chemistry tells us how it is constituted, a sort of explanation of the creation. It is a science with ontological status.

Ribéreau-Gayon compared the use of analytical chemistry to the demolition of a house followed by the separation and analysis of its materials; but analysis is incapable of giving the plan of construction and its organization. The only reassuring part of this Poincaresque analogy, which essentially makes analytical and organic chemistry servants of the new master, is that the components have to be identified before physical chemistry can explain wine as wine, not merely analyze its components. There remains a consolation prize for the old science. "Analytical and organic chemistry have an extremely important role to play in oenology, that of determining the nature and properties of the constituents of wine." And the ordinary chemists did not need worry about their jobs, because these substances were still not well known. One of the great achievements of physical chemistry was its study of solutions, especially its explanation of the *mechanism* of precipitations, how small solid particles are produced in a liquid by "chemical" reactions. Wine came naturally under this scientific empire.

The Acid Craze in Montpellier and Bordeaux

The two major centers of oenology were the faculty of sciences in Bordeaux and the faculty of pharmacy in Montpellier. From these two centers there poured forth theses, articles, and books based on the methods of physical chemistry. In a study of the role of real acidity (pH) in the preparation and preservation of wine, Jules Ventre introduced the discoveries of Ostwald, Arrhenius, and Sørensen into the basic science on which his research was founded.[14] Ventre was professor of oenology and agricultural

---

[14] Jules Ventre, *Du rôle de l'acidité réelle dans la préparation et la conservation des vins* (Montpellier, 1925). S. P. L. Sørensen (1868–1939), head of the chemical department of the

industries at the Ecole nationale d'agriculture (La Gaillarde) in Montpellier from 1920 to 1939. The new theories of ionization cleared up a series of paradoxes and questions on the action of tartaric acid, the biology of alcoholic fermentation, chaptalization, and aging. Ventre's work also provided a veneer of science to justify deplorable Midi practices like *platrâge* and *salage* (adding calcium sulfate and salt), all explained in terms of dissociation, or the movements of the ionized constitutents of the biological system called wine. The theoretical foundations of the new oenology may not have been original, but oenology had become relentlessly scientific and resolutely practical. Seemingly esoteric, physical chemistry also made its practical mark on the new oenology.

Between 1929 and 1931 Ventre published his three-volume *Traité de vinification pratique*, perhaps the first of the manuals to be explicitly based on the preceding half-century or so of oenological research, including his own considerable contribution and that of his *maître*, Auguste Bouffard, who held the chair of agricultural technology at La Gaillarde from 1885 to 1916.[15] Pierre Viala hailed its appearance as an important event in French winemaking literature, which did not contain as many books on the art of making wine as one might expect. There was the old work that had appeared under Chaptal's name. Claude Ladrey, professor of chemistry in the faculty of sciences in Dijon, had provided guidance according to the science of the 1860s. Of course there was the work of Pasteur, necessarily hailed as "the basis of modern vinification"; this was debatable by 1932. There were specialized treatises, such as the classic by Lucien Sémichon on wine diseases. Jules Laborde of the faculty of sciences in Bordeaux had begun a major project on vinification but gotten out only the first volume before he died. There was a fair amount of German and Italian literature on vinification. As Viala said, there was nothing so broad and valuable as Ventre's work, "the most complete, the most practical, and also the most scientific." In a few years the industry of vinification had been transformed biologically and its equipment changed to accommodate the latest technology. Ventre's work appeared at the right time.[16]

The faculty of pharmacy in Montpellier was a productive center of the new oenology.[17] A leader in the old, it had a long tradition of research on

Carlsberg Laboratory, Copenhagen, was a biochemist well known for his synthesis of amino acids, work on hydrogen-ion concentrations, studies of proteins, and development of a buffer system long used "as the standard for the definition of pH." Much of his work was done in collaboration with his wife, Margrethe. See H. Holter and K. Max Moller, eds., *The Carlsberg Laboratory, 1876/1976* (Copenhagen, 1976), pp. 63–81.

[15] Bouffard is hailed in Montpellier's oenological history for pioneering work on the nature of sugars in the must and as the "guérisseur de la casse des vins." Jean-Paul Legros et Jean Argeles, *La Gaillarde à Montpellier* (Montpellier, 1986), pp. 139–40.

[16] Comments by Pierre Viala in *RV* 76 (1932): 65–6.

[17] In the 1970s, the faculty of pharmacy in Montpellier was turning out an average of a thesis a year on oenological subjects. Jaulmes counted more than thirty "thèses de pharmacien

vine and wine. Well-known men of science like Jules Planchon and Henri Fonzès-Diacon gave the faculty a respectable name, even outside France. Work centered on the tough problems of the nineteenth century, mostly diseases: oidium in the 1850s, phylloxera (1860s–1890s), downy mildew and black rot in late-century and beyond, and wine fraud.[18] In 1938, H. Baylet advertised the new approach in the title of his thesis, *Contribution à l'étude des essais physiques des vins*. His historiography of oenology presented a threefold evolution from the stage or period of empirical methods of visual examination and tasting – still important – through the stage of chemical analysis, which determined a wine's total acidity and dry extract, to the third stage of physical analysis.[19] The third stage had a minor presence in the chemical analysis determining a wine's density. Baylet gave a proper French origin to the new type of analysis: use of the stactometer in Duclaux's method for measuring surface tension. The culmination of the new procedures came only in the 1930s with the measurement of the pH and even rH or oxidoreduction potential of wine. (Oxidation-reduction, or redox, is the unifying ionic idea that the two processes refer respectively to the loss and gain of electrons.) Baylet observed that these methods had not penetrated the world of viticulture or even oenology.

It may be that the housing of oenology in the faculty of pharmacy, rather than in the faculty of sciences as in Bordeaux, limited the number of researchers and, more important, limited their working input into wine production. The faculty of pharmacy produced professors and working pharmacists; the faculty of sciences produced professors and a few working oenologists. In Bordeaux there was also the powerful driving force of Genevois and an ambitious program based squarely on the latest German novelties as well as on his own work. A unity existed in Bordeaux, whereas Montpellier's resources in oenology were spread over the faculties of pharmacy and science and the school of agriculture.

The Ecole d'agriculture in Montpellier was famous for viticultural research, an investment that had paid off handsomely with the brilliant work of the older generation in inventing the post-phylloxera vine. Momentum was in favor of viticultural research. It is tempting to try to use this scientific emphasis to explain why the vines of the Midi have been better than their wines, but the market demand for cheap wine was a much more

supérieur de Doctorat d'Université" or "Doctorat d'Etat" that had been done at Montpellier over the years. P. Jaulmes, "L'oenologie à la faculté de pharmacie de Montpellier," *Revue française d'oenologie,* no. 45 (1972), p. 10.

[18] Ibid. The Ecole spéciale de pharmacie, created in 1803, became an Ecole supérieure de pharmacie in 1840 and a faculty in 1919.

[19] "The extract content (soluble, nonsugar solids) distinguishes the heavy- and light-bodied types" of wine; "wines having an extract below 2 percent are very light or thin on the palate compared with wines having over 3 percent." M. A. Amerine and M. A. Joslyn, *Table Wines* 2d ed. (Berkeley and Los Angeles, 1990), p. 439.

important factor. Still, if the same economic factors were operating in the Midi and the southwest, it is curious that the Languedoc-Roussillon allowed its areas of quality production to disappear and Bordeaux did not. Only recently in the Midi have a few areas that were once famous for quality wines begun to approach the quality of the *crus bordelais.* As late as 1988, the *Guide Dussert-Gerber* lamented that, for the Languedoc, "Le talent est rare."

It would be going too far to blame bad wine on bad science, but even Baylet expressed a reservation concerning the new oenology that seems curious coming from a supporter. Although praising physical analysis, he noted that it did not add indisputable elements of appreciation or differentiation to the analysis of wine; it required expert and careful practitioners as well as an enormous amount of boring repetition in experiments. Not that this was terribly different from the careful but often ignored requirements of traditional analysis, which was notorious for its sloppy practitioners. The point is that we are far from the enthusiasm of Bordeaux. And Baylet made it clear that chemical analysis remained of key importance for quality wine control and for providing the legal and juridical data on wine often required by governmental agencies, which had long been immune to biochemistry. Baylet's bibliography was exclusively French, in striking contrast to the heavily German- and English-language sources of the researchers who worked for Genevois.

A thesis in the faculty of sciences at Montpellier was a more serious affair: at least it was longer and usually more profound, armed with more complex mathematics, and sometimes even original. Baylet's thesis had been supervised by Paul Jaulmes, a professor of analytical chemistry and toxicology in the faculty of pharmacy, an oenologist with an international reputation, especially in methods of analysis. Being in pharmacy seems to have been more of a drag than a boost for Jaulmes. It is difficult to judge how much a reputation is built upon achievement or on self-promotion; usually a subtle combination of both, no doubt, but institutions count for a lot in the game of fame and fortune. Though he still has a solid footing in the technical literature, Jaulmes seems unjustly forgotten in the history of oenology.[20] After obtaining a doctorate in pharmacy with a thesis on the then stylish subject of weak volatile acids in a diluted aqueous solution, Jaulmes earned a doctorate in science, pure this time. His research was on the distillation of dilute solutions, with special reference to the case of aqueous solutions of volatile acids[21] – not a case of *plus ça change*

---

[20] See the list of references to Jaulmes in Amerine and Joslyn, *Table Wines,* pp. 898–9.

[21] Paul Jaulmes, *Recherches sur l'acidité volatile des vins; nouveaux procédés de dosages* (thesis, Faculty of Pharmacy, Montpellier, 1931), and his *Contribution à l'étude de la distillation des solutions diluées. Cas des solutions aqueuses d'acides volatils* (thesis, Faculty of Sciences, 1932).

but of moving from the faculty of pharmacy to that of science. His *patron* was Jules Gay, who taught physical chemistry. The dean, Fonzès-Diacon, professor of inorganic chemistry, also supported Jaulmes's research. These are not names you find in the *Dictionary of Scientific Biography;* yet they were important scientists pushing research in the direction of work being done in the world's leading scientific centers.

In the first chapter of his thesis in science, Jaulmes, beginning in good nineteenth-century style, gave a bibliographical-historical review of the work done on the composition of the vapor emitted by a solution of volatile substances. It included a parade of biggies: van't Hoff, Nernst, Rayleigh, Duclaux, and lesser-known names set the stage. In studying the volatile mixtures of water and acids, a subject neglected in France since Emile Duclaux's work in the 1860s and 1870s, Jaulmes hoped to find some new oenological applications of the laws of electrolytic dissociation and, *science oblige,* to do a general study of the phenomena. The special interest of his work was research on wine and other biological liquids, research that would move from the typical empiricism and imprecise experimental techniques of past studies into the exactitude of the theoretical. Exit Pasteur – temporarily. This work was in the style of the new oenology being introduced by Genevois on a much larger scale at Bordeaux. Jaulmes's research was more directly in the tradition of physical chemistry than the work done at Bordeaux, which leaned more toward the tradition of biochemistry. Jaulmes's research joins that of Bordeaux at the intersection of disciplines in the quantitative search for mechanism and in a common interest in wine. In his general oenological treatise, he covered much of the same ground as Ribéreau-Gayon's classic *Traité d'oenologie (1947).* Jaulmes's work was an oenologist's oenology. In one sense there was a certain basic similarity in the schools: *L'école bordelaise* (Genevois, Ribéreau-Gayon, Espil, and Peynaud) joined Montpellier in this devotion to analysis.

It could reasonably be assumed that by 1933 the theory of ions had been incorporated into science, including teaching manuals as well as laboratory guides. Not so, sighed Genevois and Ribéreau-Gayon; hence their emphasis on the need to study ionic equilibria in musts and wines.[22] Even when one takes into account the ritualistic opening of research articles declaring that "the question has not been seriously treated in any work," it must be admitted that Genevois and Ribéreau-Gayon effectively showed that the "complete" physicochemical analysis, which von der Heide and Baragiola did for a German wine (1910, 1914) and Clemente Tarantola for an Italian wine (1932), did not exist for a French wine. A review of the

---

[22] L. Genevois and J. Ribéreau-Gayon, "Les équilibres ioniques des les moût et les vins," *Annales de la brasserie et de la distillerie* (1933), pp. 273–7, 289–94, 305–11.

scientific literature and its research techniques might indicate that oenology had long managed to maintain not so much an imperviousness to novelty as a resistance permitting only partial penetration. The simplicity of Arrhenius's theory (1887), soon falling victim to the iron law of scientific progress, became complex, esoteric, and multiconceptual, even in the decades of the 1890s.[23] In France there was early scientific interest in applying ionic theory to biological phenomena. Maillard used it in his study on the toxicity of copper sulfate for *Pencillium glaucum* (1899).[24] Victor Henri's *Cours de chimie physique* (1905) made the ionic gospel easily accessible.

By 1930 Genevois and Ribéreau-Gayon could easily cite a long list of scientists who had done outstanding research using variations on a theme by Arrhenius. The Bordelais trio of Genevois, Ribéreau-Gayon, and Peynaud were prominent members of an ionic tribe including Th. Paul, Von der Heide, Geloso, Ventre, Tarantola, and Jaulmes. No serious laboratory could operate without using the methods of physical chemistry; indeed, the methods were the basis for a definition of the seriousness of the chemical research, especially in dealing with issues of equilibrium. Wine is a liquid in a state or near state of equilibrium. The methods of classical analysis give reliable numbers, but their use is limited or useless, because what an oenologist is often interested in knowing is not the total quantity of a substance but the free and combined quantities. The problems insoluble for basic chemistry were the ones that the methods of physical chemistry "solved very elegantly." Indeed, "the problem of [establishing] the state of ions in wines, the acidity of wines," had been "completely and clearly solved." Fortunately for scientists, there remained plenty of other "obscure problems in oenology" – for example, the transformation of coloring and pectic matters and the understanding of the action of uronic acids. Wine provided "an ideal case of the application of the laws of physical chemistry" because of three of its characteristics: a liquid milieu, a state of equilibrium, and a state of dilution in which concentrations of various ions existed in minimum amounts. Scientists could measure to their heart's content, and did.

The general aim of this new scientific assault was to destroy the mystery of wine. Genevois and Ribéreau-Gayon were quite explicit on the undesirability of presenting "wine as a mysterious being." This is still a popular

[23] See *The Encyclopedia of Chemistry,* article on "Dissociation" (an illustration of the law of "tout se complique"): Franklin's *solvent system concept;* the *protonic concept of acids* and Lowry; and Lewis's electronic theory. Simplification produces complexity.

[24] L. Maillard (professor of chemistry at the faculty of medicine in Nancy), "De l'intervention des ions dans les phénomènes biologiques: Recherches sur la toxicité du sulfate de cuivre pour le *Pencillium glaucum,*" *Journal de physiologie et de pathologie générale* 1 (1899): 651–64, which includes a brief review of the literature on "dissociation électrolytique ou de l'ionisation."

approach in the marketplace. Nor are aging oenologists immune to the temptation, even if, like Peynaud, they have spent their lives identifying the substances behind the veil. Not that many of the inorganic elements in wine remained unknown in 1933, although researchers admitted that the qualitative role of the unknown substances might be considerable. The inorganic or mineral elements and the organic acids of wine were known to within 1 percent of accuracy, and, in the case of all the ionisable constituents of wine on which analyses had been done according to classical methods, to within a milli-ion. Knowledge was weak on only the wine's nonionisable substances, which existed only in traces of a centigram per liter. This is drawing close to a god's eye view of the phenomenon.

Could ionic theory explain anything of interest or even importance to the ordinary wine buff? An anomaly in correlating the results of elementary analysis and of taste, for example? Von der Heide and Baragiola had analyzed two wines of the Fuchsberg vineyard (Geisenheim): a 1909 wine, very acid in taste, titrating 102 milli-ions of acidity; and a 1910 wine, tasting much less acid but titrating 127 milli-ions of acidity. Why was taste in conflict with analytical data? Only the establishment of the concentration in ions could explain the paradox. The 1909 wine, with a pH of 3.25, was much more acid than that of the 1910, with a pH of 3.32. Real acidity, established by the analysis of physical chemistry, is much more important in the appreciation of a wine than the titration acidation established by basic chemistry. A comparison of the data on the acidity of some German and Italian wines, furnished by the research done by von der Heide and Baragiola and by Tarantola, made it clear that the difference of sourness in the taste was due to German wines having a higher proportion of free tartaric acid than Italian wines: one-third of the total was free in the German and one-fourth in Italian.[25] Lactic and acetic acid levels were the same. Citric acid, often found in French wines, was absent.

French wines clearly differed in taste from these neighbors, yet oenological literature had no complete analysis of French wines that could give an idea of the distribution of organic anions (tartaric, malic, succinic, lactic, acetic, tannic, citric) in nondissociated acids and in acid salts. Years of happy work by researchers in wine acidimetry would provide the data for French wines or, as Genevois and Ribéreau-Gayon put it in sciencese, a "bilan physico-chimique des ions." This data was judged to be an indispensable piece of knowledge, with important practical implications because of the fundamental role of different acids in the taste, flavor, and longevity of wines. L'école bordelaise had hoisted one of its major research pennants.

---

[25] The tartaric acid was in the form of cream of tartar (acid tartrate). Tartrate is a salt or ester of tartaric acid, which comes from tartar (potassium hydrogen tartrate deposits in wine vats). It is optically active. See Oxford's CSD, p. 684.

Genevois was a prolific researcher and writer. By 1932 his list of publications ran to over 40 "substantial" items. For the supplement (1933–42) he added 87 major items, many of which were on fermentation and other subjects directly related to wine. Between 1928 and 1940 17 doctoral theses and diplomas of higher studies were prepared in his laboratory of physiological chemistry. The two best-known students were Ribéreau-Gayon and Peynaud. The laboratory was a scientific factory in the great tradition.[26] Much of the work done by these three scientists was published in four high-quality journals: the *Annales des falsifications et des fraudes,*[27] the *Annales de la brasserie et de la distillerie,*[28] which was renamed the *Annales des fermentations* in 1935; and the *Revue de viticulture,* the national organ of the Montpellier school. Monographs, which were often doctoral theses, helped spread the new oenology at a high level, and by the 1940s the time was ripe for a number of manuals or textbooks on oenology, of which the most famous was the *Traité d'oenologie* (1947) by Ribéreau-Gayon and Peynaud.

Scientists sometimes seem to fall into two major working categories, those whose happiness derives from shoring up the status quo and those who take a perverse delight in making science obsolete. The second category was so active in the 1950s and 1960s that the *Traité* had to be completely revised by 1972.[29] The year of the first edition (1947) is also the year in which Genevois and Ribéreau-Gayon published their austerely analytical work *Le vin.* In recent years Genevois has faded into the background of oenological bibliography, whereas Ribéreau-Gayon and Peynaud have emerged in iconographic splendor. Given the Genevoisian contribution to modern oenology, this hardly seems fair. But ancestor worship in science is hardly ever based on fairness.

[26] "Thèses et diplômes d'études supérieures . . . ," *Notice sur les travaux scientifiques (supplément 1933–1942) de Louis Genevois, professeur de chimie biologique à la faculté des sciences de Bordeaux* (Montrouge, Seine, 1945), p. 1.

[27] The subtitle of the journal was "Recueil de travaux de chimie analytique, de législation et de jurisprudence internationales appliquées à l'expertise des marchandises." It was edited by Dr. F. Bordas, chef du Service des laboratoires du Ministère des finances, and Eugène Roux, directeur des Services sanitaires et scientifiques de la repression des fraudes au Ministère de l'agriculture. The official organ of the Société des Experts-chimistes de France, the *Annales* had a distinguished editorial committee, including, in 1925, d'Arsonval, Gabriel Bertrand, Gayon, and Kling.

[28] This journal was the *Revue des industries de fermentation,* edited by Auguste Fernbach of the Institut Pasteur in Paris. He ran a laboratory at the IPP that did analytical tests, for a fee, on industrial brewing products. He also ran a Service des fermentations offering a course in microbiology and biochemistry applied to the fermentation industries. Moïse Schoen took over the *Annales* in the 1930s, but he died in 1938.

[29] Second ed., 1958. In the third edition the two original authors were joined by Pierre Sudraud (directeur du Laboratoire interrégional du Service de la répression des fraudes et du contrôle de la qualité and directeur de la Station agronomique et oenologique de Bordeaux) and Pascal Ribéreau-Gayon, professeur de chimie-oenologie and then directeur de l'Institut d'oenologie at the University of Bordeaux II.

Jean Ribéreau-Gayon: The Structure and Subtlety of the New Oenology

Jean Ribéreau-Gayon splashed heavily onto the oenological scene in 1931 with a doctoral thesis that was a striking piece of evidence of the new way in which scientists were thinking about wine. "Men may think with a view to knowledge, or they may think with a view to action."[30] Or both. *Repenser le vin,* with potentially radical consequences for winemaking itself. This was not one of the common run of theses.[31] Two of the topics Ribéreau-Gayon's work dealt with were at the heart of the Bordelais wine industry: quality and aging. The novelty of the work was its arriving at a new understanding of the relevant processes and mechanisms of vinification through the latest science and research methods.

Experiments by Berthelot and by Pasteur, among others, had established a very general theorem concerning the gluttony of wine for oxygen. And air affects the wine during all stages of its preparation, from fermentation to consumption or its maderized death in the bottle. Oxygen "is a necessary condition of aging." Emile Duclaux's work on microbiology (circa 1900) included a modified Pasteurian model of the aging of wine. Given the contemporary ignorance of the complex of elements in wine, a point emphasized by Duclaux himself, it is not surprising that explanations were more a matter of analysis of sparse data and of reasoning about them than of scientific knowledge about the elements and their functions. These intellectual gymnastics would be replaced in the post-ionic period by knowledge of the working of the biochemical processes and physical mechanisms of the main elements in aging or developing wines. Duclaux could make a distinction between microbial actions (ferments) and "chemical actions that superimpose their effects on those of the ferments." Duclaux's model of the aging of wine was based on the idea of oxidation of soluble substances – deposits and coloring matter. Although he integrated the actions of diastases (amylases) into his model of fermentation, including a secondary fermentation, his discourse on how oxygen ages wine was essentially Pasteurian, even in language. In oxidizing acids and coloring matter, oxygen also transformed the taste of wine, producing the "fullness that makes up the flavor of old wine." Duclaux's discussion was strong on actual winemaking practices – for instance, racking – especially in Burgundy.

Duclaux's clear separation of the chemical actions from the actions of the diastases made the study of the normal aging of wine a necessary prelude to the study of its diseases. "Good bottling permits only the chem-

---

[30] Stuart Hampshire, *Thought and Action,* new ed. (Notre Dame, Ind., 1983), p. 11.

[31] *Contribution à l'étude des oxydations et réductions dans les vins.* Thèse présentée à la faculté des sciences de l'Université de Bordeaux . . . grade de Docteur ès sciences physiques en 1931. Digest by the author in *Annales de la Brasserie . . . ,* 30e année (1932), pp. 155–7.

ical phenomena to occur," the yeasts having been gotten rid of by the racking. Duclaux favored stabilizing wine by heating, which, he argued, prevented the oxidation of acids and coloring matter. In one lot of heated wine he found that the fixed acids and volatile acids had not changed over twenty years, with the aldehyde figure remaining the same because the alcohol had not oxidized. (The fixed acidity, calculated as tartaric acid, is the total acidity less the volatile acids.[32] Aldehydes are chemical compounds resulting from the oxidation of primary alcohols.) Duclaux explained how wine's fragile coloring matter – which the spectroscope had shown to have different spectra for different varieties of vines – would coagulate if oxygen transformed ferrous salts into ferric salts, although the mechanism was not known. A few scientific advances had been made since Pasteur's work, but the model had not changed much, and the prevention of diseases was dependent upon luck and care in winemaking and sterilization by heating. Things stayed pretty much the same until the 1930s.[33]

Ribéreau-Gayon's experimental strategy was to reveal "the intimate mechanism of the actions of oxygen" in order to promote favorable conditions and avoid harmful ones in making and preserving wine. The mechanism of oxidations, "essential phenomena in the development of [all] wines," is the key to "the technique of [making] wines, especially fine wines." Hence the great significance of this thesis for Bordeaux. What was the problem? Ribéreau-Gayon summed up before attacking the problem's complexity through experimental simplification. "If a wine and air are in contact in different circumstances for a more or less long period of time, there is dissolution, progressive combination with certain elements directly, or by intermediary compounds, or through catalytic influences, resulting in a wine more or less different from the original wine."[34] (Solutions to scientific problems are usually more elegant than the language required to state them.)

Of course, researchers rarely say that their problems are small and simple. Ribéreau-Gayon followed standard scientific strategy in concentrating on some parts of the problem. He made constant reference to large-scale winemaking practices, in which wine was necessarily heavily aerated without the operator of the equipment even being aware of it. The difference between laboratory work and actual winemaking had long been noted, often to the extent of questioning the value of scientific research. By the 1930s scientists had turned the tables by taking the difference into account and by establishing ways in which laboratory research could be transferred

---

[32] See Amerine and Joslyn, *Table Wines*, pp. 441–4.
[33] Emile Duclaux, *Traité de microbiologie*, Vol. 4, chap. 27, pp. 559–88 ("Vieillissement des vins").
[34] Ribéreau-Gayon, "Contribution à l'étude des oxydations . . . ," *Annales de la brasserie et de la distillerie*, 30e année (1932), p. 155.

to the workplace. Ribéreau-Gayon determined the quantities of oxygen that diffused in wine in differing situations of aeration, whether in the laboratory or in the workplace. It was not a mystery but a physical phenomenon subject to the laws of physical chemistry.

Ribéreau-Gayon next investigated the speed of oxidation for both red and white wines, with a determination of the case in which sulfurous acid has a real anti-oxygenic effect.[35] He found that a powerful role is played by the anti-oxygenic effect of tannin, which thus has an important protective function in red wine, and by ferric and copper salts (discussed as metallic ions), which are intermediary bodies in the oxidation of wine. Discovering that these ions are catalytic agents of oxidation, he established quantitative relations between the speed of oxidation and the wine's content in ferric and copper ions.

What exactly is oxidized in a wine? Tannin, coloring matter, and sulphurous acid. How does a wine maderize? "In the course of the oxidation of all wine, the free sulphurous acid . . . oxidizes, thus allowing the progressive liberation and oxidation of the fraction of the sulphurous acid that is engaged in the aldehyde combinations; the dissociation of these combinations entails at the same time the progressive liberation of the acetaldehyde (ethanal); the result is flat wine, bottle sickness, maderization." So we arrive at the heart of the matter, the aging of wine, especially the desired result of improvement through aging, the transubstantiation of red wine into *grands crus*. How does it really take place? What explains the development of the fragrant rather than the foul?[36]

The first scientific analysis of the bouquet of wine had nothing to do with wine.[37] In 1836 Liebig and Pelouze found that an essential oil obtained from a pharmacist in Paris smelled like old wine. Their analysis was "the first well demonstrated example of the existence of a true ether

---

[35] Sulfurous acid (sulfuric [IV] acid) is "a weak dibasic acid, H_2SO_3, known in the form of its salts. . . . It is considered to be formed when sulphur dioxide is dissolved in water. It is probable, however, that the molecule H_2SO_3 is not present and that the solution contains hydrated SO_2, sulphur dioxide," which, according to another definition, dissolves in water to produce sulfuric and sulfurous acids. See Oxford's *CSD*, pp. 674–5.

[36] Readers interested in the archaeology of nasal perception should read Alain Corbin, *Le miasme et la jonquille*, translated as *The Foul and the Fragrant: Odor and the French Social Imagination* (Cambridge, Mass., 1986).

[37] My remarks on the understanding of bouquet in the nineteenth century are based on the article "vin" in A. Chevallier et Er. Baudrimont, *Dictionnaire des altérations et falsifications* (Paris, 1852; 6th ed., 1882); Liebig et Pelouze, "Note sur le principe auquel est dû le bouquet des vins. Extrait de quelques recherches chimiques faites à Giessen," *Comptes rendus hebdomadaires des séances de l'Académie des sciences* 3 (1836): 418; Jean-Joseph Fauré, *Analyse chimique et comparée des vins du département de la Gironde* (Bordeaux, 1844); William Franck, *Traité sur les vins du Médoc*, 2d ed. (Bordeaux, 1845); and Edme-Jules Maumené, *Traité théorique et pratique du travail des vins*, 2d ed. (Paris, 1874). See also the *Journal des connaissances médicales pratiques et de pharmacologie*, 34e année (1867), 569–70.

formed in the process of fermentation without the intervention of a chemist." The chemists discovered that the compound was made up of an atom of sulfuric acid and an atom of a new acid given the name of oenanthic acid ($C_{14}H_{26}O_2$) to indicate that it had the characteristic odor of wine. Liebig believed that tartaric acid influenced the formation of bouquet. Wines with the strongest "winey odor" and the most pronounced bouquet were also the wines richest in tartaric acid. By the 1860s it was recognized that the connection between bouquet and tartaric acid could not be found outside Liebig's powerful imagination.

In spite of the big-name research on the sources of the bouquet in wine, nineteenth-century scientists were able only to advance toward these sources rather than find them. The work done by Liebig, Berthelot, and Pasteur was most famous, or at least most often cited by other writers. Researchers of lesser renown also did interesting work, now inevitably as little known as the scientists who did it. In the 1830s it was scientifically respectable to believe that the bouquet came from an oil, distinguished by a tendency to thicken over time, obtained by imposing on wine the procedural tortures of freezing, distillation, and combination with ether. Some thick-oil men went on to argue that the bouquet was freed as a result of the decomposition of the grape seeds during fermentation.

The experimental results published in 1844 by J. Fauré, pharmacist and minor man of science in Bordeaux, established that the bouquet came from a specific essential oil derived from the grape skins. This was the scientific origin of the bouquet-from-the-fruit dogma so prevalent in today's wine discourse. Fauré argued that the aroma of wine was made to stand out by the existence of a viscous substance (*oenanthine*), found only in the fine wines of the Médoc, which gave them a mellow or velvety quality. He was proud of having discovered this specific element mentioned by no other writer on wine. The commercial importance of science was given a boost when some of Fauré's work was reprinted by the *négociant* William Franck as a supplement to the second edition (1845) of his famous *Traité sur les vins du Médoc.* But this recognition did not do much for the scientific longevity of pharmacist Fauré, whose pioneering work in applying chemical analysis to wine was not included in the later oenological canon.

Edme-Jules Maumené, a substantial figure in the scientific iconography of champagne, smoothly combined the scientific discourses of past researchers with an emphasis on factors that the great chemists often ignored in their single-minded concentration on the origin of bouquet. In the second edition (1874) of his *Traité du travail des vins,* Maumené explained bouquet as a multiple perfume, a phenomenon comparable to the odor of flowers: ethers, wine alcohols, aldehydes, and essential oils were the elements that collectively gave rise to the bouquet of wine. In Maumené's model, factors external to the grape exerted a great influence on bouquet. He considered the soil to be one of the most important factors in accounting for bouquet, for the same variety of vine produced different wines in different soils. And Maumené did not forget to take into account the influences of the nature of the fermentation and the maturity of the grapes when pressed. The chemist had become an oenologist.

Pasteur's analysis postulated the existence of a grape's natural bouquet, which passed directly into the must and survived fermentation, and wine's acquired bouquet, which was nearly exclusively produced by the oxygen introduced on the wine's exposure to air during vinification. Berthelot's more austere chemical analysis emphasized the proportion (from 1/30,000 to 1/15,000) of acid ethers, accompanied by some aldehydes, to the bouquet. He too distinguished between the "winey odor" of wines and their specific bouquets. By the 1880s, the standard oenological gospel on bouquet emphasized its dual origin and its specificity for different wines. This explanation seems unsatisfactory to us, but it was as far as science could go, though developing analytical procedures would add further elements to the basic structure of the explanation. A new explanation would come with the penetration of oenology by ionic theory and with the use of the new analytical tools.

Emile Duclaux, heir to much of Pasteur's scientific empire, summed up existing ignorance around 1900 with a tentative recommendation that researchers look at the distillation of *eaux-de-vie* – they keep their original bouquets – and also titrate aldehydes, ethers, and higher alcohols. Duclaux admitted that knowledge of the complex elements making up the bouquet of wines did not exist. Statements on the subject could not escape being peremptory and empirical. Yet it was in many ways the beginning of modern discourse on bouquet. "There is a part that is brought by the grape from the particular variety of vine. . . . There is a part that comes from fermentation, . . . the type of yeast used." These elements interacted with one another, but the only knowledge about the process was empirical, including a quantitative knowledge of a wine's alcohol, aldehydes, ethers, higher alcohols, and volatile substances. The explanation was not very satisfactory. Duclaux found that it was much easier to explain scientifically the aging of wine than its bouquet.

In 1931 Jean Geloso published the results of some work he had done at the Institut de biologie physico-chimique in Paris on the relationship between the aging of wines and the oxido-reduction potential.[38] Geloso was particularly interested in the biological theories of Otto Warburg and Heinrich Wieland concerning the mechanism of the fixing of oxygen by living tissues, a process that was regarded as analogous to that operating

---

[38] Jean Geloso, "Relation entre le veillissement des vins et leur potentiel d'oxydo-réduction," *Annales de la brasserie et de la distillerie*, 29e année (1931), pp. 177–81, 193–7, 257–61, 273–7.

in wine the living liquid. Geloso reviewed what was known about the effect of oxygen on the development of the bouquet in aging wines, adding a few of his own experimentally based observations. One of the problems was that oxygen reacts differently to wines in different states. It was found, for example, that the fixation of oxygen in wine when it is pasteurized has a bad effect on the bouquet, but a few months after the introduction of oxygen to the cold wine the bouquet improves. At different stages in the making of wine – fermentation, aging in bottles, and, to a small degree, pasteurization – measuring the wine's oxido-reduction potential showed that oxido-reduction substances are formed which are important in the development of the wine's organoleptic properties and are capable of fixing oxygen from the air.

Geloso speculated that the most delicate bouquet corresponds to the state that is neither totally oxidized nor totally reduced, that the optimum taste is a matter of equilibrium, a matter of the proportion of the state of oxidation to the state of reduction. Not that this helps a drinker to buy a wine at its peak of development without the help of a chemistry laboratory. (Perhaps the approach to wine developed in this type of oenology has its practical consequences in the vintage charts so precisely set up in books on wine by the gurus of the wine advice business: drink now, needs keeping, can be drunk now but will improve – all rather useless information unless one knows the history of the transport and storage conditions of the wine, particularly if in traveling it has been unintentionally subjected to a type of *estufado* treatment suitable only for Madeira.)

In his analysis of the aging of wines, Ribéreau-Gayon used practically Pasteurian language, but the ions brooded over his explanations. Oxygen is indispensable in the beginning of the aging process; after the process starts, the necessary becomes superfluous, even harmful in the case of white wines in casks. Aeration in casks is a slow and, up to a point, beneficial process dependent on intermediary oxidants; a quick aeration, such as agitation of wine in the air, brings the strong oxidants into play – hence the powerful effect of the drawing-off of casks or racking and of decanting bottles. While aging in the bottle, wine undergoes little or no "oxidation by oxygen of the air" – quite the contrary in the case of the development of ethers.[39] In the case of these volatile compounds, Ribéreau-Gayon introduced another process and another actor. "It is quite possible that certain

---

[39] Ethers are organic compounds like diethyl ether or ethoxyethane (an anesthetic and organic solvent), volatile sniffs of vapor made by dehydrating alcohols with sulphuric acid. Ethers are not to be confused with esters, also "organic compounds formed by reaction between alcohols and acids" – esterification; the esters "containing simple hydrocarbon groups are volatile fragrant substances used as flavourings in the food industry. Vegetable oils and butter are mixtures of esters. Animal fats are glyceryl esters of organic acids, i.e., fatty acids, but here the purist will quite rightly demand an explanation including a hydrocarbon chain and a terminal carboxyl group, etc." Oxford's *CSD*, pp. 250, 252.

bouquets whose development is hindered by the presence of intermediate oxidants (cupric ions) are the result of reductions [loss of oxygen (or reaction with hydrogen)]." Aging has its dangers, even in wine.

In the *Traité d'oenologie* (1947), Ribéreau-Gayon put forward the idea that the development of the bouquet of a wine is "certainly the consequence of a process of reduction, for it only appears in the complete absence of oxygen and when the oxido-reduction potential reaches a sufficiently low level. On the other hand, it disappears rapidly or is profoundly modified when the wine is lightly aerated." So one should not decant old wine until immediately before drinking. The bouquet comes from oxidizable substances, whose agreeable odor or aroma exists only in their reduced form. Ribéreau-Gayon catches himself speaking too generally here and points out that these substances more accurately constitute "systèmes rédox." Sulfur dioxide, a key player in this electron exchange, has a powerful effect on bouquet. An increase in sulfurous acid causes the bouquet to develop faster in red wines, but "it tends to produce a characteristic analogous to that of white wines."[40]

Temperature also has a speeding-up effect on the development of bouquet, a slightly higher than "normal" temperature can develop a bouquet fully in a few months instead of several years. Optimum temperature for the rapid development of reds is 20°C (67°F), and 25°C (77°F) for whites. But it has to be a natural process, not artificially induced. Only *vin ordinaire* can be developed faster by aeration and high temperature, mimicking the "natural" process at cooler temperature. Temperature-changing experiments with quality wines have not produced what Peynaud calls the desired gustative evolution.[41] Temperature is as important a factor in the preservation of wine as it is in its fermentation.[42] The real world of a wine's bouquet turns out to be more complex and intriguing than its mystique.

Part of Ribéreau-Gayon's thesis dealt with the oxidation-reduction of excessive metallic salts, whose tendency to produce cloudiness is a threat to limpidity, traditionally regarded as an important characteristic of high-quality wine.[43] Containing substances hungry for oxygen, wine tends to

[40] On the ionic reactions of oxygen and SO_2, with their oxidation numbers (referring to number of electrons), $+4$ and -2, respectively, see Oxford's *CSD*, pp. 499–500.

[41] Jean Ribéreau-Gayon, *Traité d'oenologie. Transformations et traitements des vins* (Paris and Liège: Polytechnique Librarie Ch. Béranger, 1947), p. 218; Emile Peynaud, *Connaissance et travail du vin* (Paris, 1981), p. 235.

[42] The *Guide Dussert-Gerber* recommends an ideal dark, clean, odorless cellar with a temperature of 10 to 12°C. Pretentious sommeliers should be first attacked with a question concerning the temperature of the restaurant's cellar, especially in summer. At too high a temperature, Dussert-Gerber fears too rapid a development from active yeasts and bacteria.

[43] A salt is "a compound formed by reaction of an acid with a base, in which the hydrogen of the acid has been replaced by metal or other positive ions" (Oxford's *CSD*, p. 611). A base is the hydroxide of a metal or metallic radical. It is the presence of an organic (alkyl)

keep its metallic salts in the least oxidized state, ferrous and cuprous salts. But the many aerations wine undergoes during its making and existence bring a dangerous dialectic into play that opposes the reduction and tends to bring the metallic salts into their most oxidized state, ferric and cupric salts.[44] Being acidic, wine easily picks up iron and copper from metals with which it comes in contact. "Iron at more than about ten parts per million and copper at one part per million or less are likely to promote oxidative and other reactions and produce changes in color, flavor, and clarity in wine."[45] Turbidity can also be produced by calcium picked up from concrete storage tanks and even diatomaceous earth filter-aids.

Iron and copper in wine pass from one valence to another according to the presence or absence of air. ("Valence" is a term referring to the relative combining power of an atom of a given element; one of the useful fictions or oversimplifications of science, the valence of a given element is not always the same in different compounds.) These shifts of valence give the metals the roles of intermediary oxidants and catalyzers of oxidation. In excess, iron and copper are responsible for certain specific precipitations that result in cloudiness: too much iron leads to iron or ferric *casse,* produced by oxidation in aerated wines; and too much copper leads to copper or cupric *casse,* produced by reduction in white or rosé wines after some time in the bottle.[46] Some factors are so important that they alone can explain why some wines very rich in iron do not undergo *la casse* and some wines with less iron do undergo it. These factors can be the copper content, the transformation of a precipitable ferric ion into a complex nonprecipitable form, along with the effects of the transformation of copper, sulfurous acid, tartaric acid, or basification. Ribéreau-Gayon concluded that the essential element in cupric *casse* is copper sulfate. Reduction of the cupric ion produces a flocculation of colloidal copper sulfide. As heating the wine eliminates the copper, the analysis of white wines is important in determining their copper content and their salvation. The novelty of Ribéreau-Gayon's work in this matter of chemical turbidity was to show that ferric (iron) and cupric (copper) salts have a catalytic role in certain reduction reactions that are closely related to their function of oxidation catalyst rather than being opposed to it.

Although there was a certain amount of debate over the significance of metallic salts in wine, more excitement was generated by the proposal to follow Germany in removing metal ions by blue fining. This process in-

---

radical that differentiates an ester from a salt, which, unlike an ester, also ionizes in a water solution.

[44] Jean Ribéreau-Gayon, "Le fer et le cuivre dans les vins blancs," *Annales des falsifications et des fraudes* (1930), pp. 535–44.

[45] Amerine and Singleton, *Wine,* pp. 113–14.

[46] See Peynaud, *Connaissance et travail du vin,* chap. 27, "Stabilisation vis-à-vis des casses métalliques."

volves the addition of carefully controlled amounts of potassium ferrocyanide, which forms a precipitate of intense blue color, Prussian blue. In use in Germany since 1918, it has only been permitted for treating white and rosé wines in France since 1955. It came close to being approved, with strong reservations, by the academy of medicine in 1936, when the academy was asked by the ministry of public health for advice on the desirability of adding potassium ferrocyanide to the arsenal of chemicals already legally approved for the treatment of wine. The academy fell back on the old principle that all substances not present in the natural composition of food should be forbidden. Besides, it became known that in Germany the additive was used only in clarifying the *petits vins*. Scientists might have believed that understanding mechanisms can lead to the production of healthy wine through more chemistry, but more conservative forces safeguarded the national drink from yet another chemical, for a short time at least.

In a summing up of the chief results of his work, Ribéreau-Gayon brought together experiment and production. He had determined exactly the conditions prevailing during oxidations that occur during the making, treatment, conservation, and aging of wine. Most significant were the quantity of oxygen introduced and the intervention of the intermediary oxidants. Ribéreau-Gayon indulged in no false modesty in noting that knowledge of these conditions had the potential of making great contributions to winemaking techniques. Second, he had established the primordial theoretical and practical importance of the natural presence of traces of metallic salts in wines and the very different but coordinated normal and accidental effects related to their being reversible oxidizing systems. Many of the phenomena exposed by Ribéreau-Gayon's techniques were part of the framework of laws and general theories. This emphasis on a close link between science in general and oenology in particular, though it was not new, was bound to bring radical results when the basis of science itself changed and oenologists like Ribéreau-Gayon and Peynaud made the new knowledge the basis of their work on wine. Begun by Genevois, this approach became a distinguishing characteristic of the school of Bordeaux in the 1930s.

The thesis's complex and even alarming part on iron, with the complication of the two series of compounds – ferrous, with a valency of $+2$, and ferric, with a valency of $+3$ – and the reduction and oxidation of iron salts, was too much science for the older oenologists. Louis Ferré, director of the oenological station in Beaune, denied the need to retain the two compounds of iron in discussing *la casse blanche* (iron or milky ferric phosphate *casse* in white wine) and pointed out errors in Ribéreau-Gayon's research. This was a provocation to the ionic establishment in Bordeaux. Ferré argued that the old way of direct testing of the wine by the sulfocyanometric method was sufficiently precise without the newfangled ionic equations.[47] This looked like Beaune taking on Bordeaux. Ribéreau-Gayon's refutation of Ferré was a masterly execution, showing Ferré's defective reasoning, errors, and inaccuracies. Worse for Ferré was Ribéreau-Gayon's skillful use of the latest novelties in analytical chemistry, including the work of French chemists like Georges Urbain, to show that Ferré was using Stone Age science, and even ignoring the work done by Moreau and Vinet and by Fonzès-Diacon.

If Beaune fared badly in this exchange, official oenology in Bordeaux fared even worse, for the abbé Dubaquié was convincingly shown to be guilty of "totally erroneous reasoning" – a very serious crime in France – on the same subject. Dubaquié did not know what Ribéreau-Gayon was talking about in the new approach. Even more embarrassing, as well as wrong, was the accusation that Ribéreau-Gayon had stolen Dubaquié's ideas on testing for iron in wine. Genevois thought it necessary to use his professorial power to defend Ribéreau-Gayon and his own science. The opposition was accused of ignorance of recent work in biochemistry. The *Journal of Biological Chemistry* was invoked, along with the authority of Smythe and Schmidt and Michaelis, to support the idea of the combination of iron with certain proteins and other stuff. There could be no doubt that iron compounds played a considerable role in biological milieux, in the living cell, and certainly in wine.[48] Perhaps there was less excitement about iron in Burgundy because there was less iron in its wine (about 10 milligrams per liter, half that in the white wines of the Gironde), although still enough to be dangerous in the right circumstances, and certainly enough to justify a more passionate interest in iron compounds.[49]

It should not be concluded that Burgundy was totally benighted. In 1935 the *Revue de viticulture* published an article by two scientists at the University of Dijon approving of and using ionic theory in measuring the acidity of wines in terms of their concentration of hydrogen ions. The authors, Augustin Boutaric and Jean Bouchard, worked in the university's

---

[47] L. Ferré et A. Michel (chef de travaux de la Station oenologique de Beaune), "Dosage colorimétrique des sels ferreux et ferriques dans les vins blancs. Méthode ferrocyanométrique ou méthode sulfocyanométrique?" *Annales des falsifications et des fraudes* (1933), pp. 18–36. Ferré and Michel had done their homework by 1934: "contribution à l'étude du mécanisme chimique de la casse blanche," ibid. (1934), pp. 197–211. This article contains a short history from Chuard (1895) through Ribéreau-Gayon and Dubaquié (1934), but it concludes that "Les lois générales de la chimie nous permettent encore d'expliquer le mécanisme de ces constations, sans avoir à faire intervenir des hypothèses plus ou moins plausibles et non-vérifiées."

[48] L. Genevois (professeur de chimie biologique à la faculté des sciences de Bordeaux), "Recherches récentes sur les complexes du fer. Application à l'étude du vin," *Annales de la brasserie et de la distillerie* (1933), pp. 188–92, 205–8.

[49] J. Ribéreau-Gayon, "Dosage rapide du fer dans les vins blancs," *Annales des falsifications et des fraudes* 22 (1929): 522.

laboratory of physical chemistry. The university was also offering a Diplôme supérieur d'oenologie and graduated six students in 1937. What was missing was the developing scientific-economic infrastructure that was present in Bordeaux.[50]

Peynaud's Research Passion for Acids: Context and Dominance

Research on acids, including their formation and development in the grape as well as in wine, continued to be the rage in oenology throughout the 1930s. Emile Peynaud was the leader of the acid pack. Other researchers also did their bit in this area, for there were enough problems in acid research for everyone; and a real leader in research usually developed either a new method of titration or modified an old one. Those who did neither could content themselves with pointing out that their colleagues' "new" methods were basically ones that had long been in use or, better, had fatal flaws in them. As in viticultural research during the phylloxera crisis, there was a good deal of quarreling, often polite, over methods. Older scientists were touchy and quick to defend their turf. As both Ribéreau-Gayon and Peynaud discovered, even minor criticisms of classic scientific procedures could provoke indignant counterattacks from scientists who had spent their lives using and perfecting the older methods.[51]

One of the best pieces of work, an analytical and physicochemical study, was done on Algerian wines in 1937 by Ernest Brémond for his doctorate in sciences at the Université d'Alger. Genevois was quick to point out that Brémond's research was in the style of the school of Bordeaux. Most analyses outside the laboratory were concerned with determining the total and volatile acidities of wine. Existing analytical methods gave an accuracy of 2–3 percent in determining the quantities of acids and bases in a wine. The total acidity could be given as grams of sulfuric acid (the old favorite method of most French researchers) or, as some modernists preferred, as cubic centimeters per liter of wine.[52] What was measured was the total of all the molecules of hydrogen replaceable by a base, the total of acid molecules present. A desire for greater precision required a determination of the totals of tartaric acid, of potash, of sulfates, and the total alkalinity of the ashes. Brémond noted that the malic, lactic, succinic, and

---

[50] Augustin Boutaric and Jean Bouchard, "Sur la mesure de l'acidité des vins exprimée par leur concentration en ions hydrogène," *Revue de viticulture,* 83 (1935): 37–43.

[51] See the critical reaction by André Kling, longtime head of the Laboratoire municipal de Paris, to Peynaud's criticism of Kling's methods for working with tartaric acid: "Au sujet du dosage de l'acide tartrique à l'état de racémate de chaux," *Annales des falsifications,* 29e année (1936), pp. 409–11. Kling notes that Peynaud addressed him in a familiar way to which chemists of his generation were not yet accustomed.

[52] For technical comments on ways of measuring acids and expressing "ratios between alcohol, acid, extract, glycerin, or potassium contents," see Amerine and Joslyn, *Table Wines,* pp. 473–6.

phosphoric acids were generally neglected, although they made up an important part of the fixed acidity of a wine. Part of the originality of the school of Bordeaux was that it carried out what, for want of a better phrase, we call definitive studies of some of the neglected acids.

Brémond's fame is less for his study of Algerian wines – hardly an elite topic – than for the development of a method of measuring the pH of wine through using a quinhydrone electrode.[53] Curiously, he seems to have played down the practical importance of this measurement of ionic or real acidity. Brémond showed the importance of the pH of wine in the development of the bacterial disease *la tourne* (tartaric fermentation) and the "accident" of iron *casse*. But he merely emphasized the pH as a subject having a key place in the chemistry of wine. This contrasts rather strikingly with the strident advocacy of the utility of pH by the school of Bordeaux. And certainly Peynaud received much more recognition for his studies of the wines of Bordeaux than Brémond did for the wines of Algeria. This may be difficult to justify on a strictly scientific level. Real acidity (pH) rather than total acidity played the key role in the preparation, taste, limpidity, and conservation of wines, but this was a much more important issue for the production of the *crus* of the Bordelais than for the wines of Algeria. The topic can make the scientist as well as the historian. Brémond's *Cours d'oenologie* (1946) has no significant history like the *Traité* of Ribéreau-Gayon and Peynaud. Brémond's subject, Algerian wine, disappeared as a research area and as a product with some quality potential when the French North African empire crumbled – a sad contrast to the success story for Bordeaux in wine as research subject and quality product.

Research on malic acid had been going on in Europe for decades when Ribéreau-Gayon, and particularly Peynaud, turned their attention to it in the 1930s and gave Bordeaux a certain notoriety for research on malolactic fermentation (MLF), which it has kept to the present day.[54] (MLF is not now regarded as a true fermentation but "an enzymatic reaction that

---

53 E. Brémond, "Un nouveau dispositif de mesure du pH des vins (l'iconomètre à quinhydrone)," *Annales agronomiques* 8 (1938): 371–9. The pH scale is a "logarithmic scale for expressing the acidity or alkalinity of a solution. . . . A pH below 7 indicates an acid solution; one above 7 indicates an alkaline solution. More accurately, the pH depends not on the concentration of hydrogens but on their activity, which cannot be measured experimentally. . . . The scale was introduced by S. P. L. Sørensen in 1909" (Oxford's *CSD*, p. 533). Larousse's *Wines and Vineyards of France*, p. 609, notes that "the pH value expresses the concentration of free protons"; the proton is a hydrogen ion with a charge (positive) equal to the charge (negative) of the electron.

54 D. Wibowo, R. Eschenbruch, C. R. Davis, G. H. Fleet, and T. H. Lee, "Occurrence and Growth of Lactic Acid Bacteria in Wine: A Review," *AJEV* 36, no. 4 (1985), cite 208 references on MLF. The French, German, Austrian, and Swiss research programs obviously continued, but they were joined by a considerable output from Australia and California.

is conducted by the bacterial cells after they have grown"; yet the term MLF is so solidly established in research, publishing, and industry that a convenient error is preferable to linguistic accuracy.)[55] Müller-Thurgau (also famous for his mediocre but popular low-acid, fruity wine grape variety of the same name), Osterwalder, Laborde, and Ferré were among the better-known scientists who had worked on malic acid, one of the commonest organic acids in the plant kingdom. W. Mestrezat had done some work on this acid in the wines of the Midi, and Ferré had worked on acid variations in the wines of Burgundy; but, apart from Laborde, hardly anyone had dealt with malic acid in the wines of the Gironde. As Peynaud noted in his classic study (1939), the lack of work on malolactic fermentation in Bordeaux wines was a serious theoretical and practical oenological disadvantage.

The oenological importance of MLF, apart from its giving better microbiological stability to wine, is its softening of young wines, "the first and surely essential act of aging. Not only is the acid make-up of the wine completely changed, but [malolactic fermentation] has an impact on the perfume of the wine and even diminishes the intensity of the color and changes its shade. It's not exaggerating to say that without malolactic fermentation there would hardly be any great reds of Bordeaux."[56] It pays to advertise the importance of one's research. At the end of his article Peynaud was careful to signal that the work had been carried out under the direction of Ribéreau-Gayon. He also said that his ideas for the research came from the work of Ribéreau-Gayon, who was close to Genevois, often collaborating with him in research on acids and oxido-reduction. It is tempting to see these three scientists as an oenological group whose basic ideas and techniques are harder to separate than the substances on which they worked.

In spite of the importance of malic acid, the titration of the malic anion (a negatively charged ion) in a complex milieu like must or wine had never been done satisfactorily. Peynaud was able to produce an agreement between an equation for acid formation and actual measurements that had eluded other researchers. More important, in the first systematic study since Laborde's in 1912, he illustrated the evolution of malic acid in the wines of Bordeaux. Peynaud was able to show the diminution of malic acid during the malolactic fermentation and to give measurements of total acidity, of pH, and of malic acid for 29 white and 21 red *grands vins*. In pasteurized wine the total acidity and the level of malic acid were greater than in nonpasteurized wine. As a result of this work, knowledge about malic acid in the wines of Bordeaux was not only superior to the informa-

---

[55] Ibid., p. 302.
[56] Peynaud, "L'acide malique dans les moûts et les vins de Bordeaux," *RV* 90 (1939): 27.

tion available about other French wines but had moved to a more advanced level of scientific research, comparable to what was being done in academic biochemistry. Indeed, it was biochemistry.

In 1928 Ferré had published his often cited article on malic acid, with specific reference to burgundy. Like the scientists of Bordeaux, he was armed with Mösslinger's equation (1901): $COOH$-CH_2-$CHOH$-$COOH$=CH_3-$CHOH$-$COOH$+CO_2, according to which hardworking bacteria transformed one gram of malic acid into 0 gr 671 of lactic acid and 0 gr 329 of carbon dioxide. Full of wisdom about the importance of this transformation for wine, the article included a warning that a retrogradation of malic acid during conservation can produce a disequilibrium in the composition of a wine. (An unbalanced wine inspires horror in the advanced taster.) But a great deal of emphasis was placed on methods, their deficiencies and necessary modifications, all culminating in Ferré's ingenious method for recalculating the initial acidity of the wine before the malolactic fermentation, which had considerable significance in official testing for legal purposes – fraud and all that sort of thing. Ferré's science was no longer the oenology of Pasteur, but neither was it very close to the new oenology.

French official analytical methods did not mention malic or lactic acids, but the Swiss *Manuel officiel* of 1912 gave the procedure for titration of lactic acid. There was a great deal of work on malic acid done in Switzerland, land of some wine and much pomology. Ferré noted that malic acid often disappeared in the wines of Burgundy, as for example in the Beaujolais and the Mâconnais in 1922 and 1925, and that it was not unusual for the same thing to happen in Alsace, the Champagne, and Switzerland.[57] Most wines have a trace of malic acid, whites more than reds. While in the 1930s Ferré continued to practice a highly competent, even original oenology in Burgundy, the new oenology developed in Bordeaux and, to a lesser extent, in Montpellier. No doubt a large part of the explanation for the retardation of Burgundy lies in the poverty of science as compared with science in Bordeaux and in Montpellier. And a science-based agricultural research program in these two southern cities gave oenological research a big boost. One also has to keep in mind the great economic importance of the vine in the Hérault, and the Midi generally, which led even pharmacy to be more interested in wine than it might otherwise have been. Nor can it be irrelevant to the preeminence of Bordeaux that Ribéreau-Gayon and Peynaud directly related their research to the *grands vins de Bordeaux*.

Ribéreau-Gayon and Peynaud never failed to publish a statement of the

[57] L. Ferré, "Indices oenologiques et rétrogradation de l'acide malique," *Annales des falsifications et des fraudes*, 21e année (1928), pp. 75–84.

results of their research that made sense to the knowledgeable reader, in their case the winemaker. They did this for problems that had been solved *grosso modo,* as well as for new research problems like their study completing the work done by Laborde and Espil on lactic acid in the wines of Bordeaux. Wines of all regions contain lactic acid in amounts equal to and often more than their amounts of tartaric acid. (Perhaps partly because of Pasteur's fascination with it, tartaric acid was a favorite topic of writers on wine in the nineteenth century.) Alcoholic fermentation produces about a gram of lactic acid per liter. In red wines the content generally goes up to 2 to 3 grams per liter as a result of the malolactic fermentation in the first summer of the wine's life. Being more sulfured, white wines keep their malic acid and therefore a degree of tartness; the lactic acid stays at the level produced in alcoholic fermentation. Yet a Sauternes without free or molecular SO_2 undergoes a "malic degradation."[58] Nothing is simple.

In 1939 the *Revue de viticulture* published a state-of-knowledge report on lactic acid by the Laboratoire Pierre Viala of the Institut national agronomique.[59] Its basic premises were Pasteurian, but the details with their practical consequences showed how it was possible to master vinification in a way that had been only a vague hope in the 1860s. Pasteur had first given bacteria the credit for lactic fermentation; they were isolated by several scientists, including Gayon, Dubourg, and Müller-Thurgau. The bacteria were soon regimented into three groups: the *Bacterium Gayoni,* which attacks sugars but not malic acid; a second group, including the gourmand *B. gracile,* which has an appetite for glucose, levulose, and citric acid in addition to having an ability to change malic into lactic acid, carbonic acid, and traces of acetic acid; and the nonparasitic *coccus* [*Pediococcus*], interested only in bringing about malolactic fermentation. The species *B. oenos* or *Leuconostoc oenos* that is referred to in French literature turns out to be strains of *Leuc. oenos* that disdain to ferment the sugars pentoses, xylose, and arabinose; it has been relegated to classificatory limbo by the International Committee on Systematic Bacteriology.

Yet there is a good word to be said for these vital life forms. Lactic acid bacteria (LAB) influence the "essential nature of wine," or, as chemists say in their economical officialese, have the ability to decarboxylate malic acid through malolactic fermentation. MLF allows the wine to keep its "totally natural taste." Because the lactic acid is weaker and less sour in taste than the malic acid, the wine softens (aging phenomenon), with a fall in both real and titrated acid. These favorable changes, including a pleasant change in the bouquet and a less violent hue to assault the eye, led some

---

[58] J. Ribéreau-Gayon and E. Peynaud, "L'acide lactique dans les vins de Bordeaux," *Annales des falsifications et des fraudes,* 30e année (1937), pp. 339–44.

[59] "Note sur l'acide lactique dans les vins," *RV* 91 (1939): pp. 242–8.

clever people – the quick-fix men – to think of adding lactic acid to low-acid wines. This was another custom long practiced in Germany. The Narbonne oenologist Michel Flanzy proposed that lactic acid would be a good substitute for citric acid, which, attacked by *B. gracile,* eventually disappears. The proposal was not without a drawback, for lactic acid could also be attacked by the vinegar-producing acetic (ethanoic) acid bacteria. Life was best for wine in areas where nature, under expert supervision, could guide the bacteria on the road to the *grand cru.* Given the importance of acids for flavor, color, and the low pH that keeps the bacterial content within a safe range in wine, LAB are respected and feared because of their ability to alter "the concentration of individual acids," and therefore wine quality. MLF can be regarded as "one of the most important events in vinification."[60]

Various components of wine seem to arouse the research passion of the oenologist at different historical times. In the 1930s volatile acidity certainly had its day, but then it went out of style for a while. Few wine drinkers even know what it is. "Volatile acidity is the total of fatty acids of the acetic series in wine. Excluded from volatile acidity are lactic and succinic acids, as well as carbonic acid and free sulfur dioxide."[61] Almost all the even-numbered acids from acetic to $C18$, or even higher, are produced in the fermentation of sugar by bacteria, yeasts, and fungi. The oenologist's favorites in volatility were acetic, formic, propionic, and butyl acids.[62] The leading names in research on volatile acidity in the 1930s were Jaulmes, Ventre, and Peynaud. Montpellier had a strong tradition in this type of research.[63]

Jaulmes had done his doctoral thesis (faculty of sciences, Montpellier, 1932) on the topic, and although his emphasis was not particularly practical, this did not prevent it from becoming a key work in oenology. Jaulmes is admired by specialists for his series of "beaux travaux" on volatile acidity, which range over a period of nearly forty years (1931–69). The second edition (1951) of Jaulmes's *Analyse des vins* devotes fifty-odd pages to volatile acidity, including a historical discussion of methods of measurement and a long theoretical discussion of the "méthode Jaulmes," for which he designed his own apparatus. In 1930 Ferré had adapted the acetometer of

[60] D. Wibowo, et al., *AJEV* 36, no. 4 (1985): 308.
[61] Definition from Fonzès Diacon et Jaulmes (1930), cited in Jean Ribéreau-Gayon et al., *Traité d'oenologie,* 2d ed. (1982), 1:107.
[62] Jules Ventre, *Traité de vinification,* 2d ed. (1941).
[63] For an examination of methods of quantitative analysis of volatile acids in fermented liquids, including the methods of Berthelot and Jungfleisch (1872), of Duclaux, of Sémichon and Flanzy, and of Werkman (1930), see Eugène Ketelbant (Institut national des industries de fermentation de Bruxelles), "Dosage des acides volatils dans les liquides fermentés," *Annales des fermentations* (1936), pp. 109–27.

Cazenave to measure volatile acidity in wine, but in 1935 Beaune followed Montpellier when Ferré adopted the method of Jaulmes. Since 1963 this method has been consecrated as the official method.

Back in 1925, Jules Ventre had been a pioneer in applying ionic theory to the study of acids in wine.[64] In 1937 he published an important study of volatile acidity and fermentation in which he supported the idea that the large quantity of volatile acids produced was the normal product of the breakdown of sugar by yeast.[65] Acetic acid was the most important. It could come from several sources: the acetic acid bacteria themselves produce it along with acetaldehyde; yeasts produce it during alcoholic fermentation; and lactic acid bacteria produce some during MLF. According to the science of the 1930s, disease-producing microorganisms (*la tourne, l'amer*) also produced this potentially dangerous acid. Hence the measurement of volatile acidity in oenology was considered a significant indication of the health of the wine or its pathological evolution. A figure of 5 dgr or less per liter (expressed in sulfuric acid) indicated a healthy wine. Nowadays most oenologists would express volatile acidity in grams of acetic acid per liter: levels above 1.2 to 1.4 g/l mean trouble. Duclaux had shown the mixture of volatile acids found in diseased wines: *vins amers* contain an excess of acetic and butyl acids and *vins tournés* contain an excess of acetic and propionic acids. The old results of microscopic examination were confirmed by the new analysis.

Ventre was able to make what had become a classic distinction between laboratory tests and "semi-industrial" tests, because in the experimental cellar of the Ecole nationale d'agriculture de Montpellier he was always able to work on batches of wine of over 100 liters. He found that the presence of acetic acid modifies the process of fermentation: the amount of volatile acids determines levels of alcohol and residual sugar. Too high a volatile acidity hinders the fermentation process or activity of the diastases (amylases), the enzymes that split starch or glycogen into dextrin, maltose, or glucose. It used to be thought that the several species of bacteria that oxidize alcohol to produce acetic acid are aerobic; this gave people the not completely justified belief that without air the small number found naturally in wine could not proliferate sufficiently to turn wine into vinegar. Sulfur dioxide also inhibits their growth.[66] Still, it is a potentially dangerous creature whose secretion scientists are wise to study carefully. Pasteur would have been pleased to see his battle plan unfold so nicely.

Peynaud's work on volatile acids used the work done by Charles Bertin,

---

[64] Jules Ventre, *Du rôle de l'acidité réelle dans la préparation et la conservation des vins* (Montpellier, 1925).

[65] Jules Ventre, "Acidité volatile et fermentation," *Bulletin international du vin* (December 1937), pp. 447–65.

[66] Amerine and Singleton, *Wine*, pp. 60–1.

by G. Mathieu, and especially by Ventre as a point of departure.[67] Shortly after Ventre published his admirably precise experimental data, Peynaud used them to support "the hypothesis that the volatile fatty acids are reduced to corresponding alcohols." Ventre noted that the presence of acetic acid resulted in more alcohol from less sugar and carefully leapt to the conclusion that this acid, in promoting better use of sugar, modified the process of fermentation. Peynaud assumed that the acid was reduced to ethanol (ethyl alcohol). The correspondence of measurements between sugar fermented and alcohol produced gave him a better explanation for the disappearance of the acid, indeed a general hypothesis for the disappearance of all the volatile fatty acids.

Ventre had lost the acids, and Peynaud, in a first-rate piece of scientific detection, had found them metamorphosed into alcohols. So another wine mystery was solved. And it was done within the context of elucidating "the mechanism of the formation of acetic acid during the alcoholic fermentation of the grape must" in anaerobic conditions. The fermentation of the acetic acid slowed down as the pH of the must rose in a reduction of the acid to ethanol. With reduction eventually consuming more volatile acids than were produced, the amount of acid in the must diminished. (The yeasts of Burgundy were known to form twice as much volatile acid as those of Algeria, but the acids diminished as well as, and even ended up less than, those in Algerian wine.)

In the end, the amount of acetic acid depends upon two sets of factors: first, the nature of the yeast, especially its richness in aldehydomutase and its reducing power; second, the milieu (pH, aeration, concentration of salts such as phosphates that increase the formation of acetic acid, and temperature). Peynaud emphasized the importance of the rH factor (oxido-reduction potential), next in importance to the pH factor, for in the absence of air the acetic or ethanoic acid is hydrogenated and in the presence of air it is dehydrogenated. Once again the fertility of the ionic scheme in explaining phenomena had enabled the school of Bordeaux to score a triumph in an important area of oenological research. Wine's volatile acid content, mostly acetic acid, a key criterion of the soundness of wine, had become an important part of the Bordelais research program.[68]

Narbonne Questions Peynaud's New Oenology

Viewed from Narbonne in the Aude, Peynaud's triumph was a little less brilliant. René de Sèze of the oenological station in Narbonne classified

---

[67] E. Peynaud, "Sur la formation et la diminution des acides volatils pendant la fermentation alcoolique en anaérobiose," *Annales des Fermentations* 5 (1939–40): 321–7, 385–401.

[68] Amerine and Joslyn, *Table Wines*, p. 440. For a modern treatment of "factors influencing higher alcohol formation," see ibid., pp. 354–9, especially table 45 on p. 357, showing the "source of alcohols in alcoholic fermentation."

the scientific coup as the discovery of a striking fact (the establishment of the acid/alcohol transformation molecule by molecule), lacking the support of a rigorous proof. Still, Sèze did recognize the work of Ventre and of Peynaud as an important oenological advance, an example of the benefits that oenology could derive from the progress of the physicochemical and chemical sciences. Yet oenology "is and must remain essentially a biological science." Sèze does not seem to have realized what had happened to biological science. Nor was he happy with the tendency of oenology to become, in certain cases, excessively rationalist and to adopt radical solutions based, in theory, on pure cultures and asepsis and, in practice, on a high dose of antisepsis. Old oenology was scorned. Sèze wondered if the old lessons should not be studied carefully to give them their place in "une oenologie rationnelle."[69] At Narbonne, Sémichon had certainly taken some hard knocks from the new oenologists. No doubt Sèze was undertaking a bit of a defense of the home front and his own status in the scientific world. Similar tensions existed at Bordeaux, because Dubaquié had also found himself in a state of ignorance about ferric ions during the course of a quarrel with Jean Ribéreau-Gayon.

Volatile acidity was also a matter of debate in Narbonne, where the importance of the issue was recognized but given a more traditional and inconclusive spin in debate. Michel Flanzy, director of the Station oenologique de Narbonne, engaged in an interesting squabble over the 1937 wines, whose increase of 20–50 percent in volatile acidity over a few days alarmed traditional thinkers. Flanzy argued that the increase was probably no more than a case of defective titration procedures rather than an actual increase. Marie-Louise Pariot disagreed with this curious remark, pointing out that the same procedures did not give elevated levels for the 1936 wines. Flanzy did not concede the point and attacked the incompetence of many analysts, some of whom did not even eliminate the wine's sulfurous acid from their calculations.

Because the issue was disease (acescence) the important thing was to detect it, and the best way of doing that, argued Flanzy, was by having an expert taste the wine. But tasting seemed to be a dying art, if one were to judge by the fact that the category of tasting was absent from many bulletins of analysis used in commercial transactions. This ignorant use of science worked against good natural wine, for volatile acidity is an element of quality, essential for bouquet. Flanzy condemned the law of June 1938 limiting volatile acidity because it reduced the margin of security for good wine. The best Corbières and Minervois averaged 0.60–0.70 grams of volatile acids per liter, not far from the legal limits of 1 gram 20 for wine sold

[69] René de Sèze (ingénieur agronome; chef de travaux à la Station régionale de recherches viticoles et oenologiques de Narbonne), "Sur l'abaissement de l'acidité volatile des vins," *RV* 68 (1938): 339–45, 357–63.

by producers and wholesalers and 1 gram 50 for retailers' wine. A return to expert tasting seemed a necessary prelude to the return of confidence in the wine market. The oenologists of Bordeaux easily agreed and, more important, brought about a scientific closure of the debate.[70]

Someone not familiar with the exotica of winemaking might wonder about all the fuss over volatile acidity. In spite of the alarm over the potential threat of volatile acidity to the health of wine, one must recognize that fresh grape juice has only traces of volatile acids, and the small quantity of acetic acid formed during the alcoholic fermentation of the sugar diminishes in amount at the end of the fermentation, peaking when half the sugar is fermented.[71] Part of the explanation for the large number of studies done on volatile acidity was the powerful stimulus of scientific curiosity about unexplained phenomena, but there was also a related strong economic incentive because volatile acidity had come to be associated with diseased wine. Pasteur had pointed out the connection, emphasizing the need to check wine for volatile acids.

What is the connection between volatile acidity and disease? It does not lie in the production of a small amount by the bacteria that produce malolactic fermentation, or in the decomposition of citric acid occurring at the same time, or in the later lactic fermentation of small quantities of sugar, especially pentoses.[72] All of these phenomena produce 0,30 to 0,40 grams of volatile acids per liter, usually of no significance for the past or future of the wine. The problem is rather the production of lactic acid by the "bad" bacteria in their attack on the wine's reducing sugars, glycerol, and tartaric acid, or, to put it simply, the development of acetic or acetobacter bacteria with the power to oxidize the wine. "So the volatile acidity of a wine gives information on its state of health, on the seriousness of the breakdowns it has undergone. Volatile acidity gives the history of a wine; it is the mark left by a disease, a failed vinification, or defective conservation. In addition, it allows one to anticipate the difficulties of conserving a wine."[73]

When the first laws were enacted in the 1930s they did not take into

---

[70] Michel Flanzy, "Remarques sur l'acidité volatile des vins de 1937," RV 88 (1938): 36–8 ("Evolution regrettable de l'oenologie"), 116–17, 223–4, 253–6 (article by J. Cros Mayrevieille), 261–4.

[71] Formed by dismutation of acetaldehyde à la Neuberg: $2\,CH_3 - CHO + H_2O = CH_3 - CH_2OH + CH_3 - COOH$.

[72] A monosaccharide is a simple sugar, "a carbohydrate that cannot be split into small units by the action of dilute acids. Monosaccharides are classified according to the number of carbon atoms they possess." A pentose is a sugar having five carbon atoms in each molecule. Oxford's CSD, pp. 454, 513.

[73] Jean Ribéreau-Gayon et al., Traité d'oenologie, 1:107–8. Amerine and Joslyn, Table Wines, p. 775, note that "The most widely distributed acid-tolerant organisms which account for the spoilage of wine are the Gram-positive bacteria, both rod and spherical forms, which produce lactic acid."

account the subtleties and complexities of the issue in the way laws do now for wines over two years old, for specially vinified wines, and for differences between white and red wines, and so on.[74] The imposition by the E.U. of a legal maximum on the volatile acid content of wines does not raise the same difficulties. In the regulation of consumption, a little science can be a dangerous thing, generally for the producer rather than the consumer, but it is often better than no regulation. The revised wisdom of the scientific community concerning volatile acidity eventually reached the regulative level, thus avoiding the absurdity of good wines with high volatile acidity being excluded from the market while wines suffering from acescence (or *la piqûre*) with low volatile acidity were admitted. As early as 1939 Lucien Grandchamp had given a lecture/demonstration/tasting to a "numerous and informed public" at the Laboratoire Pierre Viala to illustrate this fact. And scientists were not slow in using the problem to advertise the need for more research on acids, aldehydes, and esters. But the whole debate was soon made otiose by Peynaud's research on ethyl acetate, the culprit responsible for acescence in wine.

When the *Revue de viticulture* published Peynaud's results on the role of ethyl acetate in making wine acescent, the editors took the unusual step of noting that it always published with alacrity the remarkable studies of Peynaud on the physicochemistry of wines. The importance of this article was that it answered, on the basis of a set of rigorous experiments, the questions about acescence and provided the basis for changing a wrongly conceived legislation on volatile acidity. It was also part of Peynaud's general study of esterification in wine, which showed that acescence is a purely physiological phenomenon, with ethyl acetate being "a biological product formed deep in the yeast cell or in the acetic bacteria, the product of alcoholic fermentation itself."[75] Before the revelations of biochemistry it was impossible for anyone to know the mechanism involved here. Later it was found that "the acetate part of ethyl acetate is formed aerobically with the yeast cell and is not derived from the acetic acid of the medium."[76] Classic studies had been done on esters and volatile acidity by Berthelot as early as 1863, by Gayon in the early years of the twentieth century, and by Laborde during the First World War. Berthelot's work was strongly theoretical, an attempt to apply the laws of esterification to wine and vinegar. The work of Gayon and Laborde was limited by their purely chemical approach to the phenomenon. Gayon's method of titration gave inaccu-

---

[74] "Note sur l'acide lactique dans les vins," *RV* 90 (1939): 242–8.

[75] Peynaud, "Acétate d'éthyle et acescence," *RV* 90 (1939): 324.

[76] Amerine and Joslyn, *Table Wines,* p. 363. Peynaud's article "Acétate d'éthyle et acescence," *RV* 90, (1939): 321–7, is a follow-up of the basic "note pratique" on "L'acétate d'éthyle dans les vins atteints d'acescence," in *Annales des fermentations* 2 (1936): 367–84. For a brief discussion of "the mechanism of the formation of ethyl acetate by *Hansenula* and other aerobic yeasts," see Amerine and Joslyn, pp. 363–5.

rate results on the high side in the case of wines spoiled by *piqûre*. The subject was not new, but Peynaud's method and conclusions were.

Peynaud noted that the independence of *piqûre* and acetic acid had long been known but he had established it as an absolute rule and correlated it with tasting. Readers of Peynaud's recent works will have noted that he has always assigned a great role to tasting along with his emphasis on the key role of science in winemaking: he insists on the harmony of science and art, a distinguishing characteristic of the new oenology from the beginning. Indeed, Jean Ribéreau-Gayon's famous *Traité d'oenologie* of 1947 begins with one of the most profound analyses ever made of *la dégustation*. And in this apparently esoteric subject of ethyl acetate or ethyl ethanoate (combination of acetic or ethanoic acid and alcohol), we are never far from the main issue of the bouquet and taste of the wine.

Like many ambitious researchers, Peynaud was eager to point out the errors of his predecessors or even contemporaries; it is part of the scheme of justification, of both research and its egoistic inspiration. In 1939, several years after he had shown that ethyl acetate rather than acetic acid is responsible for the odor of *vins piqués* or acescent wines, nine out of ten manuals were stating the contrary. Peynaud might have noted that scientists who write textbooks, like their colleagues in the humanities, usually copy other books rather than indulge in the far slower if more desirable technique of plundering research articles. Even adding high levels of acetic acid to wine does not produce acescence, with its odor of ethyl acetate. The difference in odors can be vividly illustrated by comparing the odor of good vinegar to the bad odor of a *vin piqué*. Tasting the wine as well as smelling it enables one to perceive the two sets of gustative characteristics as totally different.

Peynaud emphasized the agreement between the aesthetic of taste and the evidence of science and even referred to "the special taste of acetic anions." In some *vins piqués* only tasting can reveal that the wine is sick, for the odor is not noticeable: nondissociated molecules of organic acids of wine have an acid taste in themselves – this is the ionic view. A healthy wine with a high level of volatile acidity would still have a low level of ethyl acetate; the reverse would be true for a *vin piqué*. The level of esters does not rise in wines suffering from the now rare disease *la tourne,* in which lactic acid bacteria, taking advantage of low acidity, multiply and gobble up all the tartaric acid to produce lactic, acetic, and carbonic acids, making the wine undrinkable, with a bouquet of mouse urine rather than ethyl acetate[77] (more evidence of the guilt of the acetic acid bacteria). In spite of the dangerous nature of ethyl acetate, it usually does not threaten

---

[77] Amerine and Joslyn, *Table Wines,* p. 784, in their excellent chart, refer to a "mousy" taste in the case of a wine suffering from *tourne.* The French judgment would seem more accurate, unless oenologists really know how mouse tastes.

wine. It exists in all wines, for its origin is in alcoholic fermentation itself, a product of the yeast cell or of the acetic bacteria. It must be at a low level. Good wines of the Gironde contained forty to 180 mgr per liter, but the odor is masked, mixed up in the complex phenomenon called the bouquet.[78] As in all living matter, virtue and vice in wine are the result of proportion and balance.

Bordeaux's success in oenology was based on high-level scientific research and its wine of high quality, and especially the wine's potential for aging while keeping its quality – the market claims that it improves with aging. Montpellier was great in viticulture and had a modest success with its oenological program in the faculty of pharmacy. Burgundy had oenologists but no oenology, although Louis Ferré, its most famous oenologist, wrote a book with the parochial title *L'oenologie bourguignonne.* The concept of the *terroir* and the cult of the pinot noir encouraged an exaggerated oenological specificity. (The contemporary work of Michel Feuillat and others in Dijon indicates that this is no longer the case.) A Bordeaux school of oenology, a Montpellier school of viticulture, and Burgundian oenologists may be not an unfair way of summing up the historical situation.

[78] Peynaud, *RV* 90 (1939): 324–6.

12

The Institute of Oenology

After the Second World War, the new oenology led to striking and largely beneficial changes in the wine of Bordeaux. The research program there evolved as an institution for producing oenologists and has also developed a fruitful interaction with the Bordelais winemakers appropriate for the "world capital of wine."[1] Emile Peynaud, one of the key figures in the transformation, has noted the Bordelais *conjoncture* of research, teaching, and capitalistic production and consumption. The basic work of Jean Ribéreau-Gayon and Peynaud, spread through their texts and articles and through the educational machine, eventually was applied by winegrowers to production and, to some extent, by business to the handling of wine from producer to consumer. The role of the university in this transformation was consecrated by Ribéreau-Gayon's appointment in 1949 as head of the Bordeaux Station agronomique et oenologique. In Peynaud's opinion, Jean Ribéreau-Gayon presided over the birth of a new science. This interpretation flatters Peynaud by making him present at the creation, but one cannot deny that oenology was now firmly based on several scientific disciplines. It was directly affected by what happened in analytical chemistry, physical chemistry, plant physiology, biochemistry dealing with plants and with fermentations, and microbiology. Finally, oenology was solidly established in higher education. The long reign of Ribéreau-Gayon and Peynaud in oenology at Bordeaux had such significance for the region that when they retired in 1976 it was a departmental event, celebrated with all the sentimental pomp French ceremonial genius brings to such occasions.

At the level of production, the immediate impact of scientific research was less visible than the changes wrought by a "real technical revolution" – the phrase is from Peynaud, who sometimes accepts the epistemology of discontinuity.[2] Whatever the discourse, the fact is that methods

---

[1] "*Burdeos-Aquitania, Capital mundial del vino,*" the Spanish title of a trilingual article on "Bordeaux-Aquitaine, the World's Wine Capital," by Pascal Ribéreau-Gayon in *Régions d'Europe: Spécial Aquitaine* (1991), pp. 144–9.

[2] See the excellent study by Leo A. Loubère, *The Wine Revolution in France* (Princeton, 1990), especially chaps. 2 and 3.

of vinification were being transformed by better techniques. There was also evidence of a new artistic subtlety in the entire process of transforming the grape juice into wine: this precise and limited process is what vinification means for a professional winemaker, nothing more. This simple process of fermentation, however, is a shorthand way of referring to a set of operations: for red wine, these would include the mechanical preparation of the grapes, alcoholic fermentation, the maceration and specific dissolution of certain elements of the grape, and malolactic fermentation. The operation of these mechanisms is determined by certain factors: temperature, aeration, pH, malaxation of the *marc* or presscake, length of fermentation, and the use of sulfur dioxide.

As the vinification process must be carried out within the context of a knowledge of general principles, one can say that technology, science, and art form the *Gesamtkunstwerk* of winemaking. Empiricism alone is a formula for disaster. As Peynaud never tired of repeating, empiricism, with its element of the unknown, leaves too much to chance, is not capable of producing the best possible wine each year. The more precise the winemaker's knowledge, the better he can control the maturity and composition of the wine being created, including the changing and guiding of bacterial activity. Although procedures used in manipulating wine have been codified, there is still no universal science of winemaking. The factors that account for this variation are the importance of adaptation of vinification to geographical and climatic conditions, the acidity of the grape, the type of wine being made, and whether the wine is for immediate or deferred gratification.[3]

The new mastery of vinification soon changed the wine-selling business. Different types of wine were adapted to commercial requirements and the tastes of consumers. The wine we have come to recognize as the bordeaux of today came into existence. Its scientific and technical requirements had been established while the men of the high aesthetic line were busy fabricating the new criteria of taste. Not that oenologists would admit that their science is now finished, with all that needs to be known clearly enunciated in texts and manuals. A complete science is a dead science. Oenology can still hold out the promise of the endless frontier to attract new minds to research, and the prospect of better wine and more profits to recruit technicians to its practice.

In the 1950s and 1960s Jean Ribéreau-Gayon took the lead in developing a school of oenology at Bordeaux with an international reputation. Institutional transformation eventually produced a "wineversity" within the old university structure. Cooperation continued with the Station agro-

---

[3] Emile Peynaud, *Connaissance et travail du vin,* 2d ed. (1981), pt. 4, "Les vinifications."

nomique et oenologique, one of whose basic functions was the detection of fraud and the regulation of quality through the Laboratoire interrégional du service de la repression des fraudes et du contrôle de la qualité. The Laboratoire interrégional later metaphorphosized into the Direction de la consommation et de la repression des fraudes, bureaucratic control emigrating from the ministry of agriculture to the ministry of finance, where this new solicitude for the consumer even produced a Secrétariat d'état for the budget and consumption. The role of this type of laboratory was essentially the same for nearly identical institutions in various key locations serving different agricultural regions. No one believed that winemakers were more virtuous than the other food producers, whose activities had to be regulated in the interest of public health. Another function of the Station was to work on the application and popularization of research through its Service d'application de la recherche et de vulgarisation. This service was a section of the Institut d'oenologie, which also supplied one of the codirectors of the laboratory. In 1951 a laboratory of oenology and agricultural chemistry was created in the faculty of sciences. This gave the subject a double dose of prestige: a laboratory is an important status symbol in science, and location in a faculty confers an academic, if stuffy, respectability.

The Ecole supérieure d'oenologie, created in 1957, flowered into an Institut d'oenologie with the status of a university institute in 1963. The advantage of a certain financial autonomy was enhanced by a contract signed in 1964 with INRA (Institut national de la recherche agronomique) with the aim of promoting research on quality wines. The agreement with INRA made it possible for the institute to borrow scientific and technical personnel from the national research organization, which had a strong regional presence, as well as to enjoy cost sharing for their scientific material. Profiting from the revolution of 1968, which destroyed the old university organization, the Institut d'oenologie consolidated all of its functions under the umbrella of a Unité d'enseignement et de recherche (UER) de l'université de Bordeaux II. Terminology had at last caught up with reality, for the UER was an ersatz faculty, a sort of wine U, which is now an official faculty.

In addition to carrying out research that could reasonably be classified as theoretical (having no clear immediate implication for the wine business), funded by the scientific council of the university (Bordeaux II), the Institut d'oenologie undertook to do research viewed as being immediately useful for winemaking. Public and private organizations requested research on precise problems or subjects. The key organization in supporting the basic level of applied research is the Conseil interprofessionnel du vin de Bordeaux (CIVB), created in 1948. It signs regular agreements with the Institute for research on viticultural and oenological projects. The CIVB and the Institute recently cooperated in setting up a unit for the regulation of procedures necessary to put quality wine on the market.

The part of the Institute's program dealing with professional training is particularly successful because of private support. Early in the twentieth century, Gayon had offered a course, under the aegis of the Station agronomique, on agricultural chemistry. It dealt with viticulture (chiefly plant nutrition and diseases) and oenology (limited to fermentation and wine diseases, including their prevention, diagnosis, and cure). In 1949 Jean Ribéreau-Gayon revived the tradition of transmitting scientific knowledge to people at the level of production – namely, technicians and knowledgeable winemakers. This *Cours et exercices pratiques d'oenologie,* although slanted toward practical oenology, was based on the latest theoretical oenological data. Taken over by Emile Peynaud, the course became world famous, and by 1971 had attracted a total of 2500 students over the years. In 1955 the economic importance of academic oenology received a sort of official recognition by governmental statute.

Like Bordeaux, Montpellier quickly organized courses leading to the Diplôme national d'oenologie. In Montpellier the faculty of pharmacy took the lead, but the faculty of sciences gave the first year of basic science and the Station oenologique later played a role.[4] The law of 1971 on professional formation, which gave an impetus to the type of course that emphasizes the spread of knowledge, led to the transformation of the course through its integration into a wide-net course on all aspects of oenology of interest to practitioners, technicians, oenologists, engineers who worked with wine, and keen amateurs who could keep up with the professionals. In addition to the standard fare of such a course – basic knowledge concerning making, preserving, and finishing wine, recent oenological advances, and production in the Bordeaux area – there was a pedagogical novelty, wine tasting, with the possibility of a university diploma in an area most people might erroneously regard as a variation on the art of the cocktail party.

The program for the Diplôme universitaire d'aptitude à la dégustation des vins recognized that wine tasting is partly an art but placed its emphasis on the part that is science, the part that can be learned from competent teachers, the part that can give successful students an occupation or better prepare them for a profession in the sectors of the economy related to

---

[4] The Station oenologique de Montpellier is the interregional laboratory for the repression of frauds and the control of quality of wine. In 1972 it was analyzing more than 4000 samples on request. It cooperates with the Institut national des appellations d'origine and the "Syndicats de défense" of various *crus* to protect the labels of wines like Clairette du Languedoc and Muscat de Rivesaltes. The station's chemical analysis takes place after the artistic judgments rendered by specially designated tasting commissions. E. Portal, "La Station oenologique de Montpellier," *Revue française d'oenologie,* no. 45 (1972), pp. 13–14.

consumption. A sommelier may be a salesman, but his performance is considerably improved, and perhaps even his usefulness to the customer increased, if he possesses solid oenological knowledge. This scientific education dealing with human faculties of taste and smell begins with the physiological mechanisms of the senses and goes on to the study of the relations between the composition of wines and the characteristics of their taste. Tasting exercises are divided into the three categories of theoretical, analytical, and applied. Bordelais novelists and critics like Mauriac, and especially Sollers, are notoriously chauvinistic about the superiority of bordeaux to burgundy. Science and commerce cannot afford this amusing bigotry; so wines of other regions and even of other countries are included in the program. Given the keen interest that the Institute has in California and Australia, this is not surprising, although the typical French consumer is more likely to test his taste on Italian and Spanish plonk rather than on the better foreign *crus.*

The Institute has taught wine tasting since 1949 and granted a university diploma since 1973. Demand for this gustatory knowledge was so high that it was necessary to create a minicourse of a week's duration for people wanting a serious approach to the subject. The announcement of a *dîner-dégustation* at domains such as Château Bouscaut (1985) did not reduce the number of registrants, usually about 40, for this course. Although professional tasting has become increasingly subject to the empire of instrumental analysis, it is still an important complement of technical analysis. Gas chromatography with statistical analysis has been found useful in eliminating from tastings certain types of defective cognacs in order to avoid modifying the perception of the taster. More conventional defects, like desirable qualities, still have to be identified by the taster.[5]

Recently it became possible to do higher academic-professional degrees at the Institut d'oenologie.[6] Because of the success of the Diplôme national d'oenologue and other similar degrees, higher-level professional formation became possible and even desirable. Between 1955 and 1977 there were 530 oenologists graduated from the institute.[7] The Diplôme d'études approfondies oenologie-ampélologie (DEA) is a degree for people headed for high-level research organizations as well as public and professional organizations. A maximum of 30 students is admitted each year. The DEA

[5] R. Cantagrel and J. P. Vidal, "L'analyse chimique et son intérpretation statistique . . . ," in Ribéreau-Gayon and Lonvaud, *Actualités oenologiques, 89* (Paris, 1990): pp 533–7; readers belonging to the rival camp of Armagnac drinkers can find consolation in the same volume: M. C. Segur, J. Pagès, and A. Bertrand, "Interprétation d'analyses et de dégustations d'Armagnacs," pp. 538–42.

[6] I thank Pascal Ribéreau-Gayon for sending me information on the programs and degrees of the Institut d'oenologie, now a faculty.

[7] Emile Peynaud, "L'oenologie à l'Université de Bordeaux: Le passé récent (1949–1977)," in Pascal Ribéreau-Gayon and Pierre Sudraud, eds., *Actualités oenologiques et viticoles,* p. 16.

also serves as the first year in the preparation for the doctorate in oenology, conferred by the Université de Bordeaux II. The DEA, having been created to produce nationally competent personnel, is based on a program connected with many research and teaching bodies, including the Institut de la vigne, Villenave d'Ornon (INRA), the universities of Reims and Bourgogne (Institut universitaire de la vigne et du vin), and the Institut national polytechnique de Toulouse (Ecole nationale supérieure d'agronomie). The justification for this vast effort is that France, seeing itself as the number one country in world viticulture, must maintain a high level of research in this area. In spite of programs in other countries, France sees itself as the world leader in viticultural and oenological research. Few would disagree.

The striking success of the programs in oenology has to be explained by intimately connected internal and external factors. The tasting program, with its original pedagogy and good reputation, certainly played a role in ensuring the success of the Institute. But perhaps Pascal Ribéreau-Gayon is correct in arguing that the essential reason is probably the active participation of serious researchers who also know the practical side of winemaking: this combination gives their teaching both originality and scientific rigor. Some of the more interesting parts of the program come from a mix of public and private funds. Under the aegis of the *Etablissement public régional,* the *Fonds de la formation professionnelle* shared costs for constructing a building containing a fully equipped tasting room of 80 places. The installation of the places was paid for by the châteaux of the Bordelais. The Institute has an air-conditioned cellar containing about 20,000 bottles of wine, some of which were gifts and others of which bought at reduced prices. Montpellier's ENA had perhaps pioneered in establishing a large experimental wine cellar, including gifts of foreign wines, in the late nineteenth century.[8] This cooperation shows that in Bordeaux the concepts of *Gesellschaft* and *Gemeinschaft* are not mutually exclusive sources of inspiration.

Food processors have been forced into listing ingredients on food packages, even if in France the information is in a neo-Pythagorean code. Winemakers have so far successfully resisted giving the consumer this information, except in the case of warnings about sulfites and fetal alcohol syndrome for wines sold in the United States, which pleases bureaucrats and enemies of any kind of alcohol and may be useful to the few Americans who both drink wine and read labels. The institute in Bordeaux believes that "to go naked is the best disguise," that there is nothing to hide from the consumer, who should be integrated into its programs. Scientific

---

[8] B. Fallot, *Le laboratoire de technologie et d'oenologie à l'Ecole d'agriculture de Montpellier* (Montpellier, 1886), pp. 7–10.

knowledge is assumed to lead to consumption. Certain cycles of the program are conceived with the aim of spreading information about wines, especially those of Bordeaux.

Within this propagandistic framework, a special place is reserved for the consumer: "Le consommateur face au vin." The consumer is promoted to a new role, that of the "privileged interlocutor." But he has a lot to learn about wines, especially in identifying their qualities and faults, before he can better appreciate them. According to this hedonized Socratic theory, it is believed that the more knowledge one has, the more pleasure one gets. The commercial advantage of this approach is that it is a good way of promoting quality wines. A final bonus is that a cultural need is being addressed. The university is fulfilling one of its missions by opening its doors to the nonprofessional world. In fine French style, Dionysus is sent on the notorious *mission civilisatrice.*[9]

In September 1995 the institute was officially recognized as a *faculté d'oenologie* and its set of new buildings inaugurated at the same time. This higher mission is more appropriate for an ambitious institution with seven professors and five *maîtres de conférences.* The "présentation officielle" of the faculty and the inauguration were the responsibility of Alain Juppé, the prime minister who happens to be mayor of Bordeaux. God approved of it all with a *grand millésime* in 1995.

---

[9] J.-N. Boidron, G. Guimberteau, and D. Dubourdieu, "Les cycles de formation professionnelle continue," *Regards sur l'Institut oenologique,* pp. 47–9; see also "L'enseignement de la dégustation," in ibid., pp. 51–3, by the same authors, with Pascal Ribéreau-Gayon.

Conclusion: Mopping-up Operations or Contemporary Oenology as Normal Science

It is tempting to see scientific winemaking today as being in a position similar to that of modern medicine: the prisoner of its own high-tech success, placed in a situation in which there are small increments of progress.[1] In the case of winemaking, not all of this progress improves the wine or increases the drinker's pleasure.[2] In the recent edition of his *Wine Buyer's Guide,* Robert Parker scolds oenologists because they "rate security and stability over the consumer's goal of finding joy in wine."[3] It is not a bad thing that gurus of the oenophilic scene like Kermit Lynch, Robert Parker, and Matt Kramer denounce the abuse of high-tech winemaking aimed more at extending the shelf life of the product rather than at the gustatory qualities preferred by consumers, or at least by Lynch and Parker and their numerous followers. Still, it is highly unlikely that scientific viticulture and oenology are going to fade away any time soon, and much of the research has had a good effect on winemaking. Whether the results deserve praise or condemnation, it is a good idea to know what exactly we are talking about when we refer to oenological research. This final chapter will indicate its main directions.

---

[1] On normal science and paradigms, see Thomas. S. Kuhn, *The Structure of Scientific Revolutions,* 2d ed. (Chicago, 1970), and *The Essential Tension: Selected Studies in Scientific Tradition and Change* (Chicago, 1977).

[2] For a view of medicine as "the prisoner of its own success," unsure of its aims except to pursue more high-tech scientific medicine and the medicalization of normal life events, see the superb essay by Roy Porter in the *Times Literary Supplement,* Jan. 14, 1994, pp. 3–4.

[3] Robert M. Parker, Jr., *Parker's Wine Buyer's Guide,* 3d ed. (New York, 1993), pp. 27–9, section on "destroying the joy of wine by excessive acidification and filtration." Matt Kramer, *Making Sense of California Wine* (New York, 1992), p. 54, aims harsh criticism at the school of viticulture and enology of the University of California at Davis for requiring students to specialize in one or the other on the ground that the two subjects are based on different scientific disciplines. See also Kramer's two chapters (no numbers) on "The Machine in the Mind" and "The Price of Success" for general critical assessments of viticulture and winemaking in California. *Parker's Wine Buyer's Guide,* p. 766, is merciless in denouncing the technologically achieved mediocrity of California's wines. Compare the praise of the department of viticulture and enology at UCD by Hugh Johnson, *Wine* (New York, 1974), p. 151.

The general subjects of oenological research that became popular in Bordeaux after the 1930s are to a large extent those which still dominate the field today: organic acids, wine pH, malolactic fermentation, aging, yeasts, enzymes, regulation of fermentation, phenols (cinnamates, anthocyanins, tannins, and so on), and color. Older topics like sulfur dioxide are still important, but generally in connection with another phenomenon, and of course knowledge of the subjects has become excruciatingly detailed and precise. Not that institutions in other parts of France and other parts of the world – especially Italy, Germany, Austria, California, and Australia – were unimportant in oenological research. Many factors contributed to the preeminence of Bordelais oenology: the size of the research program, the significance of the results obtained and published in classic texts, the pioneering research on subjects like malolactic fermentation with use of the knowledge obtained for the regulation of fermentation, the reputation of the program with its consequent attraction for foreigners, and the eventual fame of some of the oenologists who emerged. The fame of its wine was not without significance, but more important for the program was the input of oenological research into the improvement of that wine.

Today's research programs on vine and wine, whether done in institutions in Dijon or Davis, Stellenbosch or Karlsruhe, Glen Osmond or Asti, seem to have a general common core, although local material and techniques may be essential to the specificity of the research. And oenologists from nearly all countries often publish in the same journals. In an age when the American scientific patois is dominant, French oenologists publish a substantial amount of their research in American journals. In a citation index of world oenological literature, the research done in Bordeaux would occupy a prominent place.

The work of Pascal Ribéreau-Gayon on anthocyanins shows the strength of the tradition of oenology as part of other sciences, rather than an exclusively applied science – the applications came easily enough. His doctoral research was in the tradition of work done on plant anthocyanins since the early twentieth century by Richard Willstätter, whose research on plant pigments earned him the Nobel prize in chemistry in 1915, and by the equally notorious Paul Karrer of the University of Zurich. Jean Ribéreau-Gayon directed his son's thesis, with Genevois still around giving advice and encouragement. The method of paper chromatography for the study of phenolic compounds, especially anthocyanins, came from the laboratory of Bate-Smith at the University of Cambridge, where Pascal Ribéreau-Gayon spent a short time. Research contacts with the University of California (Davis), the University of Connecticut, and Georgia (U.S.A.), among others, moved oenology in Bordeaux into the international research orbit characteristic of places where the "first-rate stuff" is done.[4]

Since 1963 oenologists have been enlightening and enjoying themselves at international meetings, given Platonic status as symposia. Pascal Ribéreau-Gayon has established a convenient catalog of the themes dominating the meetings. The hot topic in 1963 was the preservation of wine, with emphasis on treatments for preventing cloudiness and deposits. A worthy obsession, but the road from virtue to vice is paved with good intentions, and some misguided producers were led to ruin their wines by excessive filtering with the latest technical marvel in filtering systems that should have been restricted to providing data for learned articles in oenological journals. The continuing importance of the chemistry of wine was reflected in the attention paid to the chemistry of colloids, one of Jean Ribéreau-Gayon's earlier research passions. The consumer of quality wines has been less appreciative of the anal character of the apostles of filtering, which may soon be applauded only by drinkers of bad beaujolais. Readers of Parker will know that the sinners are slowly returning to gentler clarification processes, or just returning to the wisdom of their fathers: racking and fining with nature's best clarifying product, egg whites. More important in 1963 was the obvious strength of microbiology and biochemistry in oenology. At the second symposium in 1967, these two disciplines showed their star status in the two dominant themes, fermentation and vinification. Pasteur and the biochemical pioneers of the interwar period would have been well pleased with their intellectual sons and a daughter or two.

A synthesis of technology (human and machine) and nature seemed to have been needed by the 1970s, perhaps to appease a neglected Ceres. At the symposium of 1977, the emphasis was on "the essential complementarity of origin and technique" in producing a quality product, as Pascal Ribéreau-Gayon delicately put it. This meant that speakers emphasized the importance of natural factors like soil, climate, and *cru,* and human factors like cultivation and techniques of vinification and conservation. The quality of the harvest had become a major preoccupation; so Peynaud's prophecy was fulfilled, the unhistorically minded might say. But the concern for obtaining quality wine grapes began in antiquity, continued in the Middle Ages, and was a major part of Dom Pérignon's program of quality control. So, in the fourth symposium (1989), the first theme was the quality and constitution of wine grapes, a subject taking up over one-fifth of the proceedings. All countries with an interest in producing high-

---

[4] Pascal Ribéreau-Gayon, *Recherches sur les anthocyannes des végétaux. Application au genre vitis* (thesis, Doctorat ès sciences physiques, University of Paris, 1959).

quality wine support substantial research on vines and grapes. In nineteenth-century France, Montpellier ruled this research empire; it is still strong there, but Bordeaux has also become a major force, with the Institut d'oenologie, INRA Bordeaux, and the Station agronomique et oenologique often cooperating to stay on the endless frontier of plant science.[5]

If God reveals himself in the details of the phenomena, modern science is divine. Instrumentation, statistical analysis, and comparisons of various methods of analysis have become much more important than they were in the old oenology. Analysis of anthocyanins and color are done by HPLC (high-pressure liquid chromatography) and spectrophotometry.[6] Oenology is done with high-tech research tools. Much of the new analysis is based on physicochemical methods. Molecular genetics makes it possible to study the plasmids of lactic acid bacteria (LAB), thus complementing research based on the taxonomy, physiology, and metabolism of the bacteria.[7] Atomic absorption methods are used to analyze a wine's phosphorus content, which if it combines with iron could produce ferric phosphate *casse* or cloudiness.[8] The complexity of enzyme activity has led to the delimiting of an area of scientific research with its own ugly name, enzymology. It is an area of biochemistry of key importance to oenology, especially from the viewpoint of developing new methods to understand and control yeast activity, malolactic fermentation, ripening of grapes, and the development of the *Botrytis cinerea* in its activity as noble rot.

Although infrared and vibrational spectroscopy have seduced oenological researchers into methodological novelties, nothing has captured gen-

---

[5] Work is also done at the Laboratoire de physiologie et ampélologie of the Université de Bordeaux I; see, e.g., M. Broquedis and J. Bouard, "Les polyamines et leurs relations avec l'acide abscissique au cours du développement de la baie du raisin," in Ribéreau-Gayon and Lonvaud, *Actualités oenologiques 89,* pp. 117–21 (see note 6).

[6] See, e.g., Leo P. McCloskey and L. S. Yengoyan (Ridge Vineyards and Department of Chemistry, San Jose State University), "Analysis of Anthocyanins in *Vitis vinifera* wines and Red Color versus Aging by HPLC and Spectrophotometry," *AJEV* 32, no. 4 (1981). There are two convenient collections of recent oenological research. First, Pascal Ribéreau-Gayon and Pierre Sudraud, ed., *Actualités oenologiques et viticoles* (1981), the proceedings of a symposium (1980) held on the 100th anniversary of the founding of the Station agronomique in Bordeaux in 1880, attended by 42 Bordelais researchers. (Maynard A. Amerine has written a benign review in *Wines and Vines,* 63 [1982].) Second, Pascal Ribéreau-Gayon and Aline Lonvaud, ed., *Actualités oenologiques 89* (1990), a volume of 567 pages of the proceedings of the 4th International Oenological Symposium held in Bordeaux in 1989, with papers in French and English by researchers in most wine-producing countries.

[7] A. Lonvaud-Funel, Ch. Frémaux, N. Bitau (all of the Institut d'oenologie, Bordeaux II), and M. Aigle (Laboratoire de génétique, Bordeaux II), "Application de la biologie moléculaire et de ses techniques à la connaissance des bactéries lactiques du vin," in Ribéreau-Gayon and Lonvaud, *Actualités oenologiques 89,* pp. 304–9.

[8] B. H. Gump and H. Chow, "Improvement in the Analysis of Phosphorus in Wine: Atomic absorption Methods," in Ribéreau-Gayon and Lonvaud, eds., *Actualités oenologiques 89,* pp. 462–6.

eral interest like imaging by low-resolution pulsed nuclear magnetic resonance, which allows the fingerprinting of sugars and other compounds in wine. With "the analysis of the different isotopes of the different atoms constituting a molecule," the origin of a molecule of sugar can be identified and the type of sugar used to chaptalize a must can be identified. The lies of producers and the mistakes of tasters can be discovered by the curious rich or an inquisitorial governmental agency.[9]

A couple of years ago, the *New York Times* reported on the curious phenomenon of the disappearance of chemistry as a result of its success in adopting the methodologies of other disciplines, especially physics, an aggressive imperialist. The success of an ecumenical oenology may lead to the same intellectual consequence. A consolation to chemists – oenologists take note – is that intellectual incoherence does not prevent their empire from expanding.

No subject seems more important in oenological research than the secondary or malolactic fermentation (MLF), which results from the growth of the acid and alcohol tolerant LAB in wine. Nowadays researchers recognize the LAB responsible for MLF as belonging to the taxonomic genera *Leuconostoc, Pediococcus,* and *Lactobacillus,* the last of which you can buy at a high price in health-food stores if you need to replace intestinal flora that have been murdered by antibiotics or the evil by-products of bad living. Among the *Pediococcus* only the discriminating species *Ped. damnosus, Ped. parvulus,* and *Ped. pentosaceus* enjoy living in wine.[10]

These bacteria are usually described as anaerobic, but they seem to require some dissolved oxygen. The growth and even survival of LAB depend on several factors, the most important of which are the wine's pH, the level of sulfur dioxide in the wine, and the bacteria's interaction with other organisms. "The metabolism of malic acid by LAB is encouraged by low pH values," and when wine sugars are metabolized by LAB at higher pH values, there is a higher level of volatile acidity because more acetic acid is produced. The effect of sulfur dioxide on LAB is to slow up or inhibit their metabolic activity, with the degree of effect depending on the species. Alcohol levels above 10 percent affect LAB adversely, but fortunately the key species producing MLF can tolerate 12–14 percent.[11] It has

[9] See A. Rapp and A. Markowetz, "Application de la résonance magnétique nucléaire (RMN) de l'isotope de carbone 13 à l'analyse du vin," in Ribéreau-Gayon and Lonvaud, *Actualités oenologiques 89,* pp. 449–55; see also Ribéreau-Gayon's conclusion to the volume, p. 565. The Laboratoire de résonance magnétique nucléaire et de réactivité chimique of the Université de Nantes has acquired a notoriety in NMR research, attracting notice by magazines and newspapers that like to inform the public on the activities of the oenological Sherlock Holmeses.

[10] D. Wibowo et al., "Occurrence and Growth of Lactic Acid Bacteria in Wine: A Review," *AJEV* 36, no. 4 (1985): 302.

[11] Ibid.; and C. Delfini and M. G. Morsiani, "Study on the Resistance to Sulfur Dioxide of Malolactic Strains of *Leuconostoc oenos* and *Lactobacillus sp.* Isolated from Wines," in

also been shown that "differences in the malolactic fermentability of wine are a reflection of the yeast involved in the alcoholic fermentation," with the "degree of inhibition" depending "upon the fatty acid concentration." Yeasts also produce amino acids and peptides, which are "growth factors for LAB, whose activity therefore changes the concentration of amino acids during MLF."[12] After all the studies of the difficulties faced by LAB in wine, one might marvel at this miracle of secondary fermentation, and the fact is that the MLF process is made much easier by technology and rigorous controls based on science.

Researchers recently tested 166 strains of lactic acid bacteria "for properties of oenological significance." Interest is turning from simply finding the bacteria and charting their growth to more fundamental issues. These issues include "physiological and biochemical properties that determine the ability of these bacteria to grow in wines and . . . the biochemical mechanisms by which these bacteria affect wine quality." The need for this basic knowledge is recognized by the wine industry as vital in determining "the selection of strains for use and commercialization as malolactic starter cultures" and indicating "the conditions under which these bacteria will be most effectively used in the winery."[13] The industry has already anticipated the protest of drinkers who want wine that has undergone a natural MLF rather than an "inoculated wine." Sensory tests by 18 qualified tasters ranked the wine produced by natural MLF as lower in quality. In any case, it would be difficult to classify the inoculated wine as unnatural. Skeptics might be tempted to drink only French wine, especially bordeaux, which, generally having a pH lower than 3.5, allows *Leuc. oenos* to fulfill the task for which nature intended it. Bordeaux is more likely to have undergone MLF through stimulation of natural flora, but the risks appear too great and the French follow pretty much the same process of inoculation as other producing countries.[14]

Obviously a lot of things can go wrong in vinification, especially in the

Ribéreau-Gayon and Lonvaud, *Actualités oenologiques 89,* pp. 331–7, which refers to the literature (Boidron, Fornachon, Lafon-Lafourcade, etc.) on the "poor resistance of *Leuconostoc oenos* species to both free sulfurous acid (H_2SO_3) molecules (especially for the undissociated ion, referred to as molecular SO_2) and to those bound with acetaldehyde and pyruvic acid. . . ."

[12] Aline Lonvaud-Funel, Annick Joyeux, and Catherine Desens, "Inhibition of Malolactic Fermentation of Wines by Products of Yeast Metabolism," *Journal of the Science of Food and Agriculture* 44 (1988): 183–91 (hereafter *JSFA*); and D. Wibowo et al., *AJEV* 36, no. 4 (1985): 307. These growth factors or promoters are substances of low molecular weight that contain sugar and nitrogen groupings.

[13] Craig R. Davis, Djoko Wibowo, Graham H. Fleet, and Terry H. Lee, "Properties of Wine Lactic Acid Bacteria: Their Potential Enological Significance," *AJEV* 39 no. 2 (1988): 137.

[14] C. R. Davis, D. Wibowo, R. Eschenbruch, T. H. Lee, and G. H. Fleet, "Practical Implications of Malolactic Fermentation: A Review," *AJEV* 36, no. 4 (1985): 294, and Pascal Ribéreau-Gayon, *Le vin* (1991; 2d rev. ed., 1994), in the famous encyclopedic collection *Que sais-je?,* pp. 36–40. See also C. Prahl et al., "The Decarboxylation of L-malic acid in

making of southern wines, low in acid and tannins and also containing unfermented sugar. Pressing dirty grapes increases the probability that the wine will be subjected to the accident of lactic fermentation. High temperature, including too fast a fermentation at high temperature – the favorite climate of LAB being 37–40°C – encourages the overpopulation of wine by these bacteria, whose unguided activity can be dangerous, even causing spoilage. The bacteria remove some elements from wine and produce others; the end products are not only lactic and acetic acids and carbon dioxide. They may degrade tartaric acid in wines with a pH above 3.5. They may produce amines, including histamine, thus giving a sensitive person an allergic-type reaction to wine. If the wine contains sorbic acid, sometimes added to protect the wine against unwanted activity by yeasts, the LAB could convert it to 2-ethoxyhexa-3, 5-diene, which gives an unpleasant geranium-like odor.[15] Since the use of sulfur dioxide to inhibit MLF became increasingly open to "controversy in the health and safety area, alternative technologies are being proposed." Researchers at the Institute of Food Technology of the University of Udine (Italy) have proposed using the enzyme lysozyme, which stops MLF without any evident damage to the wine. Lysozyme "is very diffuse in nature and has been used for a long time in the medical field" with no toxicological problems.[16] No changes were detected in the organoleptic properties and stability of wines stored for nine months at 30°C. We are reassured to learn that oenology has not become iatrogenic.

A reader of contemporary oenological journals may admire the quality of the research, the *Sitzfleisch* of the researchers, the ingenuity in the use of instrumentation and methods, the cogency of the reasoning, and the genuine concern for improving the quality of wine. The reader may also be uneasy about the almost nonchalant acceptance of the use of chemical and biological products, although optimists may point out that a science expert at detecting the rare catastrophic effect they have on vine and wine has not found comparable effect on consumers. The classic problem case in viticulture was the use of the systemic insecticide orthene to kill grape caterpillars, which used to be eaten by birds before insecticides killed the

must by direct inoculation with homofermentative lactobacilli," in Ribéreau-Gayon and Lonvaud, *Actualités oenologiques 89*, pp. 314–19. The biological decarboxylation of L-malic acid is the scientist's simple explanatory phrase for MLF. Inoculation with a homofermentative *Lactobacillus plantarum* in highly concentrated freeze-dried form is "an efficient alternative to the traditional methods of inducing MLF"; it avoids the lag phase in the starting of MLF, especially in wines with low pH values: the gospel from Prahl and co. of Chr. Hansen's Laboratorium A/S in Horsholm, Denmark, where they have always been in the vanguard of yeast research. Wine is another matter.

[15] See D. Wibowo et al., *AJEV* 36, no. 4 (1985): 307–8; and D. F. Splittstoesser and B. O. Stoyla, "Lactic Acid Spoilage in Wine," *Wines and Vines* 68 (November 1987): 65–6.
[16] A. Amati et al., "Lysozyme: A Proposal for Malolactic Fermentation Control," in Ribéreau-Gayon and Lonvaud, *Actualités oenologiques 89*, pp. 362–6.

birds – science has its own macabre theater of the absurd. The poison produced "sulphurous off-flavours in wines, reminding one of rotten eggs and cabbage or cheese." There was great economic loss in wine-producing countries, probably more than would have resulted from the loss of grapes to insect-eating birds. Everyone knows about Château Phélan-Ségur recalling its 1983 and not selling its 1984 and 1985 vintages, but this may have been only the tip of the European iceberg.[17] We may despair that issues once seemingly settled are still being actively researched because previous knowledge is now considered incomplete or downright wrong. And methods that were the last word in scientific rigor in their day are now recognized as having produced obvious errors and nonsense. The historian pontificates that this is the history of science; the philosopher argues over its meaning for scientific progress.

It may not be fashionable to argue that fallible scientists using shaky scientific knowledge are able to produce an increase in understanding and better techniques that lead to at least the possibility of better wine, but it seems as certain as most historical arguments about science. By the end of the 1930s it looked as if volatile acidity was pretty much understood, in large part as a result of the work done in Bordeaux, and not in need of much further research. Yet the understanding of the activity of acetic acid bacteria is still a hot research area. This must be yet another illustration of science the endless frontier.

While bacteriologists were fighting over the taxonomy of acetic acid bacteria, eventually recognized as one family (*Acetobacteraceae*) rather than two genera (*Acetobacter* and *Gluconobacter*), oenologists were busy trying to limit the fertility of this dangerous creature. The reason for the revival of interest in the aerobic microorganisms we call acetic acid or vinegar bacteria is the "speculation that these bacteria may survive and grow . . . [in] semi-anaerobic to aerobic conditions," thus posing a threat to wine, one of their favorite habitats. The threat begins with the harvest itself because acetic acid bacteria can grow on grapes, enter into the must, and affect vinification. The bacteria produce acetaldehyde and acetic acid from ethanol. A small amount of acetic acid "is produced . . . by yeasts during alcoholic fermentations and by LAB during the malolactic fermentation."[18] There is no way to avoid taking it into account without putting the quality of the wine into danger. There is also the legal limit imposed for volatile acidity, which is stated in grams of acetic acid per liter, mea-

[17] Details on the chemical mechanism of the appearance in bottled wine of the stinking culprit, dimethyl disulfide, can be found in D. Rahut, "Trace Analysis of Sulphurous Off-Flavours in Wine Caused by Extremely Volatile S-containing Metabolites of Pesticides E. G. Orthene," in Ribéreau-Gayon and Lonvaud, *Actualités oenologiques 89*, pp. 482–7; see also David Peppercorn, *Bordeaux*, (London, 1982), p. 216.

[18] G. S. Drysdale and G. H. Fleet, "Acetic Acid Bacteria in Winemaking: A Review," *AJEV* 39, no. 2 (1988): 143.

sured "using steam distillation followed by titration of the distillate with standard base (NaOH)."[19]

It all seems simple enough, a standard procedure that should give reliable results, and it would be if wine were simple. It turns out that the procedure is difficult because "Significant interferences in volatile acidity analyses can be caused by SO_2 (as sulfurous acid), CO_2 (as carbonic acid) and sorbic acid." Even though one can correct for the interferences, "the methods themselves are subject to errors" due to the interference of minor volatile acids (formic, propionic, butyric, valeric, and lactic, for example). In this situation, if an expert sensory evaluation conflicts with a scientific measurement of volatile acidity, the winemaker can depend on his own senses or escape from confusion by obtaining a more sophisticated scientific procedure, quick and precise analysis by gas chromatography, with the reasonable hope that he won't need another piece of science to correct this one.[20] The only way to escape from the confusion of science is by more science.

Nothing is more important than wine color, which in red wine is "due principally to anthocyanins derived from the fruit (monomeric or 'free' anthocyanins) and polymeric pigments formed from the anthocyanins by condensation with other flavonoid compounds and probably aldehydes [chemical compounds formed by the oxidation of primary alcohols] during wine aging."[21] Unlike dragons, anthocyanins do not last forever and "are important for color for . . . [only about] three years after fermentation," although they are "present for up to five years" and for reasons unknown (in 1981) may stop disappearing in some wines.[22] Temperature seems to be the main factor in the rapid evolution of the color of a wine toward orange. A moderate amount of oxygen (from the air) gives color its maximum potential for surviving the process of aging in the bottle.[23] A

[19] S. A. Kupina, J. L. Kutschinski, R. D. Williams, and R. T. DeSoto, "A Refined Gas Chromatographic Procedure for the Measurement of Acetic Acid . . . ," *AJEV* 33, no. 2 (1982): 67.

[20] Ibid.

[21] J. Bakker, N. W. Preston, and C. F. Timberlake, "The Determination of Anthocyanins in Aging Red Wines: Comparison of HPLC and Spectral Methods," *AJEV* 37, no. 2 (1986). Anthocyanins are flavonoid pigments; a flavonoid is "one of a group of naturally occurring phenolic compounds, many of which are plant pigments. They include anthocyanins, flavonols, and flavones" (Oxford's *CSD*, p. 271). The Bordeaux orthodoxy is slightly different: "the color of red wines essentially results from anthocyanins as well as from their combinations with tannins." See Y. Glories and C. Galvin, "Les complexes tanins-anthocyanes en présence d'ethanal – conditions de leur formation," in Ribéreau-Gayon and Lonvaud, *Actualités oenologiques 89,* pp. 408–13.

[22] McCloskey and Yengoyan, "Analysis of Anthocyanins in *Vitis vinifera* wines and Red Color versus Aging by HPLC and Spectrophotometry," *AJEV* 32, no. 4 (1981).

[23] Y. Glories and C. Bondet de la Bernardine, "Rôle joué par l'oxygène et la température sur l'évolution du contenu phénolique du vin rouge – mécanismes mis en oeuvre," in Ribéreau-Gayon and Lonvaud, *Actualités oenologiques 89,* pp. 398–402.

beautiful color satisfies the neurons of visual aesthetics in our brain. Color gives the drinker a sign of the health of the wine with intimations of the pleasure or pain to come. The longevity of color components in the wine is thus a matter of the highest importance, a guide to the past as well as the future of the wine.

Galileo defined wine as "light held together by moisture." Oenologists are less poetical and more precise. By the 1960s, plant phenolics and specific information on *Vitis vinifera* were integrated into oenological research. With the older instrumentation it was not even possible to define wine color precisely, for the colorimeter's quantitative analysis estimated the color of solutions by comparison with the colors of standard solutions. A new type of photometer introduced greater precision and the possibility of measuring the density of hue and color.[24] Scientifically defined reality is the offspring of instrumentation and method.

But nature usually has a nasty surprise or two in store for researchers, even if they use widely accepted practices. Hence the possibility of scientific progress. In language reminiscent of Herbert Butterfield's remarks on the scientific revolution, Somers and Ziemelis argued in 1985 that new analytical concepts were necessary in establishing the often minimal phenolic content of white wine. "The most widely used procedure . . . for [the] analysis of total phenols in wines . . . is generally [grossly] invalid when applied to white wines . . . because of a strong synergistic reaction between sulphur dioxide and . . . [one type of] phenols in response to the . . . reagent." So it turns out that the measurements of total phenols may have been "magnified several-fold."[25] So far so good, but revisionists can be revised.

It turned out that the direct "spectral method of estimating the extent of polymeric pigment formation in young red wine" was not so valid as Somers and Evans had argued in their ambitious article in 1977, which pointed to the serious deficiency of contemporary oenology in measuring wine color, as seen in errors in texts and technical literature. Bakker, Preston, and Timberlake showed, nearly a decade later, that total free anthocyanins in young red wine "measured by high performance liquid chromatography were lower than those measured by the spectral method of Somers and Evans."[26] The reader may again be tempted to wonder if

[24] T. Chris Somers and Michael E. Evans (The Australian Wine Research Institute, Glen Osmond, South Australia), "Spectral Evaluation of Young Red Wines: Anthocyanin Equilibria, Total Phenolics, Free and Molecular So₂, 'Chemical Age,'" *JSFA* 28 (1977): 279–87. The method devised by Sudraud (published in 1958) made it possible to establish the composition of wine color. Nine of the 30 references in Somers and Evans are to French work, especially that done by Pascal Ribéreau-Gayon, Institut d'oenologie, Université de Bordeaux II.

[25] T. Chris Somers and Guntis Ziemelis, "Spectral Evaluation of Total Phenolic Components in *Vitis vinifera:* Grapes and Wines," *JSFA* 36 (1985): 1275–84.

[26] J. Bakker, N. W. Preston, and C. F. Timberlake, "The Determination of Anthocyanins in Aging Red Wine: Comparison of HPLC and Spectral Methods," *AJEV* 37, no. 2 (1986).

method defines reality and even to wonder if this is not a monk's quarrel. Probably not, for as Somers and Ziemelis pointed out in 1985, "the phenolic composition of reds is extremely dynamic, with irreversible progression from monomeric to polymeric pigment forms."[27] What is at stake here is the health and quality of the wine.

Pascal Ribéreau-Gayon and two colleagues made clear the significance of the process in a broader chemical context. "During the maturation phase of red wines, extending from the end of vinification until bottling, the presence of oxygen provokes the chemical transformation of pigments, essential for aging." Traces of acetaldehyde (the liquid aldehyde ethanal) are produced by the mild autoxidation of the ethanol or ethyl alcohol "in the presence of phenolic compounds. Simultaneously, this acetaldehyde provokes a copolymerisation of anthocyanins and tannins. . . . As condensation reaches a certain level, the large molecules precipitate and colour decreases. The same oxidation produces an increase in the level of tannin concentration and consequently a decrease in astringency, favorable from the point of view of quality."[28] V. L. Singleton points out the need for care in aging if one of its principal aims – adding to the complexity of the wine – is to be achieved. It is all too easy for the process to go awry, with the result that simple flavors are produced "rather than complex mixtures of flavors." A lot can go wrong. "Ethanol and three different higher primary alcohols are oxidized in the course of aging to four different aldehydes, which in turn oxidise to four different carboxylic acids [many of these are fatty acids], which in turn esterify to produce a total of 28 different substances which have arisen from the original four. Obviously, the extent and rate of aging will produce unique complexes of these 28 different substances and this is only an example of one of the many types of components – chemical components – of wine."[29] Pasteur would be pleased to learn that there are even things to be discovered about tartaric acid. The "esterification of tartaric acid [conversion of part of the acid to ethyl esters] during wine aging accounts for appreciable mellowing by diminishing the effective sensory acidity."[30]

[27] A polymer is "a substance having large molecules consisting of repeated units (the monomers)." Polymerization is "a chemical reaction in which molecules join together to form a polymer." In condensation polymerization, "a small molecule is eliminated during the reaction." Oxford's *CSD*, p. 550.

[28] Pascal Ribéreau-Gayon, Paul Pontallier, and Yves Glories, "Some Interpretations of Colour Changes in Young Red Wines during Their Conservation," *JSFA* 34 (1983): 505–16.

[29] "Dr. Vernon L. Singleton on the Anomaly of Aging," a speech reported in *Wines and Vines* 58 (1977): 31–2. Primary alcohols are those having two hydrogen atoms on the carbon joined to the OH group they contain; secondary alcohols have one hydrogen atom on their carbon; and tertiary alcohols have no hydrogen. Oxford's *CSD*, p. 17.

[30] Theodore L. Edwards, Vernon L. Singleton, and Roger Boulton, "Formation of Ethyl Esters of Tartaric Acid during Wine Aging: Chemical and Sensory Effects," *AJEV* 36 (1985): 118. University of California, Davis, study; after his M.S. thesis, Edwards carried the gospel to the Rutherford Hill Winery in St. Helena.

Oenologists make a clear distinction between anaerobic aging in nearly hermetically sealed bottles and the maturation process that occurs during storage of wine in tanks and barrels, where the wine traditionally undergoes several oxygen saturation cycles. Red wine may improve up to even ten cycles.[31] The practice of repeated rackings, now described by Dussert-Gerber as maturing a wine in the old manner, used an empirical version of this knowledge; some producers still engage in this laudable luxury. In Saint-Estèphe, Guy Delon racks the wine of his estate of six hectares, Château Ségur de Cabanac, seven times and fines with egg whites, with storage in oak barrels (25 percent new). The hieratic status of the oak barrel is rarely questioned. During its storage in oak barrels, wine extracts part of the lignin of the wood, thus setting off a series of complex chemical reactions, which are poorly understood but nevertheless are believed to improve the taste of the wine. Pascal Ribéreau-Gayon and his colleagues were surprised to find that aerated "wine stored in large vats . . . will be, from an analytical and organoleptical point of view, more similar to a wine stored in oak barrels than a wine stored in tanks not aerated" – indeed, a discovery "of great value," even if it does not convert worshipers of the oak tree.[32] At present one of the great research efforts in Bordeaux is concentrating on what happens in wine during the long period between fermentation and bottling, "with the study of the evolution of color and the evolution of tannins and their softening."[33] Whether it will provide satisfactory answers without scientists putting on what Butterfield called a new thinking cap is another matter. And it is the nature of scientific revolutions that we can have no idea of which haberdashery will provide the cap, nor what its style will be.

One of the rewards of research is that it keeps on creating new research problems, which may be a trivial corollary to Peer's Law: "The solution to a problem changes the nature of the problem."[34] If the high quality of contemporary wine is the long-term result of modern oenology, originat-

---

[31] Steven J. Carnevale, "Oxygen Scavengers in Wine Packaging," *Wines and Vines* 69 (1988).

[32] Pascal Ribéreau-Gayon et al., *JSFA* 34 (1983): 514. For the macromolecular chemistry of wine in oak, see J. L. Puech et al., "Délignification du bois de chêne par percolation en milieu hydroalcoolique," in Ribéreau-Gayon and Lonvaud, *Actualités oenologiques 89*, pp. 388–91. The American oak (*Quercus alba*) is poor in tannin; the European oaks *Quercus robur* and *Quercus petraea* are rich in tannin. P. Chatonnet and J. N. Boidron, "Dosage de composés volatils issus du bois de chêne par chromatographie en phase gazeuse – Application à l'étude de l'élevage des vins en fûts de chêne," in ibid., pp. 477–81, emphasize the role of wood, especially charred wood, in increasing concentrations of substances already present in red wine, as well as adding new substances. Different vats require different treatment. The oaks of the Allier give the woodiest character to a wine. P. Pontallier, "Les conditions d'élevage des vins rouges en fûts et en cuves," in Ribéreau-Gayon and Sudraud, eds., *Actualités oenologiques et viticoles* (1981), p. 316.

[33] "Les types de vins du Bordelais. Entretien avec le Pr. Pascal Ribéreau-Gayon," in Guy Jacquemont and Antoine Hernandez, eds., *Le Bordeaux* (1990), p. 38.

[34] Arthur Bloch, *Murphy's Law* (Los Angeles, 1977), p. 48.

ing somewhere in the nineteenth century – several dates are possible, beginning with Pasteur – but especially with the oenology of the 1930s, what about the work done since?[35] Does the research done after the 1930s have the same significance for improving wine? Probably not. Although many individual improvements can result from a playing out of the extremely fertile theory and derivative techniques of this approach to wine, that is about all we can expect from basically steady state science. Whenever the next radical approach to wine emerges from a very different science, next week or in a few hundred years, winemaking practices will eventually be seriously affected. Meanwhile, old science produces a great deal of research of modest originality and helps produce a lot of good wine – sometimes, with the help of a quirky god and a good winemaker, a great wine.

[35] Some curmudgeons may want to admit that we only enjoy nondiseased wines, not high-quality wines. And Kermit Lynch's *Adventures on the Route: A Wine Buyer's Tour of France* (New York, 1988), a book with some good examples of science gone greedy in the wine business, is the place for them to find reinforcement for their mistaken ideas as well as recommendations for quality wines and some notable mediocrities.

Select Bibliography

Archives

Archives de l'Académie des sciences, Paris. Papiers de J.-B. Dumas. Documents relevant to phylloxera are in cartons no. 1634 to no. 1644.

Archives départementales de la Côte-d'Or, série M: 13, J (Station oenologique de Bourgogne. 1. Laboratoires de Beaune, correspondance, 1886–1901; 2. Laboratoires de Dijon, rapport de 1918); M: 13, L (Institut régional oeonologique et agronomique de Bourgogne).

Archives départementales de la Gironde, série M: 7 M 130 (projet de création d'une école nationale de viticulture et d'oenologie, 1899); 7 M 167 (législation, brochures, journaux, publications d'Ulysse Gayon); 7 M 184–94 (Commission de délimitation de la région des vins de Bordeaux, 1907–14); 7 M 195–205 (appellations d'origine, especially 198, documentation, 1889–1930: châteaux, crus classés et crus bourgeois du Médoc); 7 M 207 (*includes* protestation des viticulteurs contre le sucrage des vendanges, 1930–5); 7 M 210 (maladies de la vigne: correspondance, recherche, 1851–7); 7 M 211–29 (phylloxéra, 1875–1938).

Archives du département de l'Hérault, série 4 N liasses 161 à 168 (Ecole d'agriculture); 147 M 1 à 86 (viticulture).

California Wine Industry Oral History Project. Regional Oral History Office, The Bancroft Library, University of California, Berkeley. Interview transcripts: Maynard A. Amerine, William V. Cruess, Maynard A. Joslyn, Harold P. Olmo, and A. J. Winkler.

Journals

Amerine, M. A., and Joslyn, M. A., *Table Wines: The Technology of Their Production* (Berkeley and Los Angeles, 1951; 2d ed. 1970), pp. 829–37, provides a list of journals organized alphabetically by country of publication. French journals are listed on pp. 830–2.

Bibliographies

Amerine, M. A., and Joslyn, M. A. *Table Wines: The Technology of Their Production*, pp. 811–965 (Berkeley and Los Angeles, 1951). See especially appendices A ("A selected, classified, and annotated list of references") and B ("Journals with material on grapes and wine").

Bibliography of Publications by the Faculty, Staff, and Students of the University of California, 1876–1980, on Grapes, Wines, and Related Subjects. Compiled by Maynard A. Amerine and Herman Phaff. Berkeley, 1986.

Catalog der oenologischen Bibliothek von Dr Adolph Blankenhorn. Heidelberg, n.d. 33 pages.

Catalogue matières, Bibliothèque nationale, Paris. *See under* "mots typiques": *vigne, vignoble, vins,* etc.

Gabler, James M. *Wine into Words: A History and Bibliography of Wine Books in the English Language* (Baltimore, 1985).

Grapes, Viticulture, Wine and Winemaking. A subject bibliography of books and periodicals in the Peter J. Shields Library, University of California, Davis. Sacramento, Calif., 1975. 804 pages of xeroxed catalog cards.

Schoene, Renate. *Bibliographie zur Geschichte des Weines.* Munich, New York, etc., 1988. 14,713 items listed.

Some Key Secondary Works on Vine and Wine

Amerine, Maynard A., and Singleton, Vernon L. *Wine: An Introduction.* Berkeley and Los Angeles, 1965, 1977.

Amerine, Maynard A., and Roessler, Edward B. *Wines: Their Sensory Evaluation.* New York, 1976, 1983.

Briggs, Asa. *Haut-Brion: An Illustrious Lineage.* London, 1994.

Dion, Roger. *Histoire de la vigne et du vin en France des origines au XIXe siècle.* Paris, 1977, 1959.

Galet, Pierre. *Cépages et vignobles de France.* Vol. 1: *Les vignes américaines.* 2d ed. Montpellier, 1988. Vol. 2: *L'ampélographie française.* Montpellier, 1988.

Garrier, Gilbert. *Le phylloxéra: Une guerre de trente ans 1870–1900.* Paris, 1989.

Geison, Gerald L. *The Private Science of Louis Pasteur.* Princeton, 1995.

Huetz de Lemps, Alain. *Vignobles et vins d'Espagne.* Bordeaux, 1993.

Johnson, Hugh. *Vintage: The Story of Wine.* New York, 1989.

Lachiver, Marcel. *Vins, vignes et vignerons: Histoire du vignoble français.* Paris, 1988.

Loubère, Leo A. *The Red and the White: The History of Wine in France and Italy in the Nineteenth Century.* Albany, N.Y., 1978.

Loubère, Leo A. *The Wine Revolution in France: The Twentieth Century.* Princeton, 1990.

Mullins, Michael G., et al. *Biology of the Grapevine.* Cambridge, 1992.

Ordish, George. *The Great Wine Blight*. London, 1972.

Pasteur, Louis. *Etudes sur le vin. Ses maladies et les causes qui les provoquent. Procédés nouveaux pour le conserver et pour le vieillir.* 1866; 2d ed., 1873.

Pouget, Roger. *Histoire de la lutte contre le phylloxéra de la vigne en France.* Paris, 1990.

Ribéreau-Gayon, Pascal. *Le vin.* Paris, 1991.

Ribéreau-Gayon, Jean, and Peynaud, Emile. *Traité d'oenologie.* Paris, 1947, etc. Pierre Sudraud and Pascal Ribéreau-Gayon are also authors in the 3d edition.

Ribéreau-Gayon, Pascal, ed. *The Wines and Vineyards of France: A Complete Atlas and Guide.* New York, 1990.

Warner, Charles K. *The Winegrowers of France and Government since 1875.* New York, 1960; Greenwich, Conn., 1975.

Winkler, A. J. *General Viticulture: Contemporary Research, Practices and Progress in Present-Day Care of Grapevines.* Berkeley and Los Angeles, 1962, 1973.

Index